Practical Handbook on the 3Rs in the Context of the Directive 2010/63/EU

Practical Handbook on the 3Rs in the Context of the Directive 2010/63/EU

Edited by

Gianni Dal Negro
DVM ERT, Macerata Feltria, Urbino, Italy

Silvia Sabbioni
Department of Life Sciences and Biotechnology,
University of Ferrara,
Ferrara, Italy

ELSEVIER

ACADEMIC PRESS
An imprint of Elsevier

Academic Press is an imprint of Elsevier
125 London Wall, London EC2Y 5AS, United Kingdom
525 B Street, Suite 1650, San Diego, CA 92101, United States
50 Hampshire Street, 5th Floor, Cambridge, MA 02139, United States
The Boulevard, Langford Lane, Kidlington, Oxford OX5 1GB, United Kingdom

Library of Congress Cataloging-in-Publication Data
A catalog record for this book is available from the Library of Congress

British Library Cataloguing-in-Publication Data
A catalogue record for this book is available from the British Library

ISBN: 978-0-12-821180-9

For information on all Academic Press publications visit our website at https://www.elsevier.com/books-and-journals

Publisher: Stacy Masucci
Acquisitions Editor: Linda Versteeg-Buschman
Editorial Project Manager: Franchezca A. Cabural
Production Project Manager: Sreejith Viswanathan
Cover Designer: Victoria Pearson Esser

Typeset by TNQ Technologies

Dedicated to patients, ultimate beneficiaries of the 3Rs.

Quidquid praecipies, esto brevis. **(Whatever advice you give, be brief.)**

Horatius, Ars Poetica

Contents

Contributors

Chiara Attanasio, Department of Veterinary Medicine and Animal Production, University of Naples Federico II, Naples, Italy

Simon Bate, GlaxoSmithKline, Stevenage, United Kingdom

Olga Boruc, EMBL Laboratory Animal Resources, Epigenetics and Neurobiology Unit, EMBL, Monterotondo, Italy

Aurora Brønstad, Department of Clinical Medicine, University of Bergen, Bergen, Norway

Marino Campagnol, Laboratory Animal Consultant, Italy

Luisa Corsi, Laboratory Animal Veterinarian Consultant, Parma, Italy

Livia D'Angelo, Department of Veterinary Medicine and Animal Production, University of Naples Federico II, Naples, Italy

Paolo de Girolamo, Department of Veterinary Medicine and Animal Production, University of Naples Federico II, Naples, Italy

Chantra Eskes, Services & Consultation on Alternative Methods (SeCAM), Magliaso, Switzerland

Derek Fry, School of Biological Sciences, University of Manchester, Manchester, United Kingdom

Helmut Fuchs, German Mouse Clinic, Institute of Experimental Genetics, Helmholtz Zentrum München, Neuherberg, Germany

Sara Fuochi, University of Bern, Bern, Switzerland

Viola Galligioni, Comparative Medicine Unit, Trinity College Dublin, Dublin, Ireland

Martin Hrabe de Angelis, German Mouse Clinic, Institute of Experimental Genetics, Helmholtz Zentrum München, Neuherberg, Germany

Laura Lossi, Department of Veterinary Sciences, University of Turin, Italy

Enrico Maestri, Department of Law, University of Ferrara, Ferrara, Italy

Adalberto Merighi, Department of Veterinary Sciences, University of Turin, Italy

Lluis Montoliu, National Centre for Biotechnology (CNB-CSIC) and Biomedical Research Networking Center on Rare Diseases (CIBERER-ISCIII), Madrid, Spain

Cholawat Pacharinsak, Department of Comparative Medicine, Stanford University, Stanford, CA, United States

Marta Piscitelli, ENEA, Italian National Agency for New Technologies, Energy and Sustainable Economic Development, Rome, Italy

Marcello Raspa, Institute of Biochemistry and Cell Biology, Italian National Research Council (CNR), Rome, Italy

Merel Ritskes-Hoitinga, SYRCLE, Health Evidence, Radboud University Medical Center, Nijmegen, The Netherlands; AUGUST, Clinical Medicine, Aarhus University, Aarhus, Denmark

Ferdinando Scavizzi, Institute of Biochemistry and Cell Biology, Italian National Research Council (CNR), Rome, Italy

Patrick Sharp, University of California, Merced, CA, United States

Adrian Smith, Norecopa, c/o Norwegian Veterinary Institute, Ås, Norway

Joanne Storey, GlaxoSmithKline Medicines Research Centre, Stevenage, Hertfordshire, United Kingdom

Valentina Vasina, University of Bologna, Bologna, Italy

Annarita Wirz, Santa Lucia Foundation IRCCS, Rome, Italy

Paola Zarattini, Department of Life Sciences, University of Trieste, Trieste, Italy

Anne Zintzsch, Project on Severity Assessment of Genetically Altered Animals, University of Basel, Basel, Switzerland

Editors' preface

The Directive 2010/63/EU firmly anchors the 3Rs (replacement, reduction, and refinement) principles into European Legislation. The interpretation and the application of these principles pose some challenges from a number of factors. First, the rapidly changing global landscape makes it increasingly urgent to address gaps in traditional research, development, and manufacturing processes, especially where the translation of laboratory animal results to human studies has had a low success rate. Second, the widely diverse fields in which laboratory animals are used needs to be considered; according to the 2019 report on the statistics on the use of animals for scientific purposes in the Member States of the European Union in 2015−17 (https://eur-lex.europa.eu/legal-content/EN/TXT/?qid=1581689520921&uri=CELEX:52020DC0016), animals used for basic and applied research account for 68% in total, whereas animals used for regulatory activities (i.e., regulatory toxicology and batch quality control testing for human and veterinary medicines), account for 23%. The remaining 9% is shared between higher education and training, preservation of species, protection of the natural environment, forensic inquiries, and routine biological products (e.g., monoclonal antibodies, blood-based products, and other product types). Such diversity entails different challenges in predicting how the results of animal work will translate to effects in the target species. Third, sector-specific legislations call for different restriction levels in the use of laboratory animals, Regulation 1223/2009 on Cosmetic Products being the most restrictive of all with a ban for testing and marketing of cosmetic products tested on animals. Fourth, the stakeholders (patients, consumers, policy makers, and funding bodies) increasingly demand evidence of the value, safety, and increased cost−benefit ratio of new products; 3Rs-aligned advances in animal and nonanimal methods are expected to at least match present quality, safety, ethical and economic standards. Fifth, science and technology evolve at an incredibly fast pace. This pace, however, is not regular, which makes extrapolations and predictions of possible timelines for new achievements in the application of the 3Rs (especially concerning *replacement*) extremely hard to make. Finally, we are witnessing an increasing pressure by the society and policy makers toward phasing out laboratory animal use.

This handbook aims to provide a roadmap for navigating the multifaceted universe of the 3Rs in the context of the Directive 2010/63/EU; although the focus is clearly on the European landscape, it also aims to accomplish the ambition to extend its relevance to those outside Europe who keep to the 3Rs principles and to researchers seeking to publish in the many journals that hold to the 3Rs principles.

This handbook should help professionals for whom Laboratory Animal Science is relevant to their day-to-day work, in particular, basic and applied scientists working with laboratory animals who will be developing and supervising projects which could come within the scope of the EU Directive. Other professionals would include, but not be limited to, animal facility technicians, laboratory animal scientists, veterinarians, and regulatory officers. The publisher and the editors believe that this practical guidance will fill an existing gap.

This handbook brings together contributors who are experts in their fields and aims to provide not only practical information but also key references so that readers can extend their knowledge and understanding of each 3Rs-relevant aspect treated. Some aspects that are not usually treated elsewhere have been included, such as legal aspects relevant to the application of the 3Rs in Europe, and nonstatistical elements considered key for good experimental design. Furthermore, contributors may give interpretations of the Directive text that are colored by their own experiences in an attempt to help the reader's understanding of certain topics; as such, some interpretations might be subject to controversy, but if this handbook instigates debate and opinion exchange on the interpretation and applicability of the Directive, we will consider this a success.

Finally, with the aim to anchor each chapter to the Consensus Document on Education and Training endorsed by the EU Member States developed by the National Competent Authorities for the implementation of Directive 2010/63/EU on the protection of animals used for scientific purposes (https://ec.europa.eu/environment/chemicals/lab_animals/pdf/Endorsed_E-T.pdf), authors have been asked to refer to the respective relevant learning outcomes contained in the Document; by doing so, this handbook is consistent with the common education and training framework for the EU developed by an Expert Working Group (EWG) established by the Commission to fulfill the requirements under Articles 23, and 24 of Directive 2010/63/EU.

The Editors
Gianni Dal Negro and Silvia Sabbioni

Acknowledgment

The editors would like to thank their respective family members for the support, encouragement, and patience during this undertaking.

Chapter 1

Ethics and law on human–animal relationship

Enrico Maestri[1], Marta Piscitelli[2]
[1]*Department of Law, University of Ferrara, Ferrara, Italy;* [2]*ENEA, Italian National Agency for New Technologies, Energy and Sustainable Economic Development, Rome, Italy*

Ethical aspects (Marta Piscitelli)

Introduction

The editors' choice to begin this manual with a chapter on ethics and legislation is aligned with the spirit of the Directive 2010/63/EU suggesting the reader the right approach to the use of animals in laboratory research.

The use of laboratory animals should, in fact, be preceded by the recognition that the use of a living being always requires an ethical choice and each decision is inherently linked with the full understanding of the norms that regulate such use.

Intrinsic value of life and arguments for and against the use of animal for scientific purposes

Short story of animal experimentation and animalism

The life and welfare of animals are values that, differently from others, are relatively recent additions to our moral spheres and, as is the case for other values orienting our behaviors such as justice, peace, and honesty, must be introduced through a formative process, or rather, it must be taught. We are born with emotions and instincts but without some feelings or values, such as respect for animals, that are instead tied to an educational process (Declaration of Animals' Rights of 1978, "considering that education must teach from infancy how to observe, understand, respect, and love animals."). In fact, until the middle of the last century, laboratory animals were considered simple means through which humans could achieve their own objectives; in particular, among the personnel involved in experimentation animal welfare was not

Practical Handbook on the 3Rs in the Context of the Directive 2010/63/EU
https://doi.org/10.1016/B978-0-12-821180-9.00004-0

taken into high consideration as it is today. Young readers may find some of the reflections on the ethics of experimentation trivial, as they live in an era of great sensitivity for animals.

It is important to bear in mind that cultural changes usually require time and, most importantly, they are triggered by disruptive events. To understand the process that led to the cultural change with regards to animals, a brief overview of the history of animal experimentation and animalism will be useful.

In ancient Greece, philosophers used animals as tools to understand the relationship between human kind and nature; by studying animal bodies, they tried to find answers on the meaning and function of life. In the following table, some of the milestones in the history of animal experimentation and some of the most significant literature sources are listed (Franco, 2013).

Aristotle	384–322 BCE	*Historia animalium*	Observation of the common parts of animals and classification based on observation of common traits.
Thomas Aquinas	1221–1274 CE		Dominion of humankind over other creatures.
Galen	129–200 CE		Dissection of animals. Observation of organs and study of their functions.
Andrea Vesalio	1514–1564 CE	*De Humani Corporis Fabrica Libri Septem*	
Descartes	1596–1650	*Discourse on method*	Animals are machines (*res extensa*)
Bacon	1561–1626	*Novum Organum*	"Science is power"
Claude Bernard	1813–78		Dissection of animals in front of vast audience

Claude Bernard, a French physiologist of the second half of the 19th century known by the scientific world for his numerous and important discoveries on the function of pancreas, liver, central nervous system, and other organs and systems is also known as one of the most cruel live dissectors. In fact, his lessons generated dissent and criticism that progressively grew (Silverman, 1996).

In the following table, some of the names of those who mostly contributed to the debate that has brought animals to be no longer considered as objects but living beings worthy of moral consideration are listed. The modern animal right movement has its origins in the work of these people.

Jeremy Bentham	1748–1832	Utilitarianism	Animals possess the capacity to reason and think
David Hume	1711–66		Differences with humankind are qualitative, not quantitative
Henry Salt	1851–1939	Animal rights	Animal rights considered in relation to social progress
Charles Darwin	1856		The origin of species
	1876		The Animal Cruelty Act
Peter Singer	1946	Utilitarianism	Animal liberation
Tom Regan	1938–2017	Animal rights	Animal rights
Ruth Harrison	1920–2000	Against intensive breeding	Animal machines

The British parliament was pioneer in enacting laws regarding the use of animals in research. In 1876, the Cruelty to Animals Act was introduced, which amended the previous 1849 Act, and included a proper regulation of animal experimentation. The act called for three important obligations:

1. Animal experiments should only be carried out when there is absolute need of knowledge for the purpose of saving or prolonging life or alleviate suffering.
2. Animals must be anesthetized.
3. Animals must be killed immediately after the experimental procedure if they are injured or in pain as a result of the experiment.

This important document remained valid for more than a century until its replacement by the Animals (Scientific Procedures) Act 1986, issued in the same year as the first European Directive 86/609 EEC.

Although the origins of the animal rights movement can be found in the 18th century, to understand the origins of the progressive cultural shift in the 19th century in Europe, it is important to remember Darwin's evolutionary theory which revolutionized the concept of the centrality of humans in respect to other living beings, increasing the perception of continuity between animal species and man; later, in the 20th century, the various scientific discoveries of the 70s added knowledge on the nervous structure of different species; at the same time, studies of ethology allowed to verify the similarities between animal and human emotions; finally, there has been an increasing awareness of the public by the animal right movements (Petetta and Ciccocioppo, 2020). In the same period, the publication of books reporting animal suffering in intensive animal farms and in research (Ruesch, 1978; Singer, 1975; Harrison, 1964) by animal rights movements and the activity of environment associations like Green Peace and World Wildlife Fund, contributed to raise the awareness of the public opinion. Two books, "Animal Liberation"

(Singer, 1975), and "The Case for Animal Rights" (Regan, 1983), have become the ideological manifesto of the modern animal rights movement driven by an ideology of egalitarianism among sentient beings to the point of considering animal and human rights equal, including the right to life.

The cultural change that has taken place has led to the inclusion of animals in our moral sphere in such a way that choosing to use them for our own purposes is now an ethical choice.

From the Disney imaginary to intensive breeding, selective pressure in zootechnics and in pet animal breeds, the animal issue is multifaceted. One of the most debated issues is the experimentation in the biomedical field.

Although from the 70s onwards the legislation relating to the protection of animals used for scientific purposes has become progressively more stringent, it has been subject to criticism by animal rights activists who stand up for the total abolition of the animal use.

Ethical evaluation: a necessary clarification

It is critically important to understand what making an ethical evaluation means, entails, and what its impact on experimentation might be.

Too often, the "ethical evaluation" is limited to a superficial reading of rules with no deeper understanding of how an "ethical issue" should be addressed. In the ethical evaluation, deontological codes, social values, and individual values are compared and questioned; with all these elements involved, ethical issues never have a simple solution.

First of all, the etymological meaning of ethics helps us in understanding the ethical evaluation process. Ethics can be defined as "the branch of philosophy that deals with any form of human behavior (gr. ἦθος), political, legal or moral; strictly speaking, however, ethics is to be distinguished from both politics and law, being a branch of philosophy that specifically deals with the sphere of "good" or "bad" actions and not of those legally permitted or prohibited or those politically more appropriate" (Treccani Dictionary).

Therefore, an ethical evaluation consists of balancing the values at stake and consequently making a decision in the respect of the law that defines the boundaries of that evaluation (Assessment of ethical aspects, 2013 www.has-sante.fr).

Why ethical evaluation is important?

Using an animal model for research is a moral act for which a responsible approach is necessary.

The ethical evaluation bridges knowledge and action, being the moment of reflection that leads to the right decision-making. Beyond acting in a methodologically correct manner, an ethical evaluation also includes the capacity to argue the choices made (Foex, 2007).

Since the ethic evaluation is a multidisciplinary process that takes into account all the relevant elements at stake in a particular context, a list of possible solutions to different ethical challenges cannot be compiled. The existing guidelines and relevant bibliography can help making the decision faster and easier but each ethic evaluation needs to be tailored on a case-by-case basis. For example, the utilitarian vision mentioned earlier in this chapter admits the use of animals and accepts certain levels of pain and suffering provided a clear benefit for many. The correct harm-benefit assessment is one of the fundamental steps of the ethical evaluation because it entails the balance of the relevant values at stake in the decision-making in that context. This assessment also encompasses unbiased critical and self-critical capacity.

Although they were formulated to protect animals in intensive farming, the five fundamental freedoms reported in the Brambell Report of 1965 (lately formalized in the 1979 press statement by the UK Farm Animal Welfare Council and applied by the World Organization for Animal Health, the Royal Society for the Prevention of Cruelty to Animals, and the American Society for the Prevention of Cruelty to Animals) are equally valid and applicable in research. Unless deprivations are applied in experimentation as elements of the experimental protocol, the five fundamental freedoms are the following:

1. Freedom from hunger and thirst
2. Freedom from discomfort
3. Freedom from pain, injury, or disease
4. Freedom to express normal behavior
5. Freedom from fear and distress.

The five freedoms are the basic needs of every sentient being and they should be considered anytime animals are used in research.

Meaning of the intrinsic value

"All animals have an intrinsic worth, a worth that is independent of its value or use to man. This means that the interests of the animal, for example to avoid pain and suffering, must be respected. The animal deserves this respect because of what it is, because it is a living being that is able to experience pain and suffering" (https://www.uu.nl/en/research/ethics-institute).

In the introductory part of Article 12 of the Directive 2020/63/EU, it is stated that "Animals have an intrinsic value which must be respected." Intrinsic value means that the animal has value in itself, regardless of the benefits that the researcher may achieve. In stating "which must be respected" the Directive requires that the respect for animal welfare must be the guiding principle anytime animals are used for scientific purposes; this guiding principle is the driver for applying the replacement, reduction, and refinement (3Rs).

Reflections of animal suffering

Why animal suffering in science is more difficult to accept by many than that inflicted for meat production? As said above, the answer is likely because the benefits of experimentation are seldom immediately tangible. Showing the public the line of sight (sometimes very long!) between the laboratory work and the benefit for patients, consumers, etc., is challenging as it often entails explaining very complex science in simplest layman language. Therefore, it is the writer opinion that more efforts need to be spent in communicating science; only by doing so, acceptance and perception of research as a "common value" from the broader societal perspective will be possible (Rollin, 1989).

Legal aspects (Enrico Maestri)

The creation of an adequate legal framework for the conduct of experiments on animals requires a careful and responsible consideration of various interests at stake.

The conflicts between the obligation to maintain and improve human health, the valuable gains made possible by the freedom of science and the deep concern about preventing animal pain and suffering cannot be resolved through general judgments. Animal protection legislation in Europe calls for a careful examination and consideration of every single experiment on animals: this is a central element of the legislation (Olsson et al., 2016).

Freedom of scientific research is provided for in the Charter of Fundamental Rights of the European Union which states "The arts and scientific research are free" (Article 13). The freedom of science is an essential requirement for research and innovation to become an engine of development and well-being. At the same time, the freedom of research is not absolute since, as stated in the European Charter for Researchers (2005), they are required to "adhere to recognized ethical practices and fundamental ethical principles." On the other hand, the relationship between research and ethics raises complex questions, many of which are related to the specificity of the scientific field in which it operates.

On this basis, experiments on animals are expressly provided for in legislation on pharmaceutical products and international guidelines applicable to pharmaceutical testing (OECD guidelines or European Pharmacopeia in the case of alternative methods of quality control for batches of vaccines). The possible benefit for patients receiving a new treatment and the burden on the animals used in animal experimentation must thus be balanced against each other, taking into account that at present there is no existing replacement for animals in basic and translational research or in the development of new therapies (Yu, 2011).

The relationship between humans and animals is more contradictory than ever and lies in an area of tension between affection and personal interests. Up to now, animals have often been essential to research as a model organism, since experiments on animals provide important information about how drugs work and on the human toxicity of individual chemical products. This has given rise to a dilemma between health and safety requirements on the one hand and the moral need to protect laboratory animals on the other (Paul and Paul, 2001).

In this context, animal experimentation is subject to strict legal limits. Research projects are based—mandatorily—on compliance with the method of the 3Rs. The objective of the 3Rs is to assure that greater care is taken to avoid making experimental tests intolerably stressful for animals and to use or develop alternative methods to in vivo experiments (Kroeger, 2006).

Law and morals

When one speaks of bioethical issues related to animals, and when one feels a need to justify some form of legal legitimization of a subjective interest, it is important to distinguish between morals and law (Hart, 1961).

Distinguishing law from morals is essential in order to guarantee moral criticism and individual freedom (Hart, 1958).

The judgment of all individuals, based on which we decide what "the right thing to do" is, derives from the conscience of each one of us vis-à-vis that of others (we all answer to our own conscience): subjectivity becomes morality when it guarantees the existence of a number of conditions such as intersubjectivity, reciprocity, universalizability, and impartiality (Rawls, 1971).

Some, of course, argue that morality is a question of taste and thus strictly personal (emotivism), but even if one were to admit an argument of this kind, it would be precluded from being used (or considered viable) in the public sphere due to its absolute controvertibility.

The law, by contrast, regards a jurisdiction that is external to every individual (each one of us is answerable to an impartial judge): it is not positivized morality, and thus does not coincide with the moral standards of the majority; on the contrary, its jurisdiction is heteronomous and has nothing to do with what is right, but rather with what is valid, i.e., what is lawful based on a decision of an external entity (parliament, court) that is entitled to produce, promulgate, and apply law.

If law coincided with morality, we would end up with a totalitarian rule of the majority and there would no longer be any possibility of engaging in moral criticism against the law, given that law and morals would come to coincide (Mill, 1859).

Therefore, when we discuss bioethical questions in general, we must focus on the requirements intrinsic to law; a general and abstract nature first of all but—no less importantly—social pressure, effectiveness, and a collective sense of compulsoriness and joint responsibility.

If we failed to do so, we would risk making a serious epistemological error, that of mistaking every one of our desires for a right, every one of our demands, interests, or needs for a legal reason, wrongly identifying the "law that is" with the "law that should be" (Kramer, 1998).

The principle of the 3Rs and law: a legal genealogy

In general, the adjective "experimental" is used to refer to any method that enables a cause and effect relationship to be established between two phenomena/events on the basis of a hypothesis supported by already acquired knowledge and ultimately verified through experimental tests.

Experimentation should begin only if there are unfulfilled needs, i.e., when no good treatment yet exists or when the treatments currently used are poorly effective or cause many side effects (Rothman, 2008, 87–88).

A fundamental event marking the history of experimentation took place in the United States in 1937: a diethylene glycol (DEG) antifreeze caused the death of 107 people. An American pharmaceutical company had prepared a formula of sulfanilamide, a medicine used to treat streptococcal infections, using DEG as a solvent. DEG was poisonous to humans, but this fact was unknown to the pharmaceutical company, which added raspberry flavoring and marketed the product as Elixir Sulfanilamide. Unfortunately, the formula caused mass poisoning, resulting, precisely, in the death of over 100 people. After this tragedy, scientists administered the drug to some animals, which also died. The episode sent shock waves through the scientific community of that time and created the basis for a direct correlation between data originating from animal testing and their application in humans. In 1938, without questioning whether such experimentation could provide evidence that all animal species react in the same way to different chemical substances, the US Congress approved a law (Food, Drug, and Cosmetics Act) requiring pharmaceutical companies to test the safety of their products by conducting trials on animals. Since then, for the rest of the 20th century, research was largely founded on in vivo testing, then considered a model of reference for reliable predictive analysis (Orlans, 1998).

However, studies on animals should be avoided unless there is judged to be a real benefit for humans, and unless they are predictive for humans. Provided that animal protection laws are complied with and all possible forms of analgesia and anesthesia are used to spare animals from pointless cruelty, experimentation on animals can and must be ethically accepted (Moberg and Mench, 2000). Today this acceptance is fundamentally based on precise justificatory grounds: a very widely held theoretical notion that has been defined as the "priority of the human being" (Garattini and Grignaschi, 2017). According to this notion, animals are beings deserving of moral attention (moral patients), even though their interests are nonetheless deemed secondary compared to any competing interests of human beings (moral agents).

It is difficult to briefly outline the various moral reasons that may be put forward justify such an attitude in favor of human beings and their interests; nonetheless, the fact remains that, alongside the priority accorded to human interests, there is today also a widespread concern for animal welfare (Sunstein and Nussbaum, 2004). Consequently, those who follow this position in respect of experimentation recommend minimizing animal suffering within the framework of experimental practice. This goal may be pursued through the so-called principle of the 3Rs: replacement (to the extent possible, of trials that use animals with alternative research methods), reduction (in the number of animals sacrificed for research purposes), and refinement (adoption of strategies that minimize the suffering of animals used in experiments).

International and EU legal framework

European animal legislation presently shows three trends: firstly, a *constitutionalization* of animal protection and welfare principles has taken place with the adoption of constitutional provisions (e.g., Article 20a of the German Constitution). This trend, also reflected in other national constitutions, increases animal welfare rights (Special Eurobarometer 442, 2015). Secondly, a *semantic clarification* of the concept of animal has been imposed in civil law, which has called into question the traditional legal difference between animals and things originating from Roman law. Thirdly, the *supranationalization* of animal rights in the EU due to the transfer of legislative powers to the European Union in the sectors of agriculture and fishing, trade, the environment and consumer protection. The EU itself has thus become the competent legislator here: for example, through the approximation of European legislation on animal experimentation first with Directive 86/609EEC and then with Directive 2010/63/EU.

Council of Europe

At international level, in 1986, the Council of Europe adopted the "European Convention for the Protection of Vertebrate Animals used for Experimental and other Scientific Purposes" (ETS No 123/1986), which was subsequently revised through a series of amendments. The Convention is based on the conviction that there exists a "moral obligation" for human beings to "have due consideration" for animals' "capacity for suffering" and "memory," but at the same time the right to use animals in the quest "for knowledge, health and safety" remains.

The Convention provides that experiments may be conducted on animals only for certain purposes, including the diagnosis and prevention of disease, clinical research for pharmaceutical products, environmental protection, and basic research (see Article 2). Experiments may be carried out on animals for the specified research purposes only if they are essential in order to achieve the objective of the experiment and no alternative methods are available (cf. Article 6). In accordance with the principles of the 3Rs, the member States are encouraged to promote research into the development of alternative methods (cf. Article 6), so that over time animal experimentation can be replaced by other procedures. Where the experiments on animals are likely to cause enduring pain or significant harm to the animal, a special authorization or declaration is required (see Article 9). In order to avoid unnecessary repetition of procedures required by law on health and safety of products, the contracting Parties are also obliged, where possible, to mutually recognize the test results (cf. Article 29).

In 2010 a new "Directive 2010/63/EU of the European Parliament and of the Council of September 22, 2010 on the protection of animals used for scientific purposes" was adopted. It cancels and replaces Directive 86/609/EEC and confirmed the fundamental principles of the Convention.

European Union

The EU is authorized to adopt animal welfare regulations only to a limited degree: unlike environmental protection, animal welfare is not a goal of the EU and is therefore regulated by the Member States on the basis of national legislation. The EU can adopt animal welfare legislation that is binding for all the Member States if it prevents trade barriers and distortions of competition in the European single market. If the legislation on the keeping of farm animals and experimental animals differs in the individual EU countries, traders based in countries with stricter laws and regulations would be disadvantaged due to the increase in production costs in the internal market. In order to counter such inequalities and disadvantages, the EU can adopt animal welfare regulations that are binding for all the Member States. However, not all types of animal experimentation are of "relevance for the internal market" and those that are not cannot be regulated by the EU. If, on the other hand, the animal experiments are conducted in the framework of training and higher education or in basic university research according to the standards established by the EU, the issue is a relevant one and such experimentation is thus subject to regulation by the EU.

Treaty on the functioning of the European Union

In the "Treaty on the Functioning of the European Union" (TFEU, 2007), it is stated that "[i]n formulating and implementing the Union's [...] policies, the Union and the Member States shall, since animals are sentient beings, pay full regard to the welfare requirements of animals, while respecting the legislative or administrative provisions and customs of the Member States" (Title II, Article 13 TFEU). Although animal welfare is not one of the common goals of the EU, it has the same relevance as other principles mentioned in Title II of the TFEU, such as guaranteeing an adequate protection of health or promoting gender equality.

Article 13 TFEU states:

> In formulating and implementing the Union's agriculture, fisheries, transport, internal market, research and technological development, and space policies, the Union and the Member States shall, since animals are sentient

beings, pay full regard to the welfare requirements of animals, while respecting the legislative or administrative provisions and customs of the Member States relating in particular to religious rites, cultural traditions, and regional heritage.

On the one hand, animal welfare is recognized as having the same importance as the other principles mentioned in Title II of the TFEU, such as the promotion of gender equality, the guarantee of adequate social protection, the protection of health, the combating of discrimination, the promotion of sustainable development, consumer protection or the protection of personal data; on the other hand, animal protection is not an independent objective of the Union, but it has a significant impact on the policy sectors described. The concern for animals as sentient beings thus implies an institutional commitment to preventing them from suffering and recognition of their essential innate behaviors. It remains to be seen whether this improvement in the protection of animals, which is partially disputed in the legal literature, actually leads to consequences, in particular, as regards the rulings of the Court of Justice of the European Union.

Regulation (EC) No 1223/2009 on cosmetic products

The European Convention for the Protection of Vertebrate Animals Used for Experimental or Other Scientific Purposes (123/1986) was followed by Council Directive 86/609/EEC of November 24, 1986 on the approximation of laws, regulations, and administrative provisions of the Member States regarding the protection of animals used for experimental and other scientific purposes. This directive urged the development of activities aimed at identifying new in vitro methods and modifying some existing in vivo methods to reduce the number of animals used and minimize their suffering and harm.

This approach was also reflected in Directive 2003/15/EC, which had set precise limits on the sale of cosmetics tested on animals.

Regulation (EC) No 1223 of 2009 calls for the gradual and—ultimately—total elimination of the possibility of performing tests on animals for cosmetic products in Europe.

The new Regulation prohibits, in accordance with the 3Rs principle, animal testing of both finished products and ingredients or combinations of ingredients that will make up the finished product; the Regulation further prohibits the importation and marketing, in Europe, of products whose final formulation has been tested on animals. Moreover, since animal experiments cannot be completely replaced by an alternative method, it is necessary to specify whether the alternative method replaces the animal experiments partially or entirely [Commission Regulation (EC) No 440/2008 of May 30, 2008 laying down test methods pursuant to Regulation (EC) No 1907/2006 of the European Parliament and of the Council on the Registration, Evaluation, Authorization, and Restriction of Chemicals (REACH)].

Directive 2010/63/EU on animal experiments

Directive 2010/63/EU of the European Parliament and of the Council on the protection of animals used for scientific purposes, which came into force on November 9, 2010 and replaces the previously Directive 86/609/EEC, promotes the standardization and harmonization of national legislation concerning the protection of animals in research, a subject already addressed by the former directive. The national legislation of the Member States should aim to avoid, reduce, and improve the use of the animals for research purposes (cf. Article 4). Among the novelties of Directive 2010/63/EU, there is the inclusion of cephalopods, alongside vertebrates, as being worthy of attention (cf. Article 1, paragraph 3), as well as the broadening of the allowed purposes for carrying out experiments on animals, which include basic research, education and training, and forensic inquiries (cf. Article 5).

Like its predecessor, Directive 2010/63/EU establishes that Member States may, if necessary, adopt regulations and laws that go beyond the minimum standards defined by EU legislation; these may include, among other things, potential regulations on the procedures for the approval of animal experiments or monitoring of public opinion on the issue of animal welfare (cf. Article 43, which regards the publication of nontechnical summaries of authorized projects).

Under Directive 2010/63/EU (cf. Article 4, Articles 46–49), as well as the corresponding Convention of the Council of Europe (cf. Article 6), Member States of the European Union are called upon to seek and promote alternative methods to replace in vivo animal experimentation. To that end, organizations have been set up at the European and national levels, their task being to evaluate alternative and supplemental methods: avoiding duplication of tests should also help reduce the number of tests on animals. At a European level, the European Center for the Validation of Alternative Methods (EURL-ECVAM, European Union Reference Laboratory for alternatives to animal testing. Website: http://ecvam-sis.jcr.it/) provides information about already well-established replacement approaches and validated supplementary and alternative methods.

The 3Rs principle

The 3Rs have become a common point of reference for the EU Member States (Vessier et al., 2008) and for a wide variety of organizations and committees whose goal is to avoid, to the extent possible, animal experimentation or to improve housing and use conditions for laboratory animals; the principle is enshrined in the legislation of many EU countries (Cozigou et al., 2015) through Directive 2010/63/EU on animal experiments, which calls for the application of the "principle of replacement, reduction and refinement."

It is worth quoting a part of Article 4, which is entitled "Principle of replacement, reduction, and refinement":

> Paragraph 1: Member States shall ensure that, wherever possible, a scientifically satisfactory method or testing strategy, not entailing the use of live animals, shall be used instead of a procedure.
> Paragraph 2: Member States shall ensure that the number of animals used in projects is reduced to a minimum without compromising the objectives of the project.
> Paragraph 3: Member States shall ensure refinement of breeding, accommodation, and care, and of methods used in procedures, eliminating or reducing to the minimum any possible pain, suffering, distress, or lasting harm to the animals.

Each Member State further has the obligation of identifying "single points of contact" (Article 47), which have the task of coordinating working groups at a national level and sharing their work with the respective authorities and the EU Reference Laboratory for alternatives to animal testing (EURL-ECVAM), with the aim of facilitating contacts between the parties, shortening the time frame for the validation of alternative methods, and assuring a significant approach in the adoption of 3Rs strategies. The EURL-ECVAM moreover specifies alternative methods to animal experimentation, in which the 3Rs principle is applied: an experimental procedure thus represents an alternative to animal testing only if it is capable of replacing animals, reducing the number of animals, or refining the method to minimize the stress.

In 2005 the European Commission entered into a partnership with enterprises in various industrial sectors with the aim of developing alternatives to animal experimentation [European partnership for alternative approaches to animal testing (EPAA) https://ec.europa.eu/growth/sectors/chemicals/epaa_en].

The founding document of the EPAA is the so-called "3Rs Declaration."

The 3R′ principles have been incorporated into national and international legislation regulating the use of animals in scientific procedures with the aim of implementing more humane experimental approaches.

Through the application of the 3Rs in the realm of animal experimentation for scientific research purposes, it is hoped to improve the understanding of the human–animal relationship; moreover, researchers can ask themselves whether there exists "something" that separates us from animals and whether that "something" is sufficient to determine how we should treat them during an experimental stage. The legal reinforcement of the 3Rs principles sets an obligation for researchers, irrespective of their subjective will: they "have to" respect the 3Rs, even if they do not "feel" themselves to be obliged, and despite the fact that it is concretely difficult to respect them.

An example of legal challenges in a European member State: the case of Italy

The extension of our emphatic sensitivity toward animals recorded in Italy in recent decades has only partially gone hand in hand with of the need to acquire a greater awareness of animal rights and of researchers' duty to ensure the laboratory animal welfare.

As was natural, given the highly controversial issue involved, the new Italian legislation has spurred much debate. Are the new rules better or worse than the previous ones?

According to the European Commission, in agreement with the Italian scientific community, Legislative Decree no. 29 of 2014 (which transposed the European directive) establishes excessive limitations and violates European legislation, as it adopts overly strict criteria, following the entry into force of the directive itself (cf Table 1.1 below for details). In view of the incorrect transposition of Directive 2010/63, the European Commission initiated infringement proceedings in the European courts, demanding that Italy be fined.

However, a large part of the public views the new Italian law as an improvement over the European directive because it prohibits experiments on great apes (Annex A of EU) Regulation No. 750/2013, whereas the European directive allows exceptions, prohibits experiments for the production and control of war material, prohibits the reuse of animals in experiments causing a severe level of pain, and introduces more severe administrative and criminal sanctions for those who violate the minimum standards established by law (cf article 40).

The regulatory intervention of the Italian legislator aims to guarantee a higher level of protection by establishing precise requirements and conditions for animal breeding activities and the supply and use of animals, with a view to progressively reducing their use and ultimately completely replacing animals with alternative practices and methods.

The 3Rs principle is explicitly recognized as a tool geared toward the implementation of alternative methods to the use of in vivo animal models.

In recital 6 of Directive 2010/63, it is stated that:

> New scientific knowledge is available in respect of factors influencing animal welfare as well as the capacity of animals to sense and express pain, suffering, distress, and lasting harm. It is therefore necessary to improve the welfare of animals used in scientific procedures by raising the minimum standards for their protection in line with the latest scientific developments (6).

Therefore, it is necessary to take into account two conditions, laid down in recitals 10 and 11:

> While it is desirable to replace the use of live animals in procedures by other methods not entailing the use of live animals, the use of live animals

TABLE 1.1 European Directive 63/2010 versus Italian Legislation 26/2014.

European Directive 63/2010	Italian Legislation 26/2014	Comparison
Prohibitions		
Article 2, paragraph 1 Stricter national measures 1. Member States may, while observing the general rules laid down in the TFEU, maintain provisions in force on November 9, 2010, aimed at ensuring more extensive protection of animals falling within the scope of this Directive than those contained in this Directive.	Article 13 Delegated Legislation 96/2013. Criteria for delegation to the Government for the transposition of Directive 2010/63/EU of the European Parliament and of the Council of September 22, 2010 on the protection of animals used for scientific purposes. 1. In exercising the delegation for the implementation of Directive 2010/63/EU of the European Parliament and of the Council of September 22, 2010 on the protection of animals used for scientific purposes, the Government is required to follow, in addition to the principles and guiding criteria referred to in Article 1, paragraph 1, also the following principles and specific guiding criteria: [...]	The national delegated rules, subsequent to the entry into force of the directive, provide for more restrictive measures than those set by European legislation.

Continued

TABLE 1.1 European Directive 63/2010 versus Italian Legislation 26/2014.—cont'd

European Directive 63/2010	Italian Legislation 26/2014	Comparison
Article 2, paragraph 2 2. When acting pursuant to paragraph 1, a Member State shall not prohibit or impede the supply or use of animals bred or kept in another Member State in accordance with this Directive, nor shall it prohibit or impede the placing on the market of products developed with the use of such animals in accordance with this Directive.		Legislative gap
Article 14 Anesthesia 1. Member States shall ensure that, unless it is inappropriate, procedures are carried out under general or local anesthesia, and that analgesia or another appropriate method is used to ensure that pain, suffering, and distress are kept to a minimum.	Article 14 Anesthesia 1. Procedures that do not involve anesthesia or analgesia are prohibited, if they cause intense pain following serious injury to the animal, with the exception of procedures for testing anesthetics and analgesics.	The European Directive is not respected. (In the Parliament it is currently under discussion the amendment of art. 14 of Legislative Decree no. 26/2014 to prohibit all procedures that cause intense pain following serious injury to the animal, which do not require anesthesia or analgesia. The original text provides for an exception to this prohibition: see left column).
	Article 5 Purposes of the procedures 2. Procedures cannot be authorized: a) for the production and control of war material; b) for toxicological tests with the protocols of the Lethal	The Italian legislation provides for restrictive measures not provided (or only partially) for by European legislation.

	Dose-LD50 and of the Lethal Concentration-LC50, except in cases where it is mandatory by national or international legislation or pharmacopoeia; **c)** for the production of monoclonal antibodies through the induction of ascites, if other corresponding production methods exist and are not mandatory by national or international legislation or pharmacopoeias; **d)** for research on xenotransplants (referred to in article 3, paragraph 1, letter q); **e)** for research on drugs of abuse; **f)** during the didactic exercises carried out in primary, secondary schools, and university courses, with the exception of university training in veterinary medicine as well as higher university training for doctors and veterinarians.	

Continued

TABLE 1.1 European Directive 63/2010 versus Italian Legislation 26/2014.—cont'd

European Directive 63/2010	Italian Legislation 26/2014	Comparison
Article 1 Subject matter and scope 2. This Directive shall apply where animals are used or intended to be used in procedures, or bred specifically so that their organs or tissues may be used for scientific purposes.	Article 10 Animals used in procedures 5. The breeding of dogs, cats, and nonhuman primates for the purposes referred to in this decree is prohibited.	Due to Article 10, paragraph 5 and Article 1, paragraph 2 of the European directive is not faithfully implemented. The EU Directive provides for exceptions to experimental use on nonhuman primates.
Alternative Approaches (3R)		
Article 4 Principle of replacement, reduction, and refinement 1. Member States shall ensure that, wherever possible, a scientifically satisfactory method or testing strategy, not entailing the use of live animals, shall be used instead of a procedure. 2. [...] 3. [...].	See Article 1, letter a, Article 13, Article 26, Article 31, and Article 37	The current legislation has not transported in a corresponding and explicit mode, the Article 4 of Directive 63/2010.
OGM		
Article 3 Definitions For the purposes of this Directive the following definitions shall apply:	Article 10 Animals used in procedures 4. The breeding of genetically modified animals is permitted subject to an assessment of the relationship between harm and benefit, the actual need for handling, the possible impact it could have on animal welfare and the potential risks for human, animal health and for the environment.	The Italian legislation provides for more restrictive measures than those provided by the European directive.

> continues to be necessary to protect human and animal health and the environment. However, this Directive represents an important step toward achieving the final goal of full replacement of procedures on live animals for scientific and educational purposes as soon as it is scientifically possible to do so (10).
>
> The care and use of live animals for scientific purposes is governed by internationally established principles of replacement, reduction, and refinement (11).

And Article 1, paragraph 1, letter (a) of Legislative Decree 26/2014 affirms that:

1. The present decree establishes measures relating to the protection of animals used for scientific or educational purposes; to this end, the following aspects are considered: (a) replacement, the reduction in the use of animals in procedures and the refinement in the techniques whereby animals are bred, housed, cared for, and used in procedures."

It should be highlighted that Article 2 of the directive precludes the introduction of national measures that are stricter than those contained in the directive itself. National measures providing for a higher level of protection than those in the directive could be maintained, provided that they were already in force on November 9, 2010, but in such a case, the Member States concerned would have to inform the Commission of such measures by January 1, 2013. The examination of Article 13 of Italian Law 96/2013—which delegated the implementation of the European directive to the government—and its subsequent approval gave rise to a heated debate in the competent committees in the Chamber of Deputies and Senate, as well as within the scientific community, precisely because some of the rules of delegation established therein are more restrictive than those laid down in the European legislation. The legislative decree concerned, despite literally reproducing a large part of the directive, introduces a discretionary part with a higher, more restrictive level of regulation compared to the European legislation.

More specifically, the following provisions were introduced, which diverged from the European legislative provisions: prohibition of experiments and procedures that do not provide for anesthesia or analgesia, when they cause pain to the animal, except in cases where anesthetics or analgesics are being tested (Article 14, paragraph 1); prohibition against the production and control of war material; prohibition of toxicological tests with the lethal dose-LDSO and lethal concentration-LCSO protocols, except where they are required by national or international legislation or pharmacopeias; prohibition against the production of monoclonal antibodies through the induction of ascites, where other production methods exist and where it is not mandatory under national or international legislation or pharmacopeias; prohibition of research on xenotransplants, previously described as transplants of one or more organs from one animal species to another; prohibition of research on abused substances; prohibition of certain educational activities in primary and secondary schools and in university courses, with the exception of courses held in veterinary medicine faculties, and advanced university training for physicians and veterinarians (Article 5, paragraph 2); prohibition against breeding, but not against using, dogs, cats, and nonhuman primates intended for experimentation within the national territory (Article 10, paragraph 5).

It should be noted that the prohibition against breeding dogs, cats, and nonhuman primates intended for experimentation within the national territory, as provided in Article 1 of the decree (scope of application), does not faithfully transpose the corresponding Article 1 of the directive. It does not in fact transpose paragraph 2 of the latter article, which, in specifying the cases excluded from the provision, states as follows:

> This Directive shall apply where animals are used or intended to be used in procedures, or bred specifically so that their organs or tissues may be used for scientific purposes.

It is worth highlighting that whereas Legislative Decree 116/1992 classified any use of animals apt to cause pain, suffering, distress, or lasting harm as "experimentation," the decree, like the directive, excludes from the scope of application procedures causing less pain than that caused by the introduction of a needle, thereby establishing a more objective threshold for the assessment of pain, which is at the same time one of the most complex aspects to quantify and one of the parameters mainly analyzed by the regulatory authority before issuing the necessary authorizations for carrying out experiments on animals.

It should also be stressed that the current decree did not faithfully and explicitly transpose Article 4 of Directive 63/2010 (Principle of replacement, reduction, and refinement), though in many articles (e.g., Article 1, letter a, Article 13, Article 26, Article 31, and Article 37) reference is made to the 3Rs principle and experimental researchers are called on to implement it.

In any case, the Italian legislation failed to transpose Article 4 of the Directive, which reads:

1. Member States shall ensure that, wherever possible, a scientifically satisfactory method or testing strategy, not entailing the use of live animals, shall be used instead of a procedure.
2. Member States shall ensure that the number of animals used in projects is reduced to a minimum without compromising the objectives of the project.
3. Member States shall ensure refinement of breeding, accommodation, and care, and of methods used in procedures, eliminating or reducing to the minimum any possible pain, suffering, distress, or lasting harm to the animals.

Article 24 of the Italian decree provides that every breeder, supplier, or user must have a designated veterinarian, who is trained in laboratory animal medicine and possesses specific experience and training.

As under the Directive (Article 26), Article 25 requires each breeder, supplier, or user to set up a specific animal welfare body (OPBA) including at least the person or persons responsible for the welfare and care of the animals, the veterinarian and, in the case of a user, a scientific member. The next paragraphs 3 and 4 establish that small breeders, suppliers, and users may rely on a body operating in an establishment other than their own and that if a breeding or supplier establishment has also been authorized to engage in use of the animals, it may rely on a single animal welfare body.

The discussion on the suitable composition of this body has given rise to two different interpretative trends which, for the sake of simplicity, we could define respectively as "literalist" and "contextualist" (Martini et al., 2015). Those aligned with the former advocate a strict adherence to the letter of Article 25: accordingly, the OPBA should be made up of the person responsible for the welfare of the animals, the veterinarian, and (only in the case of user establishments) also a scientific member. This interpretation must be rejected, as it neglects to consider the existence of a clear conflict of interest between the person responsible for animal welfare (and thus for the facility) and the person in charge of a research project. Those adhering to the latter trend, by contrast, focus on the multidisciplinary character of the OPBA with the aim of identifying the different issues involved (scientific, social, ethical, legal, etc.) and providing all the necessary elements for forming an opinion.

The present composition of the OPBA should be better defined and a number of scientific members should be included to ensure interdisciplinarity in the competences and the specific knowledge relating to the 3Rs and animal welfare, which a single scientific member, however expert, might not possess in relation to specific disciplines; indeed, many OPBAs rely on external expert consultants. It should be noted in any case that reliance on qualified external experts who are specialized in given disciplines offers an assurance of impartiality and independence in the assessment of experimental protocols. A good compromise could be to require that one of the scientific members be a person recruited from outside the institution/enterprise who has specific, certified training in the field of the 3Rs and animal welfare; that a number of veterinarians be included in the case of institutions with a complex organization; that an expert in statistical sciences be included for the purpose of evaluating the quality of the statistical analysis and the scientific efficiency, that is, the possibility of obtaining valid results using the smallest possible number of animals; as well as an expert with competences of an ethical-legal or bioethical nature and an expert in alternative methods.

Finally, it should be highlighted that, pursuant to Legislative Decree 26/2014, the members of the OPBA, with the exclusion of the scientific member, are members by right as persons responsible for welfare and the designated veterinarian and they are thus internal personnel of the institution/enterprise they work at.

OPBAs set up in a public institution or an institution governed by public law are to be qualified as collegial bodies of an essentially technical nature whose tasks include providing assessments and consultation and expressing reasoned opinions on research projects involving animal experimentation; such opinions are also binding for researchers who, in the event of a rejection, cannot conduct the experimentation in the facility overseen by the OPBA. Finally, they are further tasked with supervising and monitoring the establishments and following the outcome of research projects and must offer utmost guarantees of independence and absence of conflicts of interest, above all in their relations with the researchers conducting the experiments.

Among the tasks of the OPBA as per Article 26 of the decree, it is again worth mentioning: the expression of a reasoned opinion on research projects and on any subsequent modifications, of which the project manager must be notified (paragraph 1, letter d), as well as the submittal of applications for the authorization of research projects, of which the Principal Investigator must again be notified (paragraph 1, letter f). The subsequent paragraph 2 specifies the elements that must evaluated for the purpose of expressing the opinion; these include the correct application of animal protection provisions, the technical and scientific relevance of the project and the possibility of replacing one or more procedures with alternative methods.

The body records the consultation provided and the relevant decisions in specific registers, which it must keep for a period of at least six years (three years pursuant to Article 26, paragraph 2 of the directive).

The OPBAs were established as a form of public guarantee of animal protection and welfare. The main criteria guiding the activities of OPBAs in their assessment of research projects are: the expected level of suffering of the animals used in experiments and the possibility of relying on alternative methods.

In contrast with what was established by the directive, pursuant to Article 31, the animal welfare body is responsible for submitting the application by certified electronic means to the Ministry and sending a copy to the competent health authority (according to the directive, responsibility lies with the user or the person responsible for the project).

In paragraph 4, letter r, item 1 of Article 31, there is explicit reference to the application of the 3Rs principle:

> (r) of the presence of personnel with specialist competences in the following areas:
> (1) sectors of scientific application in which the animals will be used, with particular regard for the objective of replacement, reduction, and refinement.

In accordance with Community and national legislation, the OPBA thus has the primary function of implementing methods that do not use animals for the majority of biomedical research projects and, where the use of animals is necessary, of monitoring the development and application of rigorous methods of refinement and reduction.

Even where it is impossible to achieve the objective of *replacement*, it is nonetheless necessary to analyze and seek to achieve the objectives of *reduction* and *refinement*. The analysis of the OPBA focuses most closely on the technical assessment, which is thus the aspect that warrants the greatest care. A challenge arises when it comes to arguing, in objective terms, for the application of a reduction; there are two solutions in this regard: the first is to use power analysis, which demonstrates that the researcher is making a great effort to optimize the use of the experimental animals, and enables an evaluation of the *power analysis* (G* Power; www.gpower.hhu.de/), i.e., the probability of revealing a significant difference, as well as enabling the OPBA to verify the exactness of the number of animals it is planned to use. The second solution is to follow ARRIVE Guidelines (*Animal Research: Reporting of In Vivo Experiments.* https://arriveguidelines.org/) for determining the sample size:

a. Specify the total number of animals used in each experiment, and the number of animals in each experimental group;
b. Explain how the number of animals was arrived at. Provide details of any sample size calculation used;
c. Indicate the number of independent replications of each experiment, if relevant.

(Reporting animal research: Explanation and Elaboration for the ARRIVE guidelines 2019)

Network of committees

Ethics committees focused on animal experimentation still remain an isolated phenomenon: they are mostly separate entities. They are independent bodies that are not hierarchically subordinate to anyone, and precisely for this reason they can decide whether to add to their institutional aims research interests they consider to be consistent with their mandate.

In addition to performing supervisory functions, Animal Welfare Body could effectively become a multicenter, multidisciplinary research resource, undertaking a monitoring function and ensuring the compatibility of the adopted guidelines, which the existing legislation defines as a specific responsibility.

In the European context, an initiative was recently launched to start up an exchange, at the Community level, by establishing an EU platform on animal welfare. One of the key priorities of the European Commission is to promote a stronger dialog on animal welfare issues that are relevant at EU level among competent authorities, businesses, civil society, and scientists.

To achieve these priorities, the platform assists the Commission in the development and exchange of coordinated actions on animal welfare, with a particular focus on a better application of EU rules in this area, through exchanges of information and best practices and the direct involvement of stakeholders; the development and use of voluntary commitments by businesses to further improve animal welfare; and the promotion of EU welfare standards to increase the market value of the Union's products at the global level.

Furthermore, again in a European context, it is worth pointing out the follow platforms and organizations (though the list is by no means exhaustive):

The EURL-ECVAM (European center for the validation of alternative methods), which published the latest status report on the development, validation, and the regulated acceptance of alternative methods and approaches (2019); the RepRefRed Society (Society for the Promotion of Alternative Bio-Model. https://www.reprefred.eu/)—a new Three Rs organization—has been added to Norecopa (Norway's National Consensus Platform for the replacement, reduction, and refinement of animal experiments. https://norecopa.no/), a Norwegian national consensus platform for the advancement of the 3Rs in connection with animal experiments; the PREPARE guidelines (Planning Research and Experimental Procedures on Animals: Recommendations for Excellence), which are complementary to the ARRIVE guidelines, are by now a point of reference for all the organizations that focus on animal experimentation; the ETPLAS (https://www.etplas.eu/), the Education and Training Platform for Laboratory Animal Science; the EPAA; the ESLAV (European Society of Laboratory Animal Veterinarians. https://www.eslav.org/), while FELASA (European Federation of Laboratory Animal Associations. http://www.felasa.eu/) is a federation of societies. FELASA (Federation of European Laboratory Animal Science Associations) represents common interests in the furtherance of all aspects of laboratory animal science, and enjoys the greatest political influence with the European Commission, having participated in the preparatory work on Directive 2010/63. The EARA (European Animal Research Association. https://www.eara.eu/) are platforms which, thanks to the public and private bodies that support them, engage in lobbying actions at both the national and European levels. The NC3Rs (National Center for the Replacement Refinement and Reduction of Animals in Research. https://www.nc3rs.org.uk/) and the Norecopa are centers that receive government support and play an important role in providing information and training, as well as funding.

Conclusions

A diligent implementation of the 3Rs, although not sufficient on its own, is therefore a necessary prerequisite for the justification of animal experiments (Maestri, 2021). Irrespective of the degree of stress inflicted on animals, the study design must fulfill strict scientific quality requirements in terms of objectivity, validity, and repeatability. It is a fundamental responsibility of every researcher to ensure scientific quality in order to obtain the maximum scientific significance. Moreover, this goal must always be the central criterion of the scientific evaluation of research projects (Hawkins and Bertelsen, 2019).

The 3Rs principle can fulfill its role only if an appropriate research plan is available and if the reduction principle is not misconstrued so that the number of animals is reduced to such a degree as to impair the study's informational value. Should this occur, the principle of the 3Rs would be reduced to magical formula whose legal enforcement would make it mandatory without researchers being able to understand, in their "internal legal culture" (Friedman, 1975), the reasons why it must be observed.

References

Animal (Scientific Procedures) Act, 1986. https://www.legislation.gov.uk/ukpga/1986/14/contents.

Brambell, R., 1965. Report of the Technical Committee to Enquire into the Welfare of Animals Kept Under Intensive Livestock Husbandry Systems. Her Majesty's Stationary Office, London.

Cozigou, G., et al., 2015. The European partnership for alternative approaches to animal testing (EPAA): promoting alternative methods in Europe and beyond. J. Am. Assoc. Lab Anim. Sci. 54 (2), 209–213.

European Charter of Researcher, 2005. 2013 March; 3 (1), 238–273. https://euraxess.ec.europa.eu/euraxess/charter-code-researchers Animals (Basel).

Foex, B.A., 2007. The ethics of animal experimentation. Emerg. Med. J. 24 (11), 750–751.

Franco, N.H., 2013. Animal Experimentation in Biomedical Research: A Historical Perspective.

Friedman, L., 1975. The Legal System. A Social Science Perspective. Russell Sage Foundation, New York.

Garattini, S., Grignaschi, G., 2017. Animal testing is still the best way to find new treatments for patients. Eur. J. Intern. Med. 39, 32–35.

Harrison, R., 1964. Animal Machines: The New Factory Farming Industry. CABI, UK.

Hart, H.L.A., 1958. Positivism and the separation of law and morals. Harv. Law Rev. 77, 593–629.

Hart, H.L.A., 1961. The Concept of Law. Clarendon Press, Oxford.

Hawkins, P., Bertelsen, T., 2019. 3Rs-Related and objective indicators to help assess the culture of care. Animals 9, 969.

Kroeger, M., 2006. How omics technologies can contribute to the '3R' principles by introducing new strategies in animal testing. Trends Biotechnol. 24 (8), 343–346.

Kramer, M.H., 1998. A Debate Over Rights: Philosophical Enquiries. Clarendon Press, Oxford.

Maestri, E., 2021. The 3Rs principle in animal experimentation: a legal review of the state of the art in Europe and the case in Italy. BioTech 10, 9. https://doi.org/10.3390/biotech10020009.

Moberg, G., Mench, J.A. (Eds.), 2000. The Biology of Animal Stress. Basic Principles and Implications for Animal Welfare. CABI Publishing, Wallingford.

Martini, M., et al., 2015. An ethics for the living word: operation methods of Animal Ethics Committees in Italy. Ann. Ist. Super Sanita 3, 244–247.

Mill, J.S., 1859/1982. On Liberty. Penguin Random House, London.

Olsson, A.S., et al., 2016. Protecting animals and enabling research in the European Union: an overview of development and implementation of Directive 2010/63/EU. ILAR J. 57 (3), 347–357.

Orlans, F.B., 1998. History and ethical regulation of animal experimentation: an international perspective. In: Kuhse, H., Singer, P. (Eds.), A Companion to Bioethics. Blackwell, Oxford.

Paul, E.F., Paul, J., 2001. Why Animal Experimentation Matters. The Use of Animals in Medical Research. Translation Publisher, New Brunwick.
Petetta, F., Ciccocioppo, R., 2020. Public Perception of Laboratory Animal Testing: Historical Philosophical, and Ethical View. Addiction Biology. Wiley.
Rawls, J., 1971. A Theory of Justice. Harvard University Press, Cambridge.
Regan, T., 1983. The Case for Animal Rights. University of California Press, Berkeley.
Rollin, B.E., 1989. The Unheeded Cry: Animal Consciousness, Animal Pain and Science. Oxford University Press 89 e Iowa Univesity Press 1988.
Rothman, K.J., et al. (Eds.), 2008. Modern Epidemiology. Lippincott Williams & Wilkins, Philadelphia.
Ruesch, H., 1978. Slaughter of the Innocent. Bantam Books, New York.
Silverman, B.D., 1996. Claude Bernard. Clin. Cardiol. 19, 916–918.
Singer, P., 1975. Animal Liberation: A New Ethics for Our Treatment of Animals. HarperCollins, New York.
Special Eurobarometer 442, 2015. Attitudes of EU citizens towards Animal Welfare.
Sunstein, C.R., Nussbaum, M. (Eds.), 2004. Animal Rights. Current Debates and New Directions. Oxford University Press, Oxford.
Vessier, I., et al., 2008. European approaches to ensure good animal welfare. Appl. Anim. Behav. Sci. 113, 279–297. http://ec.europa.eu/COMMFrontOffice/PublicOpinion.
Yu, D., 2011. Translational research: current status, challenges and future strategies. Am. J. Transl. Res. 3 (5), 422–433. http://web.archive.org/web/20061214034848/http://homepage.tinet.ie/~pnowlan/Chapter-77.htm.

Recommended readings

Annas, G.J., Grodin, M.A. (Eds.), 1992. The Nazi Doctors and the Nuremberg Code: Human Rights in Human Experimentation. Oxford University Press, New York.
Balls, M., et al. (Eds.), 2000. Progress in the Reduction, Refinement and Replacement of Animal Experimentation. Elsevier.
Bobbio, N., 1995. The Age of Rights. Polity press, London.
De Angelis, I., et al., 2019. The 3R principle: 60 years taken well. Preface. Ann Ist Sup Sanità.
De Waal, F., 1989. Chimpanzee and Politics.
De Waal, F., 2007. Chimpanzee Politics: Power and Sex Among Apes. Johns Hopkins University press.
Dolan, K., 1999. Ethics, Animals, and Science. Blackwell, Oxford.
Francione, G.L., 2008. Animals as Persons. Essays on the Abolition of Animal Exploition. Columbia University Press, New York.
Hartung, T., 2017. Opinion versus evidence for the need to move away from animal testing. ALTEX 34, 193–200.
Hegele, R.A., Maltman, G.M., 2020. Insulin's centenary: the birth of an idea. Lancet Diabetes Endocrinol. 8 (12), 971–977.
Herrmann, K., Jayne, K., 2019. Animal Experimentation: Working Towards a Paradigm Change. Brill.
Hubrecht, R.C., Carter, E., 2019. The 3Rs and humane experimental technique: implementing change. Animals.
Kant, I., 1775-1780/1997. In: Heath, P. and Schneewind, J.B., Lectures on Ethics, Cambridge University Press, Cambridge.

Kuhse, H., Singer, P. (Eds.), 2001. A Companion to Bioethics. Blackwell, Oxford.
LaFollette, H., Shanks, N., 2016. Brute Science Dilemma on Animal Experimentation. Routledge.
Levinovitz, A.W., Ringertz, (Eds.), 2001. In a chapter of "The Nobel Prize: The First 100 Years", Imperial College Press Editor and World Scientific Publishing Co. Pte. Ltd.
Lewis, D.I., 2019. Animal experimentation: implementation and application of the 3Rs. Emerg Top Life Sci. 3 (6), 675–679. https://doi.org/10.1042/ETLS20190061.
Midgley, M., 1983. Animals and Why They Matter. A Journey Around the Species Barrier. The University of Georgia Press, Athens.
Monamy, V., 2000. Animal Experimentation. A Guide to the Issues. Cambridge University Press, New York.
Moore, G.E., 1903/1993. Principia Ethica. Cambridge University Press, Cambridge.
Niklaus, S., Albertini, S., Schnitzer, T., et al., 2020. Challenging a myth and misconception: red-light vision in rats. Animals 10 (3), 422.
Nuffield Council on Bioethics, 2005. The Ethics of Research Involving Animals, London. http://www.nuffieldbioethics.org.
Raz, J., 1999. Practical Reasons and Norms. Princeton University Press, Princeton.
Russell, W.S., Burch, R.L., 1959. The Principles of Human Experimental Technique. Methuen, London.
Schiffelers, M.-J.W.A., et al., 2012. Regulatory acceptance and use of 3R models: a multilevel perspective. ALTEX 29, 287–300.
Schuppli, C.A., Fraser, D., 2005. The interpretation and application of the three Rs by animal ethics committee members. Altern Lab Anim.
Sepahban, L., 2015. Animal Testing: Lifesaving vs. Animal Welfare. Compass Point Books, North Mankato.
Singer, P., 1979. Practical Ethics. Cambridge University Press, Cambridge.
Spaemann, R., 1997/2017. Persons. The Difference Between 'Someone' and 'Something'. OUP, Oxford.
Tannenbaum, J., Bennett, B.T., 2015. Russell and Burch's 3Rs then and now: the need for clarity in definition and purpose. Journal of the American Association for Laboratory Animal Science 120–132.
UNESCO Animal Right Declaration, 1978. http://www.esdaw.eu/unesco.html.
Van De Veer, D., 1979. Interspecific justice. Inquiry 22, 55–79.
Weiss, R., 1978. The perils of personhood. Ethics 66–75.

Chapter 2

Communication and transparency in research involving animals

Lluis Montoliu[1], Annarita Wirz[2]
[1]*National Centre for Biotechnology (CNB-CSIC) and Biomedical Research Networking Center on Rare Diseases (CIBERER-ISCIII), Madrid, Spain;* [2]*Santa Lucia Foundation IRCCS, Rome, Italy*

Introduction (Lluis Montoliu)

Historically, the animal research community has been reluctant to discuss the benefits of animal experimentation in public, knowing how sensitive this issue is for the society. However, their absence from debates has been filled by the animalist groups, who have always been very keen to communicate their views to the public. This unbalanced situation changed in 2014, when the first Concordat on Openness on Animal Research was launched in United Kingdom, followed by similar agreements in other European countries released during the following years. This chapter will summarize these efforts toward transparency and communication in animal research.

The scientific community using animal models in their research projects ignored for too many years their duty to explain to the society why animal research was relevant in scientific projects. I wish to state this clearly from the very beginning: researchers must communicate to the public their needs to use animal models in their research projects, as well as inform about the results that could be obtained thanks to involving animals in their experiments. The use of animals in research continues to be a sensitive issue for many people in our society. Therefore, the reasons why animal research is an essential part of many experiments where no alternative are available need to be provided.

There are many reasons to infer why this behavior has been adopted, including the fear for aggressive responses from animalist groups, one being the reluctancy to place themselves under an uncomfortable spotlight; another reason has been the wrong assumption that society would normally understand that animal research was a price to pay for a much greater benefit in terms of human and animal health.

Practical Handbook on the 3Rs in the Context of the Directive 2010/63/EU
https://doi.org/10.1016/B978-0-12-821180-9.00003-9

Animal research is one of the most regulated scientific activities. There are international and national laws governing these experiments, which can only be approved after a strict compliance to all requisites demanded by these norms. In the Europe Union, animal research can only occur in registered centers, with approved animal facilities, by competent and adequately trained personnel, educated in the scientific and ethical aspects of animal procedures, following the approval of Ethics Committees (depending on the EU country) and the eventual authorization of the competent administrative authority. Researchers planning to involve animals in their experiments must justify their use and, more precisely, they must investigate whether validated alternative methods are already available for those procedures. And, if there are, they must use the existing alternative methods and avoid using animals.

Any researcher/user willing to be involved with animal research in their experiments must comply with the accepted international ethical standards dictating animal welfare. All animal legislation is based on the replacement, refinement, and reduction (3Rs) principle, established by W. M. S. Russell and R. L. Burch in 1959.

The ARRIVE guidelines (Lluis Montoliu)

The Animal Research: Reporting of In Vivo Experiments (ARRIVE) guidelines had been developed by the National Center for the Replacement, Refinement, and Reduction of Animals in Research (NC3Rs, UK) to improve the design, analysis, and publication of research involving animals, aiming to maximize the reported information and to minimize redundant studies.

The updated version of the ARRIVE guidelines (Percie du Sert et al., 2020) includes a group of 10 basic recommendations to follow to optimally design and communicate the results derived from any experiment involving animals, irrespective of the results obtained.

These are (extracted from the Percie du Sert et al., 2020 publication):

1. **Study design:** For each experiment, provide brief details of study design including: the groups being compared, including control groups. If no control group has been used, the rationale should be stated. The experimental unit (e.g., a single animal, litter, or cage of animals).
2. **Sample size:** Specify the exact number of experimental units allocated to each group, and the total number in each experiment. Also indicate the total number of animals used. Explain how the sample size was decided. Provide details of any a priori sample size calculation, if done.
3. **Inclusion and exclusion criteria:** Describe any criteria used for including and excluding animals (or experimental units) during the experiment, and data points during the analysis. Specify if these criteria were established a priori. If no criteria were set, state this explicitly. For each experimental group, report any animals, experimental units, or data

points not included in the analysis and explain why. If there were no exclusions, state so. For each analysis, report the exact value of n in each experimental group.

4. **Randomization:** State whether randomization was used to allocate experimental units to control and treatment groups. If done, provide the method used to generate the randomization sequence. Describe the strategy used to minimize potential confounders such as the order of treatments and measurements, or animal/cage location. If confounders were not controlled, state this explicitly.
5. **Blinding:** Describe who was aware of the group allocation at the different stages of the experiment (during the allocation, the conduct of the experiment, the outcome assessment, and the data analysis).
6. **Outcome measures:** Clearly define all outcome measures assessed (e.g., cell death, molecular markers, or behavioral changes). For hypothesis-testing studies, specify the primary outcome measure, i.e., the outcome measure that was used to determine the sample size.
7. **Statistical methods:** Provide details of the statistical methods used for each analysis, including software used. Describe any methods used to assess whether the data met the assumptions of the statistical approach, and what was done if the assumptions were not met.
8. **Experimental animals:** Provide species-appropriate details of the animals used, including species, strain and substrain, sex, age, or developmental stage, and, if relevant, weight. Provide further relevant information on the provenance of animals, health/immune status, genetic modification status, genotype, and any previous procedures.
9. **Experimental procedures:** For each experimental group, including controls, describe the procedures in enough detail to allow others to replicate them, including: What was done, how it was done, and what was used; When and how often; Where (including detail of any acclimatization periods); Why (provide rationale for procedures).
10. **Results:** For each experiment conducted, including independent replications, report: Summary/descriptive statistics for each experimental group, with a measure of variability where applicable (e.g., mean and SD, or median and range). If applicable, the effect size with a confidence interval.

Public perception of animal research (Lluis Montoliu)

All these details described in the previous paragraphs are rarely known by the public. Rather, it is often wrongly assumed that researchers can have one day an idea, which needs to be tested in animals, and next day they can go to the animal facility and begin the experiments. This is, absolutely, not the case. Usually, several months are required for the above reviewing and approval process to be completed successfully; in addition, applications are often delayed due to Ethics Committees and/or the Authorities' procedure amendment requests. However, if all these aspects are not appropriately communicated to the society, then we should not be surprised if the public assumes that the regulation behind animal research is weak or almost nonexisting. This imbalance and failure to communication has triggered suspicion and the lack of public support to animal research (McArthur Clark et al., 2019).

Animal research is a difficult topic to explain to noneducated audiences. Trying to explain the details often will not be of much help, unless these nontechnical summaries are prepared accurately, anticipating the type of audience that would read them. In this regard, the current Directive 2010/63/EU requires the publication by Member States of anonymous nontechnical summaries of those projects, prepared by the researchers, to ensure that the public is informed.

In contrast, the obvious benefits for mankind of numerous scientific projects where animals were used are often better accepted, if communicated adequately. Most of the Nobel Prize in Physiology or Medicine have been awarded to researchers who used animal models for their pioneering experiments (Understanding Animal Research, 2012).

Probably, the best current example is the relevant use of animal models for the preclinical analyses of the different COVID-19 vaccine candidates. Mice, both genetically modified or wild type, hamsters, ferrets and, particularly, nonhuman primates have been used, successfully, by the different consortia, before the proper clinical trials with volunteers could be organized. This led to the subsequent approval of vaccines by the regulatory agencies that are recognized as safe and effective (van Doremalen et al., 2020; Voget et al., 2021). Without preclinical and quality control tests with animals for safety and efficacy, there would be no vaccines nowadays.

The Stop Vivisection initiative

A turning point in animal research public acceptance in Europe was the initiative presented in June 2013 by a group named "Stop Vivisection," supported by about 1.2 million signatures collected from European citizens, aiming to a complete ban of animal experimentation, and requesting the European Commission to abrogate Directive 2010/63/EU. That movement was contested by researchers, veterinarians, scientific societies, research centers, and institutions in a dispersed and uncoordinated manner. There were few organizations created to proactively defend responsible animal research, from the scientific community. One of those associations was "The Basel Declaration Society," founded in 2011 and later transformed into Animal Research Tomorrow (https://www.basel-declaration.org/). Due to the massive support obtained, the "Stop Vivisection" promoters gained the right to defend their proposal at the European Parliament. A response from the European

Commission came in June 2015, being the main outcome to maintain the status quo, i.e., to maintain 2010/63EU Directive with no repeal. The response, however, also contained the Commission commitment to take the following actions:

- Action 1—Building on existing activities of the Commission, relevant EU agencies and OECD, to analyze technologies, information sources, and networks from all relevant sectors with potential impact on the advancement of the three Rs, and to present by end 2016 an assessment of options to enhance knowledge sharing among all relevant parties. The assessment should have considered how to systematically accelerate knowledge exchange through communication, dissemination, education, and training.
- Action 2—To continuously support the development, validation, and implementation of alternative approaches for regulatory and research use, including close cooperation between the Commission, Member States, and international organizations with the support of EU programmes, as appropriate.
- Action 3—To actively monitor compliance with the Directive, in particular, the three Rs principle, and with the relevant obligations in sector legislation to use available alternatives. To also actively monitor the correct enforcement by all Member States with the commitment, by end 2016, to examine regulatory requirements in the relevant sector legislation mandating animal testing to assess if the legislative text enables an efficient uptake of available alternative approaches. In addition, to ensure that future proposals for relevant sector legislation will reflect the rules on the protection of animals used for scientific purposes.
- Action 4—To facilitate an efficient dialog, by end 2016 to organize a conference engaging the scientific community and relevant stakeholders in a debate on how to exploit the advances in science for the development of scientifically valid nonanimal approaches and advance toward the goal of phasing out animal testing. On that occasion, progress on actions 1, 2, and 3 would have been reported.

According to the fourth action, the EU Commission organized a scientific conference devoted to nonanimal approaches in December 2016 in Brussels (Belgium) entitled "Nonanimal approached, the way forward" where delegates from the side of the animal research users, and from the side of those organizations opposing to animal research were invited to attend.

Nonetheless, the "Stop Vivisection" campaign had some additional, probably unintended, positive effects on the community of researchers using animal models in their experiments. For the first time, representatives of the scientific community felt their work with research animals was at stake. And this threat was not because of the good communication skills of the opponents, who presented their nonevidence-based views about the scientific progress, laboratory animal uselessness, and opportunities to fully replace animals with alternative methods. The threat was mainly due to the lack of communication from the scientific community. There is a popular Spanish saying that "if you are not talking, someone will talk for you." Indeed, the lack of communication by researchers was filled by the animalist groups, which took advantage of the mass media attention, in the absence of an adequate response from the science counterpart. This was a risky situation that could have led even to the edge of banning animal research within the EU with obvious disastrous consequences for the progress of biology and biomedicine.

The British Concordat on Openness on Animal Research

Fortunately, some innovative initiatives were launched in various countries. At first, in the United Kingdom, the organization Understanding Animal Research (UAR) set a Concordat on Openness of Animal Research. Following its presentation in May 2014, the Concordat was signed by a number of institutions with a very positive response in terms of animal research public acceptance (Jarret, 2016).

This Concordat was developed starting from a previous initiative called the "Declaration on Openness on Animal Research," presented in October 2012 and involving over 40 organizations which were committed to improve how they could be more open about the ways in which animals are used in scientific, medical and veterinary research. So far (Second quarter 2021), as many as 127 institutions have signed the Concordat on Openness on Animal Research, including universities, research centers, charities, scientific and patient associations, academies, foundations, NGOs, hospitals, and companies. There is a devoted web page with all the relevant information on this Concordat (http://concordatopenness.org.uk/). The signatory institutions agreed to comply with the following four commitments:

- Commitment 1: To be clear about when, how, and why we use animals in research.
- Commitment 2: To enhance our communications with the media and the public about our research using animals.
- Commitment 3: To be proactive in providing opportunities for the public to find out about research using animals.
- Commitment 4: To report on progress annually and share our experiences.

The Concordat web site holds the corresponding annual reports about the activities and initiatives toward openness in animal research that are promoted by the institutions. In addition, "Leaders in Openness" and "Openness Awards" have been set up to recognize the outstanding work of individuals and institutions toward communication and transparency on animal research.

The Spanish COSCE initiative on communication and transparency on animal research

A second European initiative took place in Spain, also established in January 2014. The Confederation of Scientific Societies in Spain (COSCE, https://cosce.org) created a working group named as "Commission for the study of the use of animals in scientific research," shortened and often referred to as COSCE Group on Animal Experimentation. A diverse group of delegates representing relevant stakeholders on animal experimentation were invited to join this commission, including researchers, veterinarians, universities, patient associations, biotechnological companies and those providing animals, and the Spanish Society of Laboratory Animals (SECAL). The writer of this chapter had the honor to represent public research institutions, such as the Spanish National Research Council (CSIC).

The first task identified by COSCE was to prepare a white document on animal experimentation addressed to the public at large. We felt and recognized the lack of communication that had been historically associated to researchers. We decided to present in a relatively short report the reasons why animal research had been instrumental for advancing our knowledge on basic biology, on the diseases that are affecting us and the animals, and for the development of new therapies to alleviate or cure those diseases. The resulting 10-page document was released in February 2015, under the title "COSCE Document on the Use of Animals in Scientific Research" (https://cosce.org/informe-sobre-el-uso-de-animales-en-investigacion-cientifica/) which contained numerous examples of knowledge and applications achieved with the use of laboratory animals. The document was well received by the media and the public. At the time, in Spain we were aware of the British Concordat, which had been launched one year earlier and we decided to adopt the British Concordat on Openness on Animal Research as a reference with some changes in the text and commitments to tailor it to Spain and its cultural and social environment. The resulting document "Propuesta COSCE Acuerdo de transparencia sobre el uso de animales en experimentación científica en España" (https://cosce.org/acuerdo-de-transparencia/) was ready and approved by COSCE in June 2016, and the initiative was eventually launched in September 2016, with over 80 adhering institutions. There were no signatures collected but simple adhesions to the agreement by representatives of each adhering institution. This document also contained four commitments inspired to the British Concordat. All these adhered organizations committed to:

- Commitment 1: To speak with clarity on when, how, and why we use animals in research.
- Commitment 2: To provide adequate information to mass media and to lay public on the conditions in which we carry out research projects that require the use of animals and about the results we derive from them.

- Commitment 3: To promote initiatives to generate a better knowledge and understanding in the society on the use of animals in research.
- Commitment 4: To inform on the progress of the agreement and sharing experiences on annual base.

At present (October 2021), the COSCE agreement includes 151 adhering institutions, including universities, research centers, research institutions, scientific societies, patient associations, hospitals, and companies (Fig. 2.1). All the information about the COSCE agreement on transparency on animal research can be accessed at a devoted web site (https://cosce.org/acuerdo-de-transparencia/). The Spanish initiative is the largest transparency agreement in Europe, and in the world, on the subject of animal experimentation (Montoliu, 2018). This proposal has been presented in numerous countries and, together with the British Concordat, it has been used as a seed for implementing similar agreements subsequently in other countries.

In Spain, we decided to invite the European Animal Research Association (EARA, https://www.eara.eu/) with the task of monitoring, annually, the compliance of the four commitments by the adhered institutions. In this regard, an annual report is presented by the COSCE Group on Animal Experimentation to the media and the public, informing about the diverse activities

FIGURE 2.1 Cloud of institutional logos currently (October 2021) adhered to the COSCE Transparency Agreement on Animal Experimentation in Spain. *Source: https://cosce.org/entidades-adheridas/.*

proactively organized and accomplished by the adhered institutions. This information is collected by EARA through an annual survey whose participation is mandatory.

Adhering institutions commit also to include a policy statement on animal experimentation in their institutional web sites. This is also a mandatory request that all adhered organizations must fulfill.

Other transparency agreements on animal research in Europe

Besides United Kingdom and Spain, other European countries have launched similar or equivalent initiatives on communication and transparency on animal research. In contrast, not much has happened so far to foster transparency on animal research in the USA (Wadman, 2017).

The German tierversuche verstehen

A somewhat different idea originated in Germany. In 2016, the Federal Government supported the creation of a web site devoted to communicating and explaining animal research (https://www.tierversuche-verstehen.de/). In this web page, researchers using animal models share their scientific projects involving animals using an accessible language, understandable by the public, and highlight where the role of animals was instrumental to achieve the results, and the potential relevance for the increase in knowledge and/or the development of therapies for treating human diseases.

Concerning additional proper transparency agreements, there are three recent proposals launched in Portugal, Belgium, and France. Subsequently, in 2021, similar transparency agreements have been launched in Germany and New Zealand.

The Portuguese acordo de transparência sobre investigação animal em Portugal

The Transparency Agreement on Animal Research in Portugal is a collaborative project developed by EARA and the Portuguese Society of Sciences in Laboratory Animals (SPCAL) which was launched in June 2018 with the participation of 16 Portuguese research institutions (https://www.ihmt.unl.pt./wp-content/uploads/2019/09/Acordo-de-Transpar%C3%AAncia-sobre-Investiga%C3%A7%C3%A3o-Animal-em-Portugal_PT.pdf). This third European transparency agreement also included the four commitments contained in the previous British and Spanish initiatives. In particular, the phrasing used in Portugal was:

- Commitment 1: We will be clear about when, how, and why we use animals in research.
- Commitment 2: We will enhance our communications with the media and the public about our research using animals.

- Commitment 3: We will be proactive in providing opportunities for the public to find out about research using animals.
- Commitment 4: We will report on progress annually and share our experiences.

Similarly to the Spanish Agreement, EARA was chosen to monitor the annual performance.

The Belgian Transparency Agreement

This was the fourth European country deploying a transparency agreement on animal research (http://bclasorg.webhosting.be/transparency-agreement/). In Belgium, 18 universities, research centers and companies signed and launched a transparency agreement on animal research in 2019, with a commitment to communicating in a more open way about how animals are used in research. This agreement was established under the leadership of the Belgian Council for Animal Research (BCLAS), and in collaboration with the EARA.

The four commitments, as presented in this Belgian Transparency Agreement on animal research, are as follows:

- Commitment 1: To create more clarity about how, when, and why animals are used in research.
- Commitment 2: Better communication with the media and the public about implementing the alternatives and reducing or refining animal testing.
- Commitment 3: To offer the general public the opportunity to become acquainted with laboratory animal research and the regulations that apply to it, for example, via open lab days.
- Commitment 4: To report the impact of their communication and share experiences annually.

The French Transparency Agreement

The French Transparency Agreement on animal research (https://www.recherche-animale.org/transparency-charter) is the fifth agreement of this type launched in Europe. In February 2021, 30 French research institutions and companies signed a new transparency agreement on animal research. In France, this proposal was initiated by the French Ministry of Higher Education, Research and Innovation and coordinated by the animal research advocacy body GIRCOR. Similar to the rest of transparency agreements, the French version also included the four known commitments whose wording in this case replicated the terms used by the Belgian document.

All these European agreements can be accessed through a devoted EARA web page at (https://www.eara.eu/transparency-agreements).

Practicing communication and transparency in animal research

All transparency agreements show the commitment of the signatories or adhering organizations to be more open on animal research. However, beyond this commitment, they also request proactive behaviors by the participating institutions. In this regard, transparency can also be applied and become one the objectives to be achieved through practicing. A number of activities can be organized to this aim, including: guided visits by scholars, associations, journalists, or individual interested persons; workshops or symposia aimed to disseminate the benefits of animal experimentation and the strict rules under which researchers operate; videos, audios, and articles shared via the institutional networks; infographics to be shown on the walls of the institution, etc.

For completion of this chapter, a list of examples of transparency events and activities selected from the COSCE Transparency Agreement on Animal Experimentation in Spain is provided under Readings and Other Resources (Fig. 2.1).

School and research hand in hand (Annarita Wirz)

For a long time, it seemed as if citizens and scientists were traveling on parallel tracks, each being aware of the importance of the others, but still separated by a barrier of misinformation. Citizens felt the need to understand if, how, when, and why scientific research activities affected their daily lives, and scientists felt the need to have their work understood. Over time, the need to break down this barrier and bring these two sectors of society closer together has become more and more pronounced.

The best way to deal with this situation seemed to be to increase the transparency of research, with the idea that this would be beneficial for both science and society. Transparency, however, is a complex concept and it is necessary to be clear about the objectives of how to implement and promote certain initiatives and to consider the most appropriate contexts. First of all, it is necessary to understand whom one wants to address, in order to distinguish between "scientifically relevant" transparency, addressed to the scientific community, and "socially relevant" transparency, addressed to the public opinion (Eliott and Resuit, 2019); the former allows scientific data to be shared among those working in the same field and is certainly useful for increasing knowledge and speeding up research; the latter, which is aimed at the public, must, however, go through a series of necessary steps to make scientific information accessible to all. It is necessary to identify the most appropriate tools, methods, and language to explain and share the daily work of a researcher. It is precisely this sharing that contributes to the enhancement of the research work itself (Eliott and Resuit, 2019).

Situations may arise where efforts to promote one form of transparency may prove counterproductive for the other if due attention is not paid. This can happen especially when dealing with sensitive scientific issues that are also strongly felt by the public, such as the use of animal models in biomedical research. For this reason, the Directive 2010/63/EU also clearly refers to the need to increase transparency in this area.

Animal models are still crucial in biomedical research, but public opinion plays a key role in influencing the way this area of science is regulated and funded (MacArthur Clark et al., 2019).

The scientific community has been slow in communicating with the public about its work with animals using an appropriate language, thus allowing information about it to be misreported by others. This imbalance has led the public to adopt a suspicious attitude. The public still does not fully understand the need for the use of animals, how they are used and how results obtained on animals can be transferred to humans (MacArthur Clark et al., 2019). As already reported earlier in this chapter, initiatives aimed at the public are increasing in order to reverse this phenomenon. Making available nontechnical summaries of projects involving the use of animals, clearly stating their purpose, the number of animals involved and the severity of the procedures (Decision 2020/569/EU), or including references to research carried out using animal models on the websites of both public and private research institutions are important steps forward. Once the rationale and regulation on animal are explained, the public welcomes the results of the research, but still wants the animals to feel no unnecessary pain or distress (MacArthur Clark et al., 2019). It is now well recognized that this balance between the needs of science and those of the animals is a prerequisite for building public confidence.

However, this type of communication is aimed mainly at an adult audience that never before has so strongly expressed the need for reliable scientific information. For transparency to be "socially relevant," however, it is necessary to address not only the adult audience but also the younger generations. For children and young people, it is particularly important to acquire the tools to be able to listen, understand and make sense of the news they hear in the media. A correct and widely understandable dissemination of scientific information by experts in the field is the first fundamental step for citizens to have the critical ability to form their own personal and autonomous opinions on the various aspects of research, without prejudice or preconceptions. For talking to younger generations about research, communication methods and language must be adapted to the age of the target audience. With children and young people, it is necessary to use simple, playful language, with many references to their daily lives, bearing in mind that they have great ability to "move around" on the Internet; within this immense "toolbox," it can sometimes be difficult to find or choose the best tool, both for those who want to inform themselves and for those who want to inform. That is why it is important to first understand and establish the objective to be achieved (Bik and Goldstein, 2013).

In recent years, online tools (e.g., Facebook, Instagram, Twitter, various blogs, etc.) have widely increased their popularity and reach, offering a powerful way for scientists to strengthen their professional profile and act as a public voice for science. The potential benefits are numerous, but one needs to know how to make good use of these resources which allow communication to travel fast (Bik and Goldstein, 2013). Scientific (and other) news can spread like little streams from a river. The risk is that the information that reaches the final point will have lost its initial connotation: this is a risk to be avoided at all costs.

Moreover, the speed of these tools can sometimes generate "noisy confusion" for young people who find it difficult to navigate around scientific news, especially when sensitive and hotly debated topics such as, e.g., vaccines or animal research using are concerned. For these reasons, it is necessary, from primary school onwards, to provide younger generations with adequate tools to build their own critical abilities, knowledge being the first and most important one. In addition to didactic knowledge strictly linked to institutional school curricula, practical knowledge too puts them in contact with the world of research. Indeed, to understand and appreciate what scientists say, young people need to be "scientifically literate," to assess what is being said. Understanding science and its applications is important for them also to make informed electoral and political choices. In order to increase this scientific culture, it is necessary to start in schools.

Involving students, especially those in secondary schools, in adequate scientific communication/education is crucial for a society that requires a high level of scientific literacy to face the economic and social challenges of the 21st century (Miller, 2011) and to have new generations of scientists capable of innovation in a competitive knowledge-based society (Eurydice, 2006).

Collaborations between researchers and teachers is pivotal in this sense, because through these interactions, teachers are provided with up-to-date information and experience, as well as a different view especially on controversial topics such as animal research. Active collaboration between the world of research and schools can help students develop a scientific approach. Young pupils from primary and intermediate school have a simple and spontaneous way of understanding and explaining the natural phenomena that surround them (Eurydice, 2006). Starting from this, it is possible to provide them with additional information and tools to increase their capacity for in-depth analysis and reasoning, until they can provide more scientifically sound explanations and identify problem-solving strategies. It is particularly important to teach students how to do a literature search and identify reliable sources. To a large extent, the public does not have a clear picture of what researchers do on a daily basis, so it is important to start again from the scientific method and how, through the experimental phases, one goes from observation and hypothesis to results.

Furthermore, to make it easier and less confusing for children and young people to understand research, the adults around them (parents and teachers) should also have the opportunity to obtain adequate information. In this sense, considerable efforts have been made in various countries for years and in different ways. Here, we report only a few examples. Some websites, such as that of UAR (https://www.understandinganimalresearch.org.uk/), or Kids4research (https://kids4research.org/), have contents dedicated not only to students but also specifically to teachers and parents. Other initiatives put scientists in direct contact with both children and adults. In Italy, for example, federations that bring together scientific associations and societies, such as the Federazione Italiana Scienze della Vita (FISV https://www.fisv.org/) or SPERA-Le Ragioni della Ricerca (http://www.federazionespera.eu/), organize events, workshops, and courses for both students and teachers, as well as more informal initiatives where researchers talk and discuss with parents about topics addressed in class with their children.

Research is becoming more and more complex, and it is often difficult to "translate" it into layman language, especially for younger people, but these are worthy efforts. In addition, in universities, students should learn and practice skills on how to effectively communicate their research work at different levels.

Good use of scientific knowledge would be invaluable in steering decisions toward the most rational solutions, enabling decision makers to achieve the best possible outcome, avoiding wrong and inefficient solutions.

Animal research has attracted widespread society attention because of the ethical issues involved. If the communication is clear and effective, children are attentive listeners, free from prejudice and preconceptions; of course, it is important to contextualize the use of animal models in research by explaining what results can be obtained and what important objectives could not have been achieved without animals.

Contextualization also means explaining how and why data generated with laboratory animals can be transferred to humans to understand the patho/physiological bases of human diseases, the ultimate scope being to develop new and effective cures.

On a wider landscape, the COVID-19 pandemic has highlighted the need for proper scientific communication and information to the public. During the pandemics, communication to the public about the science and technology behind the vaccines being delivered to the human population have not always been so effectively provided, skepticism and antivax movements being negative consequences of this gap. For example, as early as 2011, Miller said that if someone does not understand the nature of a virus, for example, he/she may misjudge the situation and not make the right choices. Indeed, science is a way of asking questions and critically evaluating the answers by applying an experimental approach. This is precisely the sort of approach that needs to be taught the public opinion, starting with young people. They need to be educated on how research works and on why it cannot provide immediate and absolute answers.

During the COVID emergency, television confirmed its role as a great catalyst and social media showed their full potential for both storytelling and information gathering. The pandemic imposed an acceleration in dealing with the problem of fake news and news coming from unreliable sources that overlapped with the real news, creating confusion in the public. It was precisely during the pandemic, however, that the main social media platforms chose to take action to improve the quality of their content, often pointing the audience to institutional websites. Although with different strategies, all social media tried to direct their users toward reliable information and intervened on their algorithms in order to make the most reliable sources more visible. As the pandemic progressed, social media also standardized their strategies (Carelli and Vittadini, 2020). Hopefully, this steering toward reliable sources is a transformation that will be preserved even after the end of the COVID emergency.

In some respects, the Internet can be a more powerful tool than traditional means (e.g., congresses, courses, and seminars) for transferring scientific communication. The increasing use of online resources, so familiar to children and young people, can eventually transform and broaden the culture of science. Blogs and social media can provide good tools for extended scientific discussions also involving a wider audience. Online interactions have the potential to increase and improve communication between scientists and the public. An established online communication and outreach strategy can also meet the demands of some funding agencies (Bik and Goldstein, 2013).

In order to use resources efficiently, objectives and resources need to be thoroughly evaluated first. In addition, each social network has its own characteristics that can complement each other. With this in mind, the best choice can be made. For example, blog posts need to be thought of as long-term projects, as they require a greater commitment in terms of time for the creation and continuous administration of posts, which need to be constantly updated, but on the other hand, contents can be also taken up by other social networks and be disseminated more widely (Bik and Goldstein, 2013).

Internet-related technologies are increasingly ubiquitous and, therefore, there is an urgent need for scientific institutions to provide students and teachers with training in using these technologies effectively and appropriately for the dissemination of scientific knowledge and in being effectively engaged in discussions on complex topics such as research in general and animal research in particular. The COVID-19 outbreak has been a "raining ground" for models of communication between citizens and institutions, and it has helped reevaluate the role of science and research, changing not only the relationship between experts and the public but also that between scientific experts and law makers. In this regard, trust is a central issue, if not the most relevant one, to question the dynamics between the different actors of communication and the mechanisms that regulate the reliability of a source, therefore its credibility (Sfardini, 2020).

Experts are recognized as such because they know what they are talking about, they act responsibly, they have a deep knowledge and experience of the subject at hand and their analyses can guarantee reduced margins of error (Gili and Panarari, 2020). The authoritative voice of the world of science is being increasingly listened, and social platforms, institutions, and the media are increasingly interacting in an unprecedented way in the effort to stem the spread of unreliable information. This joint effort is key, and must be safeguarded and nurtured in the future, although who is writing suspects that there is still a lot to do before the best way forward is found.

When "open science" is conceptualized as "socially relevant," especially in controversial areas like, e.g., animal research, the objectives of scientific communication and the needs of the public should overlap (Eliott and Resuit, 2019). The active collaboration between many different professional figures—researchers, teachers, disseminators, and journalists—would allow to increase citizens' scientific awareness, starting with students. This awareness is crucial in the social, political, economic, and decision-making spheres.

References

Bik, H.M., Goldstein, M.C., 2013. An introduction to social media for scientists. PLoS Biol. 11 (4), e1001535. https://doi.org/10.1371/journal.pbio.1001535.

Carelli, P., Vittadini, N., 2020. Social-virus. Piattaforme, istituzioni e hashtag durante la pandemia. In: Sala, M., Scaglioni, M. (Eds.), L'altro Virus. Comunicazione e Disinformazione al tempo del Covid-19. Vita e Pensiero, Milano, pp. 18–35.

Elliott, K.C., Resnik, D.B., 2019. Making open science work for science and society. Environ. Health Perspect. 127 (7), 75002. https://doi.org/10.1289/EHP4808.

EU Decision 2020/569, 2020. Commission Implementing Decision (EU), Establishing a Common Format and Information Content for the Submission of the Information to Be Reported by Member States Pursuant to Directive 2010/63/EU of the European Parliament and of the Council on the Protection of Animals Used for Scientific Purposes and Repealing Commission Implementing Decision 2012/707/EU.

EU Directive 2010/63, 2010. Directive of the European Parliament and of the Council on the Protection of Animals for Scientific Purpose.

Eurydice, 2006. Science Teaching at School in Europe. Policies and Research. Eurydice, Brussels, p. 96.

Federazione Italiana Scienze della Vita (FISV). https://www.fisv.org/. (Accessed 31 May 2021).

Gili, G., Panarari, M., 2020. La credibilità politica. Radici, forme, prospettive di un concetto inattuale. Kids4research, Marsilio, Venezia, p. 224. https://kids4research.org/ (Accessed 31 May 2021).

Jarrett, W., 2016. The Concordat on Openness and its benefits to animal research. Lab. Anim. (NY) 45, 201–202.

MacArthur Clark, J., Clifford, P., Jarrett, W., Pekow, C., 2019. Communicating about animal research with the public. ILAR J. 60, 34–42.

Miller, J.D., 2011. To improve science literacy, researchers should run for school board. Nat. Med. 17 (1), 21.

Montoliu, L., 2018. Transparency in Animal Experimentation. FEBS Network. https://network.febs.org/posts/32348-transparency-in-animal-experimentation (Accessed 17 April 2021).

Percie du Sert, N., Hurst, V., Ahluwalia, A., Alam, S., Avey, M.T., Baker, M., et al., 2020. The ARRIVE guidelines 2.0: updated guidelines for reporting animal research. J. Physiol. 598, 3793–3801.

Sfardini, A., 2020. Come comunicare la pandemia? Credibilità e fiducia delle fonti istituzionali nell'informazione italiana sul Covid-19. In: Sala, M., Scaglioni, M. (Eds.), L'altro Virus. Comunicazione e Disinformazione al tempo del Covid-19. Vita e Pensiero, Milano, pp. 42–49.

SPERA-Le Ragioni della Ricerca. http://www.federazionespera.eu/. (Accessed 31 May 2021).

Understanding Animal Research (UAR). https://www.understandinganimalresearch.org.uk. (Accessed 31 May 2021).

Understanding Animal Research, 2012. Animals and the Nobel Prize. https://www.understandinganimalresearch.org.uk/news/research-medical-benefits/animals-and-the-nobel-prize/ (Accessed 17 April 2021).

van Doremalen, N., Lambe, T., Spencer, A., Belij-Rammerstorfer, S., Purushotham, J.N., Port, J., et al., 2020. ChAdOx1 nCoV-19 vaccine prevents SARS-CoV-2 pneumonia in rhesus macaques. Nature 586, 578–582.

Vogel, A.B., Kanevsky, I., Che, Y., Swanson, K.A., Muik, A., Vormehr, et al., 2021. BNT162b vaccines protect *Rhesus macaques* from SARS-CoV-2. Nature 592, 283–289.

Wadman, M., 2017. A trans-Atlantic transparency gap on animal experiments. Science 357, 119–120.

Recommended reading

Articles on the Value of Animal Experimentation for Scientific Colleagues (Revista SEBBM, March 2021, a Dossier on "Animal Experimentation" Coordinated by Lluís Montoliu. https://www.sebbm.es/revista/articulo.php?id=746&url=experimentacion-animal.

Interviews With Digital Press on this Subject (Lluis Montoliu Interviewed in farmacosalud.com. https://farmacosalud.com/el-acuerdo-por-la-transparencia-ha-cubierto-el-deficit-de-informacion-sobre-experimentacion-animal-que-cubrian-los-grupos-animalistas/

Further Processing the Raw Data on Animal Uses Per Year in Spain Released by Authorities and Preparing Graphics With them, Documenting the Evolution of Several Parameters Over More than 10 Years. https://cosce.org/informes-de-los-usos-de-animales-en-investigacion-y-docencia-en-espana-2009-2017/.

Launching the INVIVO Journal, by the University Miguel Hernández (Elche, Alicante, Spain) About Science and Animal Experimentation. https://issuu.com/invivo_/docs/invivo__2_.

Interview at the Public Radio Station (RNE) Within the Popular Science Program "A hombros de gigantes" to Lluis Montoliu on Transparency in Animal Experimentation. http://www.rtve.es/alacarta/audios/a-hombros-de-gigantes/hombros-gigantes-mayor-transparecencia-uso-animales-experimentacion-10-09-18/4727338//.

Interview to J.M. Orellana, Veterinarian at the Animal Facility of University of Alcalá (Madrid). http://portalcomunicacion.uah.es/diario-digital/ingles/the-uah-animalary.

Exhibition on Laboratory Animals at University of Oviedo. https://eticainvestigacion.uniovi.es/actualidad/-/asset_publisher/4bR1/content/exposicion-animales-de-laboratorio.
Podcast Lab Chat Rat, interview to Javier Guillén on animal research in the EU. https://labratchat.buzzsprout.com/772673/3961181-9-animal-research-in-the-european-union.
Quiz Game on Animal Experimentation by SECAL (Spanish Society of Laboratory Animals). https://secal.es/quiz-secal-el-juego-para-ponerte-al-dia-en-experimentacion-animal/.
Murillo-Cuesta, S., Artuch, R., Asensio, F., de la Villa, P., Dierssen, M., Enríquez, J.A., et al., 2020. The value of mouse models of rare diseases: a Spanish experience. Front. Genet. 11, 583932.
The Value of Animal Models to Defeat the Coronavirus, by Lluís Montoliu in the Conversation, March 15, 2020. https://theconversation.com/no-podremos-derrotar-al-coronavirus-sin-modelos-animales-133887.
The Relevance of Animal Models in Research on Rare Diseases, YouTube Popular Science Video by CIBERER. https://www.youtube.com/watch?v=t-WC3f4XXNs.
Animal Experimentation Provides Life, YouTube Popular Science Video by SECAL. https://www.youtube.com/watch?v=PqDs_QH0C0M.
Secondary School (High school) Students Visit the Animal Facility of the University of Cantabria. https://web.unican.es/noticias/Paginas/2018/noviembre_2018/El-Servicio-de-Estabulacion-y-Experimentacion-Animal-de-la-UC-recibe-la-visita-de-escolares-por-la-Semana-de-la-Ciencia-.aspx.
Delivering a Popular Science Conference on Transparency in Animal Research in a Pub (Skeptical in the Pub group) Lluís Montoliu. http://www.escepticos.es/node/5090/.
Introducing a Rare Disease Patient With Albinism to her Corresponding Mouse Model Carrying her Mutation at the Homologous Gene, Established by Crispr Genome Editing. https://www.youtube.com/watch?v=Sigxo3Virho.
Reporting About the Number of Uses of Animals in Spain, Distributed by Species, Aim and Severity of Animal Procedures, CCC Blog. https://culturacientifica.com/2018/01/05/cuantos-animales-se-usan-espana-experimentacion/.
Interview to the Responsible Person at the Animal Facility at the HNP Hospital in Toledo (Spain) YouTube video. https://www.youtube.com/watch?v=LJ2qDVWd8GE.
Presenting the Adhesion to the Transparency Agreement to the Entire Community of the University of Alicante. https://web.ua.es/es/actualidad-universitaria/2019/febrero19/11-17/la-universidad-de-alicante-apuesta-por-la-transparencia-en-experimentacion-cientifica-con-animales.html.
Compilation of Online Resources for Practicing Transparency Obtained From Participating Institutions in the COSCE Transparency Agreement. https://cosce.org/ejemplos-de-divulgacion-y-de-transparencia-en-experimentacion-animal-2019/.
Scaglioni, M., 2020. Il virus sullo schermo. Il "sismografo" televisivo e la mediazioni della crisi. In: Sala, M., Scaglioni, M. (Eds.), L'altro Virus. Comunicazione e Disinformazione al tempo del Covid-19. Vita e Pensiero, Milano, pp. 10–17.
Takeo, T., Sztein, J., Nakagata, N., 2019. The CARD method for mouse sperm cryopreservation and in vitro fertilization using frozen-thawed sperm. Methods Mol. Biol. 1874, 243–256.

Chapter 3

Anatomy, physiological features, genetics and genetic alterations, breeding and strain differences relevant to the choice of the model—Impact of 3Rs

Livia D'Angelo[1], Paolo de Girolamo[1], Laura Lossi[2], Adalberto Merighi[2], Marcello Raspa[3], Ferdinando Scavizzi[3]

[1]*Department of Veterinary Medicine and Animal Production, University of Naples Federico II, Naples, Italy;* [2]*Department of Veterinary Sciences, University of Turin, Italy;* [3]*Institute of Biochemistry and Cell Biology, Italian National Research Council (CNR), Rome, Italy*

Introduction and general concepts: basic anatomical features relevant to the choice of models (Paolo de Girolamo and Livia D'Angelo)

While primary and utmost effort should be given to develop and use nonanimal alternatives, the other 2Rs are essential: when conducting experiments, the final decision very often is not between animal and nonanimal; instead, it is which animal? Which strain? Thus, success and validity of research depends on the selection of the most appropriate animal model to address key scientific questions.

A laboratory animal model is *a living organism in which normative biology or behavior can be studied, or in which a spontaneous or induced pathological process can be investigated, and in which the phenomenon in one or more respects resembles the same phenomenon in humans or other species of animal* (Wessler, 1976). Commonly, the term animal model refers to modeling of humans. However, the choice of the species to use as models for humans is only one of the principal connotations for selecting a model. A model can be selected also to evolutionary reshape model species' traits in the laboratory

Practical Handbook on the 3Rs in the Context of the Directive 2010/63/EU
https://doi.org/10.1016/B978-0-12-821180-9.00009-X

environment. In both cases, the evolutionary perspective and approach guide the choice of model species for research.

Finding the most appropriate model remains a challenge in several fields of translational research. Some diseases are unique to humans and do not have an animal counterpart, which might serve as a surrogate for human patients. Moreover, critical gaps in the knowledge of human biology makes selecting nonhuman models for translational research especially difficult because we are aiming at a target that is not yet clear.

This chapter highlights the most relevant aspects of anatomy, physiology, genetics, and breeding of commonly used mammalian and fish species and in which field of biomedical research they are more suitably used. The mammalian species here considered are the mouse (*Mus musculus*), rat (*Rattus norvegicus*), Syrian hamster (*Mesocricetus auratus*), Mongolian gerbil (*Meriones unguiculatus*), guinea pig (*Cavia porcellus*), rabbit (*Oryctolagus cuniculus*), and pig (*Sus scrofa*). Our aim is to put under the spotlights the morphophysiological peculiarities of mammalian and species in comparison to human (Tables 3.1 and 3.2) and the main current uses of individual species in research according to those features (Table 3.3).

We also describe the main and relevant anatomical and physiological features of some of the most used teleostean species in research (Table 3.4) to be taken into consideration when selecting a fish model: the turquoise killifish (*Nothobranchius furzeri*), medaka (*Oryzias latipes*), zebrafish (*Danio rerio*), and goldfish (*Carassius auratus*).

Evolution of the anatomy and physiology of pain pathways and its relevance in relation to the severity of procedures (Adalberto Merighi and Laura Lossi)

Nociception and pain (Adalberto Merighi and Laura Lossi)

In sensory physiology, an adequate stimulus defines a class of environmental phenomena that entails the least amount of energy to elicit a percept mediated by a specific sensory system. Most sensory systems are specialized to detect a corresponding class of environmental phenomena (stimuli), such as, e.g., the visual system that is dedicated in detecting visual information from the surrounding space. The nociceptive system is unique in that many different stimuli may cause pain, and a common denominator for these stimuli is found in tissue damage. Thus, an adequate stimulus to elicit pain is called a *noxious stimulus* (Koerber, 2009) and can be identified with any actual or potential tissue harmful event. Noxious stimuli are collected by nerve endings found throughout the body and originating from a group of specialized peripheral sensory neurons (PSNs), referred to as *nociceptors*, which represent the first nerve cell of a polyneuronal chain ultimately responsible for the perception of pain. Notably, pain consciousness only occurs at the end of this chain when

TABLE 3.1 General physiological features of rodents (mouse, rat, hamster, gerbil, and guinea pig), rabbit, and pig used in biomedical research.

	Mouse	Rat	Hamster	Gerbil	Guinea pig	Rabbit	Pig
Adult weight	20–40 g (male)	300–500 g (male)	85–145 g (male)	65–130 g (male)	900–1000 g (male)	2–5 kg	230 kg
	18–35 g (female)	250–300 g (female)	95–120 g (female)	55–133 g (female)	700–900 g (female)		
Life span	1–years (median)	2.5–3 years (median)	2 years (median)	2–5	3–4 years	5–7 years	6–10 years
Chromosome number (diploid)	40	42	44	44	64	44	38
Body temperature	37–37.2°C	37.5°C	36.2–37.5°C	37–39	37.2–39.5°C	38.5–39.5°C	38.9°C
Puberty	28–49 days (male)	50/60 days	6–8 weeks (male 90 g)	75–85 days (male)	3–4 months (male 600–700 g)	4–5 months	3–7 months
	28–49 days (female)		8–12 weeks (female 90–100 g)	65–85 days (female)	2–3 months (female 350–450 g)		
Breeding season	None	None	None	None	None	None	
Gestation	19–21 days	21–23 days	15–18 days	24–26 days	59–72 days	30–32 days	114–115 days
Litter size	4–12 pups	8–14 pups	4–12 pups	1–12 pups	2–5 pups	1–14 pups	8–12 pups
Birth weight	1–1.5 g	5–6 g	2–3 g	No value found.	60–115 g	60–70 g	1300 g
Eyes open	12–13 days	10–12 days	15 days			6–10 days	

Continued

TABLE 3.1 General physiological features of rodents (mouse, rat, hamster, gerbil, and guinea pig), rabbit, and pig used in biomedical research.—cont'd

	Mouse	Rat	Hamster	Gerbil	Guinea pig	Rabbit	Pig
				No value found.			
Weaning	21 days	21 days	21 days (35–40 g)	21–28 days (12–18 g)	14–28 days (180 g)	5–8 weeks	3–5 weeks
Heart rate	310–840 beats/min	300–500 beats/min	280–412 beats/min	260–600 beats/min	230–380 beats/min	200–300 beats/min	70 beats/min
Blood pressure	133–160 mmHg (systolic)	116 mmHg (systolic)	No value found.	No value found.	80–94 mmHg (systolic)	90–130 mmHg (systolic)	Mean arterial blood 150 mmHg
	102–110 mmHg (diastolic)	90 mmHg (diastolic)			55–58 mmHg (diastolic)	80–90 mmHg (diastolic)	
Whole blood	5.85 mL/100 g body weight	6 mL/100 g body weight	65–80 mL/kg	60–85 mL/kg	69–75 mL/kg	58.5–585 mL/kg	
Respiratory frequency	163 breaths/min	85 breaths/min	33–127 breaths/min	70–120 breaths/min	42–104 breaths/min	32–60 breaths/min	15 breaths/min
Tidal volume	0.18 (11–36) mL/min	1.5 mL	0.6–1.4 mL	No value found.	2.3–5.3 mL/kg	4–6 mL/kg	10–15 mL/kg

TABLE 3.2 A summary of the utilization of different mammalian species in current biomedical research.

Main fields of use		Mouse	Rat	Hamster	Gerbil	Guinea pig	Rabbit	Pig
Anatomy and physiology		Basic research		Urinary apparatus (pathology)	Brain circulation	Embryogenesis Teratogenicity	Basic research	Biology of reproduction circulatory system (pathology) Renal diseases
Applied pharmacology development of biomedical devices		General use				Vitamin C deficiency Beuhler and Kligman tests		General use Xenotransplantation
Pathology	*Cancer and cancerogenesis*	General use		Oral cavity tumors Pancreatic tumors	Radiation research		General use	
	Toxicology	General use			General use		General use	General use
	Neurological disturbances, neuropathology, and neurodegeneration	General use		Epileptic syndrome Prion diseases	Audiometry Vascular pathology	Audiometry	Neurological disturbances	Neurodegeneration (ALS)
	Endocrinology and endocrine pathologies	General use		Diabetes	Adrenal glands			Diabetes

Continued

TABLE 3.2 A summary of the utilization of different mammalian species in current biomedical research.—cont'd

Main fields of use		Mouse	Rat	Hamster	Gerbil	Guinea pig	Rabbit	Pig
	Genetic pathologies	General use		Muscular dystrophy Cardiomiopathy		Asthma	Hypertrophic cardiomyopathy	Cystic fibrosis
	Others	General use		Atherosclerosis Thrombosis Caries Infectious diseases		Infectious diseases	Atherosclerosis Infectious diseases	Cutaneous regeneration
Surgery		General use (microsurgery, transplantation)				Eye (hypermetropia)	Eye	General use

TABLE 3.3 General morphophysiological features of the most used fish in biomedical research.

	Turquoise killifish	Medaka	Zebrafish	Goldfish
Skeleton		Mononucleated osteoclasts	Independent feeding after 5 days postfertilization	Independent feeding after 5 days postfertilization
Kidney		C-shaped epithelial layer of primitive podocytes		
Fins	Large dorsal, anal, and caudal *fins*	Unpaired anal, caudal, and dorsal fin, paired pelvic and pectoral fins	Unpaired dorsal, anal, and caudal tail fins; Paired pectorals and pelvic fins	Unpaired dorsal, caudal, anal fin; Paired ventral and pectoral fins
Digestive apparatus	Oral and pharyngeal teeth Stomachless	Oral and pharyngeal teeth Stomachless	Teeth attached to the fifth branchial arch. Stomachless	Pharyngeal teeth. Stomachless
Reproductive system	Sexual dimorphism	Sexual dimorphism	Sexual dimorphism	Sexual dimorphism
Skin	Highly colorful skin in male; pale female	Pigmented skin. Four types of chromatophores	Five uniformly, pigmented, horizontal stripes on the trunk	Head without scales
Lateral line		Subdivided in anterior and posterior lateral line, according to the neuromasts position	Subdivided in anterior and posterior lateral line, according to the neuromasts position	Complete

Continued

TABLE 3.3 General morphophysiological features of the most used fish in biomedical research.—cont'd

	Turquoise killifish	Medaka	Zebrafish	Goldfish
Adult body size	5–6 cm	2.5–3 cm	4–5 cm	10 cm
Life span	9 months	4 years	4–5 years	10–15 years
Chromosome number (aploid)	19	24	25	50
Temperature	26–28°C	24–28°C	24–28°C	
Feeding habits	Carnivorous	Omnivorous	Omnivorous	Omnivorous
Eggs (spawning)	30–50	40–50	200–300	3–10
Embryonic development	Hatching 12 days after fertilization, as independent feeding larva	Hatching 9 days after fertilization, as independent feeding larva	Independent feeding after 5 days postfertilization	Independent feeding after 5 days postfertilization
Sexual maturity (post fertilization)	3–5 weeks	2–3 months	3–4 months	6 months

TABLE 3.4 **A summary of the utilization of different fish species in current biomedical research. Turquoise killifish and medaka belong to sister groups, respectively. Cyprinodontiformes and Beloniformes; whereas zebrafish and goldfish, both belong to the order Cypriniformes.**

Main field of use		Turquoise killifish	Medaka	Zebrafish	Goldfish
Anatomy and physiology		Basic research (age research)	Basic research (developmental biology)	Basic research (developmental biology)	Basic research (neur-endocrine studies)
Applied pharmacology		Drug validation for aging	Anticancer drug	General use	
Pathology	*Cancer and cancerogenesis*	Spontaneous tumors in brain, liver, and genital apparatus	Spontaneous lymphoma, ovarian tumor; induced melanoma, liver tumors, and xenograft	Spontaneous and induced tumors, and through transgenesis[a] and xenograft	
	Toxicology		Ecotoxicity/toxicity tests	Ecotoxicity/drug development	Ecotoxicity
	Neurodegeneration and neuropathology	Neuro-degeneration	Retinite pigmentosa	Genetic-based pathologies	
	Endocrinology and endocrine pathologies			Diabetes. Food intake regulation	Hypothalamic −pituitary axis. Food intake regulation
	Genetic pathologies		General use	General use	
	Others			Cardiovascular diseases	

[a]*Melanoma, pancreatic tumors, T cell lymphoma or leukaemia, B cell leukaemia, rhabdomyosarcoma, neuroblastoma, lipoma, Ewing's sarcoma, myeloproliferative neoplasms, corticotroph adenoma and neoplasm, testicular germ cell tumor.*

nociceptive inputs reach the cerebral cortex. The transduction of noxious stimuli into electric signals operated by the nociceptors is commonly referred to as *nociception* (Tracey, 2017). Once activated, nociceptors in skin, muscles, joints, or viscera generate a nerve signal (action potential) that is ultimately transferred to the somatosensory cortex, the parietoinsular cortex and to the anterior cingulate cortex of the cerebral hemispheres. As they respond to stimuli of a heterogeneous nature, i.e., mechanical (e.g., high pressure), thermal (too high or too low temperatures), and chemical, nociceptors are polymodal receptors. Remarkably, their thresholds of activation are lower than the stimulus intensity needed to damage tissues and thus, at least in normal conditions, pain has a *warning function* to avoid tissue damage.

Whether emphasis is on the subjective experience of pain or the functions elicited by pain, there is consensus that pain is closely (though not always) linked to nociception. Thus, while nociception is diverse from pain (Sherrington, 1906), activity in nociceptive pathways usually leads to pain, and augmented nociceptive activity increases pain (Basbaum et al., 2009; Gold et al., 2010). This makes the understanding of the mechanisms of nociception and of nociceptive alterations across diverse species essential for a correct evaluation of the degree of pain that may be suffered by a given species during experimental procedures. However, although evidence is accumulating for a robust conservation of the basic molecular mechanisms of nociception across species (Sneddon et al., 2014), one has to keep in mind that even closely related animals such as rat and mice and even between different murine strains there appear to exist a considerable evolutionary divergence in the response to pain, as reviewed in Mogil (2019).

Mammals

In mammals, nociceptive stimuli conveyed from somatic and visceral organs to cortical centers follow different routes. In particular, the trigeminal and spinal nerves exclusively provide somatic sensory fibers to the skin and the locomotor apparatus, whereas most viscera have two distinct types of sensory nerves originating from the sympathetic (via the white communicating branches of the thoracic and lumbar nerves) and parasympathetic (via certain cranial nerves and the splanchnic pelvic nerves) autonomic nervous system.

Somatic pain pathways

Somatic pain pathways collect stimuli from the skin, muscles, joints, ligaments, and bones. Except for the head and the proximal regions of the neck, which are innervated by the peripheral projections of the trigeminal ganglion (TG) PSNs and a few specific organs in the same regions that are innervated by the proximal ganglia of the glossopharyngeal and vagus nerves, nociceptive stimuli from all other parts of the body are encoded by the PSNs of the dorsal root ganglia (DRG). The main somatic pain pathways originating from these neurons are the *trigeminothalamic* and the *spinothalamic pathways*. They are crucial for the sensory discriminative aspects of pain perception, in other words for the animal to be aware of which part of the body is suffering pain. Both pathways consist of a polysynaptic chain of three key neurons, often referred to as first, second, and third order somatic sensory neurons. First order neurons are the PSNs located in the above listed ganglia. Second order neurons lie in the spinal cord dorsal horn, or the spinal nucleus of the trigeminal nerve (SNTN) in the medulla oblongata. These projection neurons give rise to axons that cross the midline and ascend all along the spinal cord and brainstem to reach the thalamus. Finally, third order sensory neurons are situated in the ventral posterolateral nucleus of the thalamus and eventually send their axons to the cerebral cortex. Modulation of nociceptive signals can occur all along this polyneuronal chain, primarily at the level of the substantia gelatinosa of the spinal cord dorsal horn or the SNTN, which has an architecture analogous to that of the spinal cord dorsal horn, and by several descending pathways that exert inhibitory or facilitatory effects onto the trigeminothalamic or the spinothalamic neurons. Other ascending pathways are also significant for the general features of pain and pain control, as they convey stimuli related to motivational and cognitive aspects (spinoreticular tract and spinoparabrachial tract), motor responses and affectivity (spinomesencephalic tract and spinoparabrachial tract), and neuroendocrine/autonomic responses (spinohypothalamic tract) to pain. Notably, the spinoparabrachial pathway is an important route of convergence of somatic and visceral nociceptive stimuli.

First order neurons

First order neurons are pseudounipolar (Dogiel, 1908) and give rise to a single main process splitting into a peripheral and a central branch displaying some functional peculiarities common to dendrites and axons. The process of primary sensory neurons is commonly referred to as a primary afferent fiber (PAF). DRG neurons have been classified following morphological, electrophysiological, and neurochemical criteria (Willis and Coggeshall, 2004). A similar classification also holds for trigeminal neurons, and, likely, the other sensory neurons localized in the other above-mentioned ganglia. DRG neurons are thus classified into three main categories: the RT97-positive medium- to large-sized light neurons (about 40%) that are immunoreactive to the RT97 antibody recognizing the phosphorylated form of neurofilament protein NF200; the LA4-positive small dark neurons (about 30%) that are immunoreactive to the LA4 antibody, which reacts with a specific cell membrane oligosaccharide, but not to RT97. LA4-positive cells typically do express binding sites for the isolectin B4 (IB4) from *Griffonia simplicifolia*; and the peptidergic small- to medium-sized dark neurons (about 30%) that synthesize many pain-related neuropeptides, such as substance P and calcitonin gene–related peptide. Under physiological conditions, the RT97-positive cells transmit non-nociceptive information by means of large myelinated fibers (Aβ); however, some RT97-positive cells involved in nociception are supposed to have thin-myelinated fibers (Lawson and Waddell, 1991). Many of the LA4-positive neurons and of the peptidergic neurons are nociceptors and give rise to small unmyelinated C fibers or thin-myelinated Aδ fibers (Merighi, 2018).

The peripheral branch of the DRG and trigeminal neurons ends innervating different tissues and organs and gives rise to several morphological specializations according to sensory modality. In general, non-nociceptive terminals form encapsulated nerve endings displaying a few morphological arrangements (e.g., Meissner's corpuscle, Pacinian corpuscles, and Ruffini endings), whereas nociceptive fibers terminate as free nerve endings that give rise to discrete arborizations in target organs.

PAFs are classified based on diameter, presence of a myelin sheet, and conduction velocity. Large myelinated Aαβ fibers (or group I and II in muscles) transmit low intensity mechanical, tactile, proprioceptive inputs to the central nervous system (CNS), but not noxious stimuli, although a group of A fibers with intermediate Aαβ/Aδ features has been suggested to participate in nociceptive transmission. Under physiological conditions, nociceptive information is transmitted only by thin-myelinated Aδ or unmyelinated C fibers (group III and IV, respectively, in muscles).

The central branches of PAFs originating from DRG neurons enter the spinal cord via the dorsal roots of the spinal nerve. Aδ and C fibers approach the dorsal horn via the dorsolateral funiculus and, after splitting into an ascending and a descending branch, primarily terminate into the superficial laminae of the dorsal horn. High-threshold cutaneous mechanoreceptor Aδ fibers appear to terminate preferentially in lamina I and outer lamina II, but some thin-myelinated fibers are likely to reach deeper localizations. A recent detailed description of the pattern of termination of the different types of PAFs in the superficial dorsal horn can be found in Merighi (2018).

The anatomical distribution of C fiber terminals within the superficial DH is related to their neurochemical features. In fact, whereas peptidergic nociceptors mostly terminate in lamina I and in lamina II_o of the superficial dorsal horn, IB4-positive terminals are concentrated within inner lamina II (II_i), suggesting the existence of two different pathways for pain processing.

Nociceptive PAFs originating from the PSNs of TG and other encephalic ganglia terminate within the trigeminal spinal nucleus. The pattern of termination of these fibers has been mostly described for trigeminal afferents and is very similar to that of the spinal nociceptive PAFs.

Second order neurons

The spinothalamic tract (STT) is composed by the fibers of both nociceptive-specific and wide-dynamic-range (WDR) spinothalamic neurons (II order neurons). Schematically, neurons of the STT localized in lamina I and in deeper laminae (IV–VI) ascend in the ventrolateral funiculus of the spinal cord and terminate in the lateral thalamus. Other neurons located at the base of the medial aspect of dorsal horn and in the ventral horn (laminae VII–VIII) ascend in the ventral funiculus and terminate in the medial thalamus (Giesler et al., 1981). Collectively, the terminals of STT neurons exhibit a patchy distribution in three main thalamic regions: (1) the ventrobasal complex; (2) the medial thalamus including the intralaminar nuclei; and (3) the posterior complex, an important site for pain integration involved in the so called *cortical-thalamo-cortical loop*, whereby the cortex reinforces the excitation of thalamic neurons by pain stimuli and vice versa. It has been suggested that neurons in the ventrobasal complex are mostly responsible for sensory and discriminative aspects of noxious stimuli, whereas neurons in the intralaminar nuclei/posterior complex are responsible for motivational and emotional aspects of pain, as well as the escape reaction to acute pain.

Third order neurons

The thalamus is the crucial relay for the reception and processing of nociceptive information *en route* to the cortex. The thalamocortical inputs are mainly conveyed to the first somatosensory area of the postcentral cortical gyrus (S I). However, imaging studies have found other cortical areas with nociceptive or WDR neurons activated by noxious stimuli, such as: the second somatosensory area (S II; receiving inputs from the ventral posterior inferior nucleus and intralaminar nuclei); certain regions of the parietal cortex; the insular cortex; the anterior cingulate cortex; and the medial prefrontal cortex. These cortical structures are highly interconnected to each other and with limbic structures. Their activation is predominantly contralateral to the noxious stimulus, except for some bilateral activations (i.e., in the prefrontal and anterior cingulate cortex), perhaps associated with emotional-cognitive factors.

Visceral pain pathways

Visceral pain arises from the internal organs including the heart and vessels, airway structures, gastrointestinal (GI) tract, and urinary and reproductive organs. However, visceral pain is not evoked from all viscera, and it is not always linked to a genuine visceral damage (Cervero and Laird, 1999; McMahon et al., 1995). The diffuse nature and difficulty in locating visceral pain is due to a relatively low density of visceral sensory innervation and extensive divergence of visceral input within the CNS. Input from the viscera has a more diffuse and less precise topographic distribution compared to somatic structures (see above) and reaches lamina I and the deep dorsal horn (Hunt and Mantyh, 2001). In patients, pain from different visceral organs can have differing areas of presentation, e.g., bladder to perineal area or heart to left arm and neck. Therefore, visceral pain is often localized to distant structures and thus, recognized as referred pain. Development of symptoms may entail referred pain to somatic structures within the same metameric field as the affected viscera (McMahon et al., 1995). In addition, secondary hyperalgesia of superficial or deep body wall tissues might occur due to visceral-somatic convergence (Cervero and Laird, 1999; McMahon et al., 1995). Finally, visceral pain is often associated with marked motor and autonomic reflexes such as profuse sweating, nausea, vomiting, GI disturbances, and changes in body temperature, blood pressure, and heart rate (Cervero and Laird, 1999; McMahon et al., 1995).

PAFs innervating the viscera project to the CNS through autonomic sympathetic and parasympathetic nerves. The sympathetic innervation follows the hypogastric, lumbar, and splanchnic nerves which traverse both prevertebral and paravertebral ganglia, until they reach the thoracolumbar region of the spinal cord. The parasympathetic innervation is mediated by vagal and pelvic afferents that terminate in the brainstem and lumbosacral cord, respectively (Gebhart, 2000).

Visceral pain is principally signaled by spinal afferents, while vagal afferents signal nonpainful sensations such as hunger, satiety, fullness, and nausea (Sengupta, 2009). Although, vagal afferents may not signal pain directly, several studies have documented that stimulation of the vagus nerve attenuates somatic and visceral pain (Sengupta, 2009). Vagal afferents originate from PSNs in nodose ganglion that project to the nucleus tractus solitarius (NTS) in the brainstem. Spinally converging visceral afferents, instead, synapse onto the second order neurons within the dorsal horn, which, in turn, project to higher centers through the dorsal column pathway, the parabrachial pathway, and the STT. Superficial dorsal horn projections, mostly form the spinoparabrachial pathway, are associated with autonomic and affective responses to painful stimuli (Bester et al., 2000). Along with NTS projections from vagal afferents, spinoparabrachial projections are transmitted to limbic and cognitive higher relay centers including those parts of the brain involved in affectivity, such as the amygdala, hypothalamus, and periaqueductal gray. Visceral spinothalamic projections travel contralaterally from the deep dorsal horn to the main thalamic projections located in the ventroposterior medial and lateral nuclei of the thalamus. The medial thalamic nuclei eventually project to the areas of the prefrontal cortex that are correlated with visceral pain (Groenewegen, 1988; Bushnell and Duncan, 1989).

Imaging studies of pain in rodents

Human brain imaging has provided much information about pain processing and modulation. The two most used brain imaging methods in rodents are magnetic resonance imaging (MRI) and positron emission tomography (PET) (Borsook and Becerra, 2011; Thompson and Bushnell, 2012). MRI has a better spatial and temporal resolution than PET in terms, but PET allows for the imaging of neurotransmitter release and glia, in addition to functional imaging of neurons. Despite several technical difficulties, studies with functional MRI or PET to image pain processing in rodents have yielded unexpectedly consistent results, with brain regions commonly activated in humans (somatosensory cortex, cingulate cortex, and thalamus) also being activated in rats or mice. Imaging has been applied to the study of both acute and chronic pain models, as well as to study the effects of analgesic (Borsook and Becerra, 2011). Remarkably, the latter studies showed corresponding activation patterns with pain and opiate analgesics, like what was observed in humans (Thompson and Bushnell, 2012) and recent studies have focused onto preclinical imaging of anesthesia and the response to anesthetics (Vesce et al., 2017).

Fish (Paolo de Girolamo)

At present, most anatomical and electrophysiological studies on pain circuits in fish have been conducted on teleost or bony fish with relatively few on elasmobranch (cartilaginous) fish. Earlier than 2002 nociceptors were not yet identified in fish and thus it was believed that they could not perceive pain (Rose, 2002). Using histology and electrophysiology, nociceptors were initially identified in the rainbow trout, a teleost, or bony fish, although previous research left open the question of the existence of nociceptors in elasmobranchs, which, remarkably are devoid of C fibers (Sneddon, 2015). Like in mammals, trout nociceptors give rise to C and Aδ fibers, and were functionally identified in three classes: polymodal (responsive to mechanical, thermal, and chemical stimuli), mechanothermal (unresponsive to chemicals), and mechanochemical (unresponsive to temperature), with electrophysiological properties similar to those of mammals (Sneddon, 2015; Mettam et al., 2012). Anatomical and electrophysiological investigations have found that, unlike mammals, only about 4%−5% of fish nociceptors are innervated by C fibers. C fibers in mammals contribute to second dull pain whereas the faster conducting Aδ fibers signal first pain to the CNS. Therefore, according to some lines of thought, the small number of C fibers implies that fish do not feel pain (Rose et al., 2014). As Aδ fibers conduct pain at a faster rate, others claim that the fish nociceptive system is more rapid and efficient than that of mammals (Sneddon, 2015).

Tract-tracing studies in fish have demonstrated that the nociceptive neuroanatomical pathways from the periphery to the brain are highly conserved and like those described above in mammals. Dorsal horn neuroanatomical strata and synaptic organization have been reported in certain species and considered to be analogous of those in mammals (Kitchener et al., 2010). In fish too, the spinothalamic and trigeminothalamic are the main pain pathways and are structured like in mammals (Sneddon, 2004). Moreover, in teleost fish (zebrafish), there are several connections between the thalamus and the subpallial ventral telencephalon (Rink and Wullimann, 2004). Several brain areas are active during noxious stimulation as demonstrated by early gene expression, electrophysiology, or functional MRI studies. Therefore, also in fish, neuronal activity varies between noxious and innocuous stimulation and the response to nociceptive stimuli is not restricted to the hindbrain and spinal cord reflex centers (Rose, 2002). As a further proof for pain responses in fish, numerous analgesic drugs have been shown to be effective in ameliorating the pain-related changes in behavior and physiology after nociceptive stimulation.

As mentioned before, the assortment of pain-related behavioral responses in mammals is very different among species (Flecknell et al., 2007). The same occurs in fish that may, e.g., show increased or decreased swimming according to species. Other specific behaviors such as tail beating or vigorous tail fin wafting may be observed in zebrafish.

Breeding, genetics, genetic alterations, and strain differences relevant to the choice of the model (Marcello Raspa and Ferdinando Scavizzi)

Introduction (Marcello Raspa and Ferdinando Scavizzi)

Laboratory animal models have been traditionally used to study human diseases and biomedical research due to their similarity to humans in terms of genetics, anatomy, and physiology. In particular, the laboratory mouse (*Mus musculus*) represents a powerful mammalian system and model organism to analyze the genetics of human hereditary pathologies since it is the first mammalian organism that has been genetically modified (Jaenisch and Mintz, 1974). Thousands of mouse lines have been now generated by individual research groups and by international consortia like EUCOMM, KOMP, and IKMC to progress toward the goal of knocking out all the mouse genes. To better understand the function of essential genes, conditional and not constitutive knockouts are preferable, typically using the Cre/*lox*P system (Sauer and Henderson, 1988) allowing the control of the time and the tissue where the gene is turned off. The use of a tamoxifen-inducible Cre (Feil et al., 2009) which can be activated in a spatiotemporal manner can help to finely regulate gene expression. Gene editing using nucleases, especially CRISP-Cas9, has recently caused a new impulse on the creation of mouse model development with elevated speed and precision, facilitating the precision medicine approach to relate human patient data with mouse model development at nucleotide level disclosing a new biomedical era. Rat genetics also started at the beginning of the last century, but the difficulty to culture rat ES cells delayed the creation of targeted rat mutant, only in 2009 authentic rat ES cells were used for to generate a rat knockout. The use of nucleases like CRISP-Cas9 that can be used for all species has brought back the importance of rat models to study specific human diseases (Ma et al., 2014). Rat was the third mammalian organism whose genome has been completely sequenced in 2004. The possibility to study reverse genetics in mice and rats and the opportunity to sequence whole genomes with increasing easiness will help to create models to study also complex and multigenic human diseases. Another organism, whose popularity is growing in animal facilities is zebrafish. It is the second vertebrate most used in biomedicine worldwide. This is due to the high fecundity, genetic manipulability, pharmacological tractability, and transparency of embryos and larvae which allow to monitor the in vivo progression of pathologies. Furthermore, the genome has been sequenced. However, as all animal model, zebrafish is not perfect, presenting some drawbacks, i.e., inbreeding depression and genome duplication.

Fundamentals of mouse reproduction and breeding performance (Marcello Raspa and Ferdinando Scavizzi)

To properly manage a mouse colony, an understanding of the fundamentals of mouse reproductive biology is essential. Generally, laboratory mice become sexually mature between 5 and 8 weeks of age. Males of most strains usually mature sexually by 6 weeks of age (e.g., DBA/2 and C3H/He) while females can conceive when they are as young as 23 days old; however, mice bred that early generally produce small litters and usually mice breed well when they are 6 to 8 weeks old. Mice are annually polyestrous animals with spontaneous ovulation and breed continuously. The duration of the estrous cycle is 4–5 days and lasts about 12 h, occurring in the evening. Vaginal smears are useful in timed mating to determine the stage of the estrous cycle. Mating can be confirmed by the presence of a copulatory plug in the vagina up to 20–24 h postcopulation. The average gestation period is 18.5–21 days. A fertile postpartum estrus occurs 14–24 h following parturition, and simultaneous lactation and gestation prolongs gestation by 3–10 days owing to delayed implantation. Litter size varies among strains, ranging from a litter of about two to three pups in some poorly breeding substrains to 12 or more pups per litter (e.g., FVB/N). The average litter size can range between five and eight pups but is highly strain-dependent. As a general rule, *inbred* mice tend to have longer gestation periods and smaller litters than *outbred* and *hybrid* mice. The newborn pups weigh 0.5–1.5 g at birth, are hairless, and have closed eyelids and ears. In a laboratory mouse husbandry context, weaning (generally, 21–24 days) refers to removing a pup from its home pen (rather than to the time a pup stops nursing and starts eating solid food). Weaning age depends on weanling size and maturity. Pups are conventionally weaned at three weeks of age when they weigh about 10–12 g. Typically, laboratory mice can breed for about 7–8 months, producing four or more litters. However, some strains produce only one or two litters, usually because strain-specific characteristics and/or mutations affect their fertility. Reproductive life spans and fertility for each strain are average values. If a pair is breeding well beyond its expected reproductive life span, retain it until the female is not pregnant within 60 days of her previous litter's birth. Generation time in laboratory mice is about 12 weeks: ~3 weeks gestation, 3–4 weeks suckling, and 5–6 weeks until sexual maturity. There are hundreds of established inbred, outbred, and transgenic strains. A *strain*, in reference to rodents, is a group in which all members are as nearly as possible genetically identical, and this can be accomplished through the inbreeding. By having this type of population, it is possible to conduct experiments on the roles of genes or conduct experiments that exclude genetic variation as a factor. In contrast, outbred populations are used when identical genotypes are unnecessary or a population with genetic variation is required, and are usually referred to as *stocks* rather than *strains*. Laboratory mice can have a variety of coat colors, including agouti, black, and albino. The different strains are identified with specific letter-digit combinations, e.g., C57BL/6 and BALB/c. Maintaining adequate recording of the all strain generated at the genetic background level is crucial. The information on each strain should be maintained and transferred to all collaborators. Important information include: correct strain name, genetic description of the mutation, the genetic background of the animals, the genotyping protocol, and phenotype. Every mutant strain name must provide precise information on the affected gene, the type of mutation, and the genetic background. Nomenclature rules are described on the MGI webpage under "Guidelines for Nomenclature of Mouse and Rat Strains" (http://informatics.jax.org/mgihome/nomen/strains: shtml). The first inbred strains were generated in 1909 by Clarence Cook Little, who was influential in promoting the mouse as a laboratory organism. Sequencing of the laboratory mouse genome was completed in late 2002 using the C57BL/6 strain. This was the second mammalian genome to be sequenced after humans. The haploid genome is about three billion base pairs long (3000 Mb distributed over 19 autosomal chromosomes plus 1 respectively 2 sex chromosomes), therefore equal to the size of the human genome. Estimating the number of genes contained in the mouse genome is difficult, in part because the definition of a gene is still being debated and extended. The current count of primary coding genes in the laboratory mouse is 23,139, compared to an estimated 20,774 in humans.

Fundamentals of zebrafish breeding performance (Livia D'Angelo)

In parallel to the growing number of zebrafish models have been using in research over last decades, a conspicuous number of breeding protocols have been published and made available for scientific community. Majority of these methods are based on decades of experience with this species. A very recent manuscript, endorsed by the European Federation of Laboratory Animal

Sciences (FELASA), has been released (Aleström et al., 2020) and collects and reviews all available protocols to ensure the welfare of zebrafish as well as reproducibility of experimental results.

Efficient breeding of zebrafish requires a comprehensive understanding of the natural behavior of this species and knowledge of abiotic characteristics of their environment. Breeding is a complex process that is influenced by several factors, including mate choice, courtship, and other social interaction–related behaviors, and perception and processing of a variety of cues, including visual, lateral line, and olfactory cues (Harper and Lawrence, 2010). In addition, breeding is also influenced by numerous abiotic aspects of the environment, including water chemistry (e.g., overall salinity, the type and amount of dissolved materials in the water, pH), temperature, amount and type of food available, density of vegetation or availability of hiding places, the size of the breeding tank, turbidity, illumination levels, circadian and seasonal changes in the light cycle, etc (Nasiadka and Clark, 2012).

Zebrafish are highly fecund species: a female can spawn hundreds of eggs per single mating. Zebrafish are generally bred from age 3 months postfertilization to well more than 3 years postfertilization. However, the ability to spawn does decline with age, and, thus, most laboratories do not breed fish older than 12–15 months of age.

Zebrafish are oviparous organisms with external fertilization and no parent care. They are asynchronous spawners, scattering mature, nonadhesive eggs in batches. Fecundity is typically high, and a single female may release clutches of several hundred eggs in a single spawning session. The presence of males stimulates ovulation in females and oviposition (Harper and Lawrence, 2010).

The spawning should be set up including a recovery period of at least 1 week to allow for sufficient regeneration and maturation of new eggs. Similarly, adults periodically not used for embryo production should be kept as mixed sex groups to allow natural breeding behavior and avoid egg-associated inflammation of the oviduct. To avoid inbreeding depression, each new generation should be produced by an outcross, and sibling mating should be performed only when absolutely necessary (Varga, 2016).

Rearing of embryos and larvae represents a key issue in a zebrafish facility. Embryos are generally kept in embryo medium, at a stocking density of up to 100 embryos/35 mL in a Petri dish, at 28.5°C in dark-light cycle. Daily check of eggs after is necessary, and carefully removal of dead and unfertilized eggs. The embryo medium should be changed regularly, and nonviable embryos, dead eggs, or chorion remains should be removed at the same time. At around 120hpf, the digestive tract has developed, the swim bladder is inflated, and larvae are thus able to swim and feed independently (Strahle et al., 2012). Early larvae are often maintained in small tanks with the water flow turned off at up to 250 larvae/L. When reaching an appropriate size, the juveniles are transferred to aquaria with water exchange. The period to reach sexual maturity varies between two and four months depending on the environment, feeding, and husbandry conditions.

Inbred, outbred, hybrid, transgenic, and genetically altered strains (Marcello Raspa and Ferdinando Scavizzi)

Inbred strains are laboratory animals produced by inbreeding. By definition of The International Committe on Standardized Genetic Nomenclature for Mice and The Rat Genome nomenclature Committe, an inbred strain is produced by systematically mating brother to sister for 20 or more generations. Inbred strains are isogenic and the individuals may be regarded as genetically identical. The objective of inbreeding is to reduce the genetic variability making all the genomic loci, homozygous. Isogenic animals are histocompatible therefore accept tissue transplants from any individual of the same strain and sex. Furthermore inbred strains are ideal to generate targeted mutant lines on a specific genetic background making possible to study a knockout gene and compare its experimental phenotype to control animals of the same strain that are genetically identical with the exception of the gene of interest. Baseline phenotypic data for the most common mouse strains are available through The Mouse Phenome Database (http://phenome.jax.org/). Outbred stocks are large populations of laboratory animals that are genetically heterogeneous preserving the allelic frequencies generation after generation. Compared to inbred strains, the genetic composition of an individual is not known a priori. Outbred stocks are not genetically defined, but they share group characteristics such as being albino and good breeders, features that make them popular as foster mothers for assisted reproductive techniques (Benavides et al., 2020). If the population is of a small size, the "inbreeding effect" will generate homozygousity. Outbred colonies, like human populations, are heterogeneous; they are used in toxicological and pharmacological studies. When animals of two different inbred strains are crossed, the resulting progeny are called F1 hybrids. They are heterozygous at all the loci where the parental strains differed. The F1 hybrids are isogenic uniformly heterozygous therefore genetically uniform as the inbred strains. They exhibit hybrid vigor, compared to inbred strains, are robust, genetically identical and histocompatible, therefore suitable for many experiments. Tumor graft, skin, and other tissues are accepted from either the parental strains. The basic types of genetically altered (GA) rodents are listed below (Table 3.5), but before generating your specific mutant, it is important to check on databases if they have been already created. Websites to visit to search for rodent mutants are: www.jax.org, www.Infrafrontier.eu, www.mousephenotype.org, www.findmice.org, etc. Transgenic rodents are subjected to random insertion of a foreign DNA. Transgenic animals are generated by pronuclear microinjection of foreign DNA fragments directly into one of the two pronuclei of the zygote, before the pronuclear fusion. This

TABLE 3.5 Generation of genetically altered rodents: a summary of different approaches.

Mutation type	Generation of the mutation	Introduction of the mutation	Strain involved	Features
Transgenic (Tg)	Random insertion of foreign DNA by pronuclear injection.	Integration in a variable number of copies and in different genome sites	Traditionally the FVB/N strain has been the preferred strain for pronuclear injection	Each integration event create a founder. Phenotype may vary from founder to founder due also to the possible inactivation/alteration of a gene in the site of insertion
Targeted mutation (Tm)	Electroporation of pluripotent Embryonic stem (ES) cells with a linear DNA vector carrying the mutation flanked by long homology arms. Homologous recombination in ES cells	Generation of specific (Targeted) mutations like knock out (KO) and knock in (KI)	ES cells derived from C57BL/6N selected by international projects like EUCOMM and IMPC	C57BL/6N ES line allows the generation of a mouse line which is genetically identical to its wild type control with the exception of the mutated gene
Conditional knock out	Electroporation of ES cells with a linear DNA. Selected gene is flanked by two lox/P sites (floxed). Homologous recombination in ES cells	Floxed genes are conditional alleles that may be turned off in a spatiotemporal manner typically using Cre/loxP system	Two independent lines to be crossed to generate a double mutant: one expressing the Cre recombinase under a temporal or spatial promoter and the other bearing the gene of interest flanked by two lox/P sites	Cre recombines the "floxed" allele at lox/P sites generating a time or tissue-specific knock out
Endonuclease mutation (EM), e.g., CRISPR	CRISPR-based guide RNAs are designed to target Cas9 endonuclease to cut DNA in a sequence specific manner and let the host cell to repair the break. Electroporation or microinjection into either the pronuclei or the cytoplasm of 1-cell or 2-cell embryos	Generation of small insertions and deletions to generate frame shift mutations	Can be used in any genetic background in mouse rat and other species	Multiple mutations have been produced in a single step. Rapid. Mutation must be confirmed by sequencing

process bring an addition of a piece of DNA that integrates in the genome in a variable number of copies. This leads to the expression of the transgene under the control of the promoter and other regulatory elements contained in the DNA construct. Transgenic animals are represented by the symbol Tg. Each event of microinjection creates a unique animal model due to insertion site and to the number of copies inserted. Phenotype may vary from founder to founder due also to the possible inactivation/alteration of a gene in the site of insertion. The founder animals are hemizygous for the transgene and designated Tg/0 (preferable to Tg/+) since they don't have any corresponding "wild type" sequence in the genome. Generally hemizygous animals have the same phenotype of the homozygous therefore it is not necessary to generate the homozygous animal. Traditionally the FVB/N strain has been the preferred strain for pronuclear injection because mothers have very large pronuclei that facilitate the injection and embryos recover very well into live born animals, compared to other strains. Other methods for random integration transgenesis are based on vector-mediated transgenic methods of different origins. Retroviral, lentiviral vectors, and transposons are commonly used to create a transgenic line.

Pretreated spermatozoa with subsequent Intracytoplasmic sperm injection have been used to produce transgenic animals but possible long-term effects on health and behavior of the offspring are still a major concern. Historically the most important technology to specifically delete or modify a gene utilize rodent embryonic stem (ES) cell lines (Homologous recombination). ES cells are pluripotent, undifferentiated embryonic cells derived from the inner cell mass of a blastocyst. The first line to be cultured in vitro was derived from the strain 129 and first knockouts were generated on this background. Now several ES lines are available, but the most used are those from C57BL/6N because selected by international projects like EUCOMM and IMPC. This new ES line allows the generation and the following phenotypic characterization of a rodent line, on the same genetic background with great advantage on the genetic homogeneity of the line. Many constitutive gene knockout lines are lethal, subviable, or have complex phenotypes, in these cases, conditional alleles may be used to control time and tissue where the gene is turned off, typically using Cre/loxP system. This system derives from the bacteriophage P1, and it consists of the recombinase Cre that recognizes and cuts two lox/P DNA sites (of 34 basepair each) then joins the two ends generating a deletion of what was inside the lox/P. Production of conditional KO requires two independent lines to be crossed to generate a double mutant: one expressing the Cre recombinase under a temporal or spatial promoter and the other bearing the gene of interest flanked by two lox/P sites. Cre recombines the "floxed" allele at lox/P sites generating a time or tissue-specific KO. Tamoxifen-inducible Cre ^{ERT2} lines are now widely used. Several techniques based on gene-editing nucleases have been used to produce target mutations in mouse and rats, like zinc finger nuclease (ZFN) or transcription activator-like effector nuclease (TALEN), but recently the clusters of regularly interspaced short palindromic repeats (CRISPR) system has proved to be the more rapid, efficient, and precise. CRISPR/Cas9 is based on a primitive bacterial defense mechanism to protect against the infection of viruses and phages. CRISPR-based guide RNAs are designed to target Cas9 endonuclease to cut DNA in a sequence specific manner and let the host cell to repair the break. Damage is repaired by nonhomologous end joining usually generating small insertions and deletions to generate frame shift mutations. Homology-directed repair can occur if a DNA template is provided to the reaction to produce specific (point)-mutations. CRISPR/Cas9 can be electroporated or microinjected into either the pronuclei or the cytoplasm of 1-cell or 2-cell embryos. Multiple mutations have been produced in a single step (Wang et al., 2013). The advent of CRISPR/Cas9 is revolutionizing genome editing, but also this powerful technology has some drawbacks since every founder is unique and needs to be carachterised by extensive sequence analysis to ensure the presence of the desired mutation and the absence of off-target mutations. The advantage of this system is that avoid the use of ES cells, therefore can be used for gene editing in any genetic background, in mouse and rat embryos and virtually in any other species or cell type. Spontaneous mutations represent a great variety of molecular events that lead to obvious phenotype. Many of these mutations are very popular mouse strains, such as *nude* ($Foxn1^{nu}$), *scid* ($Prkdc^{scid}$), *diabetes* ($Lepr^{db}$), *obese* (Lep^{ob}), and *X-linked muscular distrophy* (Dmd^{mdx}). Chemically induced mutations as the N-ethyl-N-nitroso urea (ENU) which is an alkylating agent that has been used as a mutagen in several large mutagenesis programs to produce point mutations, generating numerous mutant alleles for protein-coding genes. ENU has been widely used for forward genetic screens. Although generating random mutations, ENU screens are still carried out since the identification of the mutated gene can be easily obtained by sequencing of the whole genome.

Genetic quality control and genetic monitoring (Marcello Raspa and Ferdinando Scavizzi)

Maintaining inbred strains genetically pure is a big issue for any animal facility, in fact changes in the genetic composition may derive from contamination by accidental mating of individuals from one strain with animals of other origin or by genetic drift.

Genetic contamination seems to be the most important source of genetic alterations in inbred strains. Genetic drift is the inevitable accumulation of spontaneous mutations, generation after generation. This lead to strain divergence when the same strain is maintained in different places and the generation of new substrain after 10 generations. There are 15 substrains of C57BL/6 including the J and N from The Jackson Laboratory (Jax) and the National Institutes of Health (NIH), respectively, with distinct phenotypic features.

Genetic quality control is essential to determine the genetic composition of an animal and to screen for uniformity and authenticity of a strain (Benavides et al., 2020). Genetic quality control of laboratory rodents relies on polymorphic genetic markers that are specific for any genetic background. Genetic markers are specific (generally noncoding) sequences that, under no selective pressure, have high variability. Currently, single nucleotide polymorphisms (SNPs) and microsatellites are used for genetic monitoring. Microsatellite markers are simple sequence length polymorphisms (SSLPs) and Short Tandem repeats (STRs) that can be easily genotyped analyzing the length of the PCR products. A comprehensive list of SSLPs, including primer sequences and size of amplicons can be found at http://informatics.jax.org/marker for several mouse inbred strains and at http://rgd.mcw.edu for rat inbred strains. SNPs are the most common genetic variations. Hundreds of SNPs have been described for mouse and rat but only 28 (Petkov et al., 2004) are sufficient to characterize most mouse inbred strains. Inbred strains should have all polymorphic loci in homozygosity, on the contrary outbred colonies cannot be tested for authenticity and they can only be screened for the level of genetic heterogeneity to avoid the "inbreeding" process.

Background characterization, genetic stability, and cryopreservation programs (Marcello Raspa and Ferdinando Scavizzi)

The enormous number of mutant lines raise up the problem of undefined "mixed genetic background" in experimental rodents, in particular when multiple mutations are needed and two or more strains have to be crossed to obtain inducible and/or conditional models. Since genetic background influences phenotype through the presence of modifier genes, the expected phenotype may vary when the mutation is brought on a different genetic background. Genetic markers selected to cover the entire genome can be used to genetically monitor the new strain to know the percentage of genome of each inbred strain. In the past, this typically occurred between C57BL/6 and 129 substrains because the ES cells used for homologous recombination were derived from 129 substrains but animal studies were carried on C57BL/6 background. It is important that each line is characterized for its genetic background and also for the selection of the corresponding control animals. In case of pure genetic background (congenic), proper control would be animals of the corresponding inbred strain, instead in case of mixed background control animals should be littermates with similar genetic composition. The problem of mixed background can now be solved using CRISPR/Cas9 into inbred embryos introducing the mutation in the strain of choice or using ES cells of the desired strain (e.g., C57BL/6). In case of an existing mutant strain has already been developed on a different or mixed background and a fully congenic is needed, then a classical or marked assisted backcrossing should be performed. Periodic backcrossing of a congenic strain to the background strain (of reputable source) also minimizes divergence and keeps the congenic close to the strain background of control animals (Benavides et al., 2020). To reduce the risk of genetic drift or genetic contamination, the number of generations of in-house breeding should be minimized and lines submitted to repositories to be archived as frozen embryos or sperm. These repositories such as the Jackson Laboratory, the European Mouse Mutant Archive (EMMA), the Mouse Mutant Resource and Research Center (MMRRC), the International Mouse Strain Resource (IMSR), or RIKEN secure the relevant mouse strains facilitating their distribution to the scientific community.

Strain differences relevant to the model selection (Marcello Raspa and Ferdinando Scavizzi)

The explosion in number of genetically modified mouse strains make the choice of the best model a difficult decision. Effects of genetic background on the phenotype of genetically engineered mice are anecdotal due to diverse allele composition. For example, even a comparison between two closely related substrains like C57BL/6J and C57BL/6N demonstrates genetic and phenotypic differences that affect physiological, biochemical, and behavioral systems (Simon et al., 2013). Specialized searching approaches are necessary to identify the best mouse model and strain. A comprehensive literature search is the first approach, but in addition to that, a visit to The Mouse Phenome Database based at the Jackson Laboratory is recommended. This database is a collection of baseline phenotypic data regarding all commonly used inbred strains and GA models collecting all the strain characteristics helping the correct strain/model choice. IMPC consortium (International Mouse Phenotyping Consortium) has a specific approach to eliminate the confounding effects of the different genetic background and should be consulted at www.mousephenotype.org. This consortium continuously update mouse model phenotypes generated on a single and pure genetic background (C57BL/6N) through a systematic and standardized phenotipization pipeline. The IMPC has created a collection of open datasets of genome/phenome information that surveys a significant proportion of the genome landscape. IMPC has the goal to complete a null mouse mutant resource for the coding genome and create a comprehensive catalog of mutant strains available to study mammalian gene function. The reproducibility of the data collected is guaranteed by a cross-center verification to achieve the highest standard.

Genetic modification of zebrafish (Livia D'Angelo)

Zebrafish are used to model human diseases through hypothesis-driven reverse approaches (functional characterization of candidate genes), forward approaches (random mutagenesis, forward genetic screens, high throughput small molecule

TABLE 3.6 Common tools used for genetic modification in zebrafish.

Reverse approach		Forward approach	
Transgenesis		Mutagenesis	N-ethyl-N-nitrosourea (ENU)
			Viral mutagenesis
			Transposon-driven insertion
			Radiation
Morpholino			
Genome editing	Zinc finger nucleases (ZFNs)		
	TAL effector nucleases (TALENs)		
	CRISPR (clustered regularly interspaced short palindromic repeats)		

screens), which contribute to understanding the etiology, phenotype, and treatment of diseases (Rafferty and Quinn, 2018). Here, the major molecular and genetic tools used in zebrafish genetics are briefly reported and summarized in Table 3.6.

Reverse approaches

Transgenesis

The transgene typically consists of a ubiquitous or cell type–specific promoter sequence that drives expression of a gene or gene variant of interest. By varying the promoter and gene, transgenic zebrafish can be engineered for a variety of functions, such as:

- spatial control of gene expression by engineering transgenes to have cell type–specific promoter sequences driving gene expression (Huang et al., 2003; Rothschild et al., 2013);
- temporal control of gene expression using heat-shock protein promoters (Shoji et al., 2003);
- overexpression: by designing a transgenic construct where the promoter drives expression of the gene of interest. Expression level can be titrated by altering transgene copy number and promoter strength;
- biosensors: transgenic reporter lines can be made to express biosensors under the control of signaling pathway-responsive elements, making useful and reliable molecular tools to decipher the activation or inhibition of distinct cell signaling events. These biosensors are valuable resources for drug screening (Parson et al., 2009).

Varshney, G.K., Huang, H., Zhang, S., Lu, J., Gildea, D.E., Yang, Z., Wolfsberg, T.G., Lin, S., Burgess, S.M., 2013. The Zebrafish Insertion Collection (ZInC): a web based, searchable collection of zebrafish mutations generated by DNA insertion. Nucleic Acids Res. 41, 861−864.

Vesce, G., Micieli, F., Chiavaccini, L., 2017. Preclinical imaging anesthesia in rodents. Quart. J. Nuclear Med. Mol. Imag. 61 (1), 1−18.

Wang, H., Yang, H., Shivalila, C.S., Dawlaty, M.M., Cheng, A.W., Zhang, F., Jaenisch, R., 2013. One-step generation of mice carrying mutations in multiple genes by CRISPR/Cas-mediated genome engineering. Cell 153 (4), 910−918.

Wessler, S., 1976. Introduction: what is a model?. In: Animal Models of Thrombosis and Hemorrhagic Diseases. National Institutes of Health, Bethesda, pp. xi−xvi.

Willis, W.D., Coggeshall, R.E., 2004. Sensory mechanism of the spinal cord. In: Primary Afferent Neurons and the Spinal Dorsal Horn, 3th ed. Plenum Press, New York.

Recommended reading

Popesku, J.T., Martyniuk, C.J., Mennigen, J., Xiong, H., Zhang, D., Xia, X., Cossins, A.R., Trudeau, V.L., 2008. The goldfish (*Carassius auratus*) as a model for neuroendocrine signaling. Mol. Cell. Endocrinol. 293 (1−2), 43−56.

Schartl, M., 2014. Beyond the zebrafish: diverse fish species for modeling human disease. Dis. Model. Mech. 7 (2), 181−192.

Sneddon, L.U., 2012. Clinical anesthesia and analgesia in fish. J. Exot. Pet Med. 21 (1), 32−43.

http://www.informatics.jax.org/greenbook/index.shtml.

http://ko.cwru.edu/info/breeding_strategies_manual.pdf.

https://www.jax.org/jax-mice-and-services/customer-support/technical-support/genetics-and-nomenclature#.

https://en.wikipedia.org/wiki/Laboratory_mouse#cite_note-Ensembl-12.

Chapter 4

Animal welfare

Viola Galligioni[1], Olga Boruc[2], Helmut Fuchs[3], Martin Hrabe de Angelis[3], Paola Zarattini[4]

[1]*Comparative Medicine Unit, Trinity College Dublin, Dublin, Ireland;* [2]*EMBL Laboratory Animal Resources, Epigenetics and Neurobiology Unit, EMBL, Monterotondo, Italy;* [3]*German Mouse Clinic, Institute of Experimental Genetics, Helmholtz Zentrum München, Neuherberg, Germany;* [4]*Department of Life Sciences, University of Trieste, Trieste, Italy*

Animal welfare and behavior (Viola Galligioni and Paola Zarattini)

Proper animal welfare (good welfare) is tightly related to good science in terms of both ethics and data reproducibility (Poole, 1997; Prescott and Lidster, 2017). Therefore, promoting well-being of laboratory animals is essential and should be included in education of new researchers' generation.

There are many different welfare indicators and assessment standards, but it has been suggested the animal welfare can be effectively assessed and, therefore, implemented, by answering two main questions: "Are the animals healthy?" and "Do animals have what they want?" (Dawkins, 2003).

Although the answer to the first question refers strictly to the animal physical status and therefore is not properly related to the topic of this chapter, the second question is fundamental in assessing and improving the animal well-being. This is true especially for animals maintained in captivity, since they are totally dependent on us for husbandry and care. For this reason, the current legislation on the protection of animal used for scientific purposes reflects the reduction, refinement, and replacement's (3R's) principle and in particular the refinement concept to improve husbandry and scientific procedures which minimize pain, suffering, distress, or lasting harm and to improve animal welfare in situation where the use of animal in unavoidable.

The potential welfare problem in the laboratory animal facilities is associated with behavioral restriction during breeding, housing, and using animals in research, therefore we have to focus our attention on the reframed question: "Do animals have what they need?" Only by answering this question correctly, we will be able to improve quality of science trough better animal welfare.

Practical Handbook on the 3Rs in the Context of the Directive 2010/63/EU
https://doi.org/10.1016/B978-0-12-821180-9.00010-6

TABLE 4.1 The Five Freedoms provide fundamental principles to pursue animal welfare. However, they are just the starting point to achieve animal welfare. If an animal experienced hunger, thirst, pain distress, suffering, fear, and diseases, these could affect the status of the animal itself. Instead, the Five Domains model includes four physical/functional domains and one "mental" domain that represents animal's overall welfare state, since all the individual negative affects identified have a cumulative impact on welfare.

Five Freedoms	Five Domains
Freedom from thirst, hunger, and malnutrition	Nutrition domain
Freedom from thermal and physical discomfort	Environment domain
Freedom from pain, injury, and disease	Health domain
Freedom from fear and distress	Behavior domain
Freedom to express normal behavior	Mental domain

The paradigm has shifted from the Five Freedoms (Farm Animal Welfare Council (FAWC), 1993) to the Five Domains model (Mellor and Reid, 1994), Table 4.1. The Five Freedoms can be seen as snapshots of animal in a specific moment in time, not reflecting causes and consequences of stresses that lead to long-term problems. The Five Domains (nutrition, environment, health, behavior, and mental state), instead, are categories of factors acting on the animal that help to assess outcome indicators.

More recently, animal welfare has been switched toward the concept of Quality of Life and positive welfare framework: instead of focusing on freedom from something, animal welfare should aim to improve quality of life, by providing opportunities for positive experiences, to experience pleasures, engage with the world, and conspecifics they want and for the time they want.

But in practice, how can we recognize the well-being of an animal? How can we define whether it is suffering, is it in pain or distress? Indeed, having animals capable to express wide range of naturalistic behaviors has an important ethological validity. Therefore, we need to know, first, what is their behavior in nature, recognize their physical, social, and psychological requirements and, secondly, how to assess the behaviors we observe. If the animal has an opportunity to express a wide range of behaviors in enriched environment, it will express a natural repertoire of social behaviors, with positive affective engagement. That will help in stress resilience and social competence, giving in turn the ability to recover from stressors. Therefore, an enhanced welfare will bring to meaningful behavioral and physiological readouts.

In the scientific literature, several approaches have been published to define behavior and welfare assessment. Below there are some examples:

- Using vocalizations and spatial distribution of animals as behavioral indicators (Dawkins, 2004)
- Leach et al. (2008) proposed a welfare assessment based on physical appearances, health status, provoked/unprovoked behaviors as well as husbandry procedures and management-based parameters (environmental conditions, staffing, and resources available in the cage).
- Spangenberg and Keeling (2016) defined criteria and parameters related to welfare principles of good feeding, good housing, good health, and appropriate behavior.

Welfare issues can be consequent or not to procedures carried out on animals, so it's important to understand what's "normal." Some examples of welfare issues are summarized in Table 4.2.

Within the same species, specific parameters could be peculiar of the strain, sex, age, or the genetic modification and before planning any experiment, these features should be clear to justify and validate the model used. By simply watching your animals, you will be able to understand how they interact with the environment, other animals, and the human presence.

Therefore to keep and improve animal welfare, the thorough understanding of animal behavior is necessary.

Laboratory animals well-being is the result of fulfillment of their physiological needs (*physical health*) and their behavioral needs (*from the affective to the aggressive states in animals*).

TABLE 4.2 Examples of welfare issues and possible causes, related or not to experimental procedures.

Welfare issues	Possible causes related to procedures	Possible causes not procedures-related
Wounds or aggressive behaviors	Intramuscular injections, postsurgical wounds	Cage change, new environment
Animals smaller than others in same cage	Pain consequent of a procedure, infection of surgical area	Malocclusion, infection
Poor grooming	Distress and pain consequent of a procedure	Stressful environment
Impaired locomotion	Consequence of surgery or experimental infection (i.e., Experimental Autoimmune Encephalomyelitis model)	Phenotype injury, infection

The physiological needs, such as the need for food, water, or thermal comfort, are common to all species whereas behavioral needs such as nest-building and social contact in rodents are the outcome of a long adaptive and evolutionary process which is specific for each species. They are influenced by the genetics of the animal (innate behaviors) and by the behavior in which they live (behaviors learned through individual experience). The inability to satisfy these needs leads to the development of behavioral anomalies (e.g., stereotypies).

Studying and identifying species-specific and individual-specific behaviors is not trivial. Definitions of behavioral categories should be clear and unambiguous before starting data collection. Behavior is not something that can be captured in a screenshot, it consists in stream of movements and events, including effects of animal behavior on the environment, other animals, or itself (Martin and Bateson, 2007). An ethogram is a catalog of species behavior patterns and the objective criteria that identify them. Clear examples of mice ethograms are reported on mousebehaviour.org. Behaviors can be classified based on the presence of movement (active behaviors), on the function (i.e., maintaining physiological stasis as in maintenance behaviors).

By knowing the normal behavior of the species, it's then easier to spot what is abnormal.

Rodents are social animals and they establish their own hierarchy in the group. Rodents are altricial animals, so the care of the litter by the mother has an enormous impact on the phenotype development (and that's why foster mother is important, when needed). They are also nocturnal animals and in the wild they occupy large spaces, traveling long distances to find food.

If we consider senses of laboratory rodents, they have the same as humans, but they differ in acuity (Olsson et al., 2003). Olfactory clues (or their absence) are more important than visual cues in communication and the interaction with other animals: for this reason, it is important to preserve nest odor cues during cage change, thus reducing aggression. Rodents have a playful behavior: pouncing, pinning, and tickling are a positive experience and are key parameter in studying the affective state in animals (Table 4.3).

Husbandry and enrichment

Laboratory animals spend the majority of their lifetime in their home cage and the cage environment needs to be adapted to the specific needs of the animals. For this reason, it is important to use animal-based benchmarking parameters to provide an appropriate environment (Table 4.4).

The assessment of husbandry practices and materials is commonly performed only on resources availability or management needs (most often budget constraints), without considering that animals of different strains, sexes, and ages will react differently to the same conditions.

TABLE 4.3 **Resources and links related to laboratory rodent behavior.**

Mice	mousebehaviour.org	Mouse-specific ethogram (list of species-specific behaviors, describing the elements and function of each behavior performed by the animal)
	phenome.jax.org	Phenotype (physiological and behavioral) data correlated to the genetics of the strain
Rats	ratbehaviour.org	Collection of information on rat behavior and biology
	https://www.youtube.com/watch?v=giu5WjUt2GA	Short documentary about what happens to lab rats when left in the wild. Explanation on why it's important to keep some features of the outside world in the lab rat cage

TABLE 4.4 **Influence of husbandry practices on animal behavior.**

Actions	Elements to be considered	Physiological and behavioral needs or animal benchmarking parameters
Caging	Dimension (spatial requirements), ventilation system, position on the rack	Exploratory behavior, natural movements, social interaction
Watering	Quality, delivery system	Growth, weaning, survival
Feeding	Formulation (various nutrient content), manufactures (pelleted, extruded, meal diet, flaked or dry or powdered diet)	Growth, reproduction, survival
Bedding	Type of material, size and manipulability, volume and depth of the litter	Spatial segregation (clean and dirty areas), animal comfort, thermal preferences, parental behavior
Changing	Frequency, preservation of nesting materials	Social interaction, parental behavior
Handling	Method of picking up the animal, duration and frequency	Social interaction, exploratory behavior,

In accordance with Bailoo et al. (2018) "the manner in which laboratory animals are housed is driven by economics (minimal use of space and resources), ergonomic (easy of handling and visibility of animal), hygiene and standardization (reduction of variation). This has resulted in housing condition that lack sensory and motor stimulation and restrict the expression of species-typical behavior. In mice, such housing conditions have been associated with indicators of impaired welfare, including abnormal repetitive behavior (stereotypies, compulsive behavior), enhanced anxiety and stress reactivity and thermal stress."

Considering the European framework, the Directive 2010/63/EU defines species-specific care and accommodation of animals in Section A and Section B of Annex III.

Refinement in the care of laboratory animals starts at the level of housing and husbandry, putting in place measures to enhance physiological and behavioral needs.

Environmental factors that have a huge impact on animal behavior and physiology are: temperature (see Table 4.5 for more details on the thermoneutral zone, TNZ), relative humidity, light and light cycle, smell and odor cues.

Today the use of individually ventilated cages is widespread in rodent facility, since it decreases effectively the accumulation of intracage ammonia and also the frequency of the cage changing routine. Nevertheless, laboratory animals are routinely handled for cleaning and scientific procedures and

TABLE 4.5 Thermoneutral zone and its impact on the animal and on research.

Thermoneutral zone definition	Pathophysiological parameters correlated:	How to increase temp, so that it's closer to the thermoneutral zone	Resources
A range of ambient temperatures where metabolic rate is at basal or resting levels. Within this range of temperatures "a stable core temperature is achieved with adjustments in insulation, posture, and skin blood flow and there are no active mechanisms of evaporative water loss."	- Morphological and physiological differences; - Immune response; - Cancer development.	Include nesting material in the cage in proper quality and quantity. Nesting is an important behavior, how mice make nests or fail to do so after new substrate is introduced in the cage could also be an indicator of their welfare.	Chen et al. (1943), Gordon et al. (2012), Gaskill et al. (2013), Maloney et al., (2014), Hylander and Repasky (2016), Hankenson et al. (2018).

humans working with them wish to avoid stress and anxiety with the implementation of nonaversive handling. For instance, picking up mice by tail can compromise their welfare and affect scientific outcomes due to aversion and high anxiety levels which can be minimized by using a tunnel or a cupped hand (Gouveia and Hurst, 2019).

One other good example of positive handling technique is mimicking biological behavior in juvenile rats by tickling that simulates the rat rough-and-tumble play of young rats. This interaction between the handler and the rat produces vocalizations of 50 kHz that are associated with positive state. A maximum of 15 s three days a week per rat has been shown to have a positive impact on the young rat: rats develop more social competence, therefore having more access to affiliate social interaction in their lifetime. It also decreases the avoid/escape behavior (avoiding experimenter hands). Tickling decreases the negative impact of social isolation and, when done prior to injection, it decreases the negative affective response.

Tickling enriches the human−animal relationship: it helps in habituation to handling and before manipulations such as injection and it makes handling and other interactions consistent and predictable.

Environmental and social enrichment

As already discussed, housing and care should be adjusted to the (natural) needs of the species and should comply with demands of the legislator and those of the investigator. However, what we can offer is at best a compromise: a cage with nesting material and enrichment (Table 4.6), to be shared with other conspecifics (in case of social species) that the animal itself does not chose.

TABLE 4.6 **Type of behavioral needs and environmental enrichment options.**

Category of enrichment	Type of enrichment	Tools
Environmental	Physical objects (see Fig. 4.1)	Burrowing substrate, nesting material, house mouse, tunnels, igloos, shepherd shacks, shelves, cage climbers
	Physical exercises	Running wheels, climbing walls, and other apparatuses, increased vertical space, gnawing materials, balls or marbles
	Nutritional enrichment	Seeds, grains, flavored pellets, gel treats or crumbles
Social	Housing	Paired- and grouped-housing within the same space; nonsocial housing in the same room

Environmental enrichment usually means any change in a barren environment. They can have a positive impact, promoting natural behavior that are usually suppressed in laboratory animals or in avoiding abnormal behavior, due to distress caused by restricted conditions that limit control of the animal on the physical and social environment. One of the main aims of enrichment is to allow the animal to perform species-typical behaviors that return control over the environment to the animal and promote homeostasis (Garner, 2005). To evaluate whether the material/object is enriching, we need to understand whether it is addressing a specific behavioral need and, therefore, it is biologically relevant.

In the past, many scientists, especially in the neurobiology area, believed that having barren environment could produce better science. That would also allow comparison with data from the past, generated in barren housing, because biological data would have been biased by enrichment and nesting in the cage. However, Andrè et al. (2018) showed that simple environmental enrichment and nesting material tested in two mouse strains of both sexes did not significantly affect data collected on 164 physiological parameters, therefore nesting and enrichment can be applied without skewing results. That means that research and animal welfare can go hand in hand.

Of course, the enrichment should be suitable to the research study, since it shouldn't confound or bias the data. One classic example is the use of nutritional enrichment in obesity research, or high dust content tissue as nesting material that can be problematic for nude mice. Another example is improving cage complexity that runs the risk of creating choke points and other ambush sites that can induce aggression, which can have consequences on stress physiology (Howerton et al., 2008).

Usually, attention is given to the physical conditions in which laboratory animals are housed, including environmental enrichment, but meeting their social needs should be considered equally important for both animal welfare and scientific reasons (Hawkins et al., 2014).

A meta-analysis of 100 research papers using mice and rats, published in 2010–11, found that 99 of them failed to take into account the social organization of these animals (Hendrie and Pickles, 2011).

To evaluate whether the material/object is enriching, we need:

1. To understand whether it is addressing a specific behavioral need and, therefore, its species-specific relevance.
2. To make sure that it is safe for the animals.
3. To ensure that it will not interfere with the scientific enquiry (Fig. 4.1).

FIGURE 4.1 Examples of physical objects that can be used as environmental enrichment. A: polycarbonate tunnels for mice; B and C: polycarbonate shelters with different colors; D: different paper strips for nesting; E: special bedding with different size and manipulability; F: example of complex structural enrichment: two-levels environment, in which the rat can climb, jump, and stand in up-right position. *Credit: Paola Zarattini.*

Design of in vivo research infrastructures and laboratory animal facilities (Helmut Fuchs and Martin Hrabe de Angelis)

In vivo research infrastructures

Research Infrastructures are designed for the benefit of the scientific community to make use of know-how, facilities, equipment, etc., that are available as central infrastructures and accessible, for example, via collaborative projects. An exact definition is given on the website of the EU commission (https://ec.europa.eu/info/research-and-innovation/strategy/european-research-infrastructures_en): "Research Infrastructures are facilities that provide resources and services for research communities to conduct research and foster innovation. They can be used beyond research, e.g., for education or public services and they may be single-sited, distributed, or virtual. They include major scientific equipment or sets of instruments, collections, archives or scientific data, computing systems and communication networks or any other research and innovation infrastructure of a unique nature which is open to external users." In vivo research infrastructures can serve different purposes as, for example, the generation and breeding, archiving or phenotyping of mutant animals, but there are also very specialized infrastructures such as gnotobiotic facilities. This means, in vivo research infrastructures are special cases of laboratory animal facilities. Examples for in vivo research infrastructures for the archiving of mutant mouse lines are the European Mouse Mutant Archive (EMMA Wilkinson et al., 2010), the Jackson Laboratory, or Riken to mention just a few of them. Research infrastructures for mouse phenotyping, and in some cases also for rat phenotyping are located in Europe, for example, at the Institut Clinique de la Souris ins Straßbourg, the Czech Center for Phenogenomics in Prague, the MRC in Harwell, or the German Mouse Clinic in Munich.

The existence and use of in vivo research infrastructures have many advantages for the scientific community as well as for animal welfare and 3Rs (Table 4.7), and large-scale collaborative research is considered as a procedure to improve research quality and reliability of results (Ioannidis, 2014). In the European context, infrafrontier (Consortium, 2015) (Raess et al., 2016) acts as a common interface and coordinates the activities of mouse research infrastructures from different countries in Europe and Canada. In this chapter, we will mainly focus on examples for in vivo research infrastructures and laboratory animal facilities for research based on mice as this is one of the major proportions of laboratory animals, and examples will be given based on our experience for mouse phenotyping units. But most information might also apply for other areas of research and other species.

Regulations from the EU Directive 63/2010 that apply on the design of in vivo research infrastructures and laboratory animal facilities

In the following part of this chapter, a brief collection of topics is presented that are regulated by the EU Directive 63/2010 and that are of special relevance for the design of in vivo research infrastructures and laboratory animal facilities (see also Table 4.8).

Article 22 of EU Directive 63/2010 regulates the requirements for installations and equipment. They have to be suited for the species and the procedures have to be carried out as effectively as possible, minimizing the number of animals and their burden but ensuring to obtain reliable results.

In article 30, the animal records that have to be collected and stored are defined. This topic might also address the design of a laboratory research

TABLE 4.7 Advantages of in vivo research infrastructures in the context of 3Rs.

Reduction
• Synergy effects by fostering collaborations between scientists • Efficiency needs less animals centralized at one facility o Broad spectrum of parameters o All analysis on same animals enables comparability • Share control animals (e.g., by windowing approach (Haselimashhadi et al., 2020)) • Data pool from other projects to compare and evaluate results • Constant improvements in statistics tools and data analysis • Avoids repetitions of archiving and distribution system o Data freely available for the scientific community
Refinement
Experience and routine of procedures Trained stuff (for procedures and animal welfare) Standardization Highest quality standards, quality management systems (e.g., ISO 9001 certification, NF X50-900 certification) International networks for constant exchange of know-how (e.g., INFRAFRONTIER, IMPC (Brown and Moore, 2012a,b; Koscielny et al., 2014; Mallon et al., 2012; Meehan et al., 2017; Munoz-Fuentes et al., 2018)) Medical relevance by involvement of collaborating physicians Meta-analysis
Replacement
Data in databases is the bases to avoid animal experiments and for the establishment of strategies without the need for animals (e.g., by technologies based on artificial intelligence)

facilities in terms of design and connectivity to database systems that are able to support the personal of the facility to collect and store the information mentioned in article 30.

The most important and detailed topics are listed in the Annex III of the Directive that regulates the requirements for the establishments and for the care and accommodation of animals. For this chapter, topics 1 and 2 are relevant. For sure, the design of the complete facility will be the basis and thus have consequences for further topics, for example, Annex II, 3.4c cleaning and sterilization of feeders.

A large proportion of the requirements mentioned in Annex III topics 1 and 2 have to be met and designed by the architecture of the facility, and will not be addressed in detail (e.g., noise insulation and absorption materials or isolation and heating systems). But for other topics, for example, the design of

TABLE 4.8 Regulations from the EU Directive 63/2010 that are relevant for the design of in vivo research infrastructures and laboratory animal facilities.

Article	Topic
Article 22	Requirements for installations and equipment
Article 30	Animal records
Annex III 1.1	Functions and general design
Annex III 1.2	Holding rooms
Annex III 1.3	General and special purpose procedure rooms
Annex III 1.4	Service rooms
Annex III 2.1	Ventilation and temperature
Annex III 2.2	Lightening
Annex III 2.3	Noise
Annex III 2.4	Alarm systems

general and special procedure rooms in the facility (topic 1.3), or the lightening controls (topic 2.2), there are additional aspects that might be worth a more detailed consideration.

General architecture and design of an in vivo research infrastructures and laboratory animal facilities

The technical planning of an animal facility requires to plan and make decision about general strategies that fit best the purposes of the planned facility. For example, a decision has to be taken about the kind of diet that will be used: Depending whether autoclaved or irradiated diets are applied, appropriate configurations for the entry of the diet have to be arranged. For example, for autoclavable diets appropriate autoclaves and autoclaving capacities have to be planned.

A decision has to be taken concerning the supply of water and food for the animals. Animals might have access to water either by an automatic watering system or via bottles. The decision will have consequences on costs for constructions of the building and costs and options for the cage system, as well as later on expenditures. Bottles are in general the simpler choice, but the consequence will be the permanent workload for the refill and change of the bottles, as well as the need for a filling station and bottle cleaning. Care has to be taken when there are fluctuations in the room pressure. This might lead to running out of the bottles (for example, at restart of the ventilation system after regular checkup) and anadipsia can be a consequence. With an automated watering system, the mentioned problems can be overcome, but it is more expensive and the animals have to learn to use the system. There is always a danger that mice play around with the nipples and open the valve by putting in some bedding material that causes a continuous water flow and the animals drown. The cage system needs to be compatible with an automated watering system. A bottled watering system might be the better choice if medication or experimental substances should be supplied with the water.

The general blueprint of a laboratory animal facility is a barrier consisting of a clean corridor for the supply of the animals with clean and fresh material (cages, bedding, food, water, etc.) and a dirty corridor where the used material is disposed (Fig. 4.2A). The rooms of the lab animal facility are in between these corridors, and have direct connection to both, the clean and the dirty corridors. The clean corridor is connected to entry points for material and staff. Material is entered via different kinds of disinfection (e.g., autoclave, hydrogen peroxide chamber, disinfection bath). The present gold standard for thc entry of personnel is a sluice with complete change of garment, sitover/stepover, and air shower. The dirty corridor is connected to a sluice to exit the material from the barrier, and sluices for exit for the staff. The used material is then either disposed or cleaned in the washing facility that can be accessed for the entry of the dirty material and is connected on the clean part to the entry points for the material.

In addition to the mentioned elements, the EU Directive specifies in the Annex III expectations concerning storage rooms (Annex III, 1.4a) as well as for the hygienic storage and safe disposal of carcasses and animal waste (Annex III, 1.4c). There is also the need for a quarantine area (Anne III, 1.3b) for new animals, to enable newly acquired animals to be isolated until their health status is determined and there is no potential health risk for the existing animals in the facility.

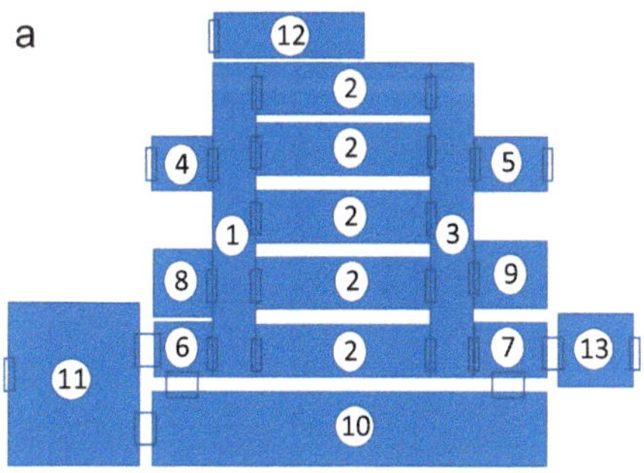

FIGURE 4.2A General design of a laboratory animal facility with a two-corridor system. *Credit: Helmut Fuchs.*

There are many possibilities for variants from this general design, depending on the specific needs and the scientific directions of the facility. There is the option that there is a subdivision in the lab rooms into different sanitary levels (Fig. 4.2B). In this case, the entry of the clean material might be shifted toward the cleanest areas, and the exit of the used material is closer to the exit points. Thus, there is a gradient of material flow from clean to dirty.

Even the deviation from the two corridors concept has been practiced in many facilities with strong success. This variant of the design can be even more advanced, for example, for the use in phenotyping facilities, where there is one corridor for the logistics of the animal rooms and another corridor to access the lab rooms (Fig. 4.2C). Thus personnel for animal care and animal procedures can be separated. This might be reasonable in cases where animal

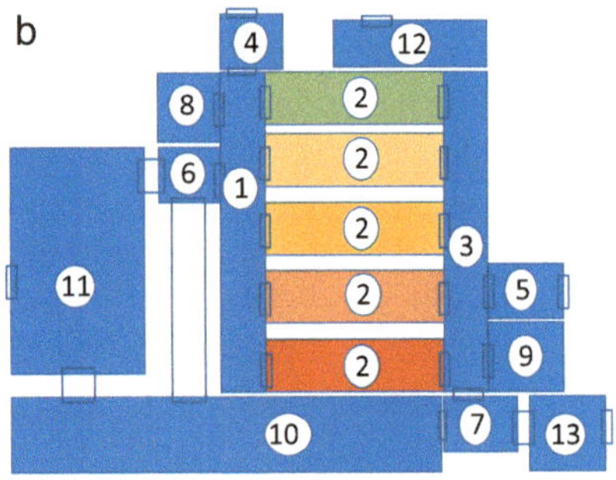

FIGURE 4.2B Design of a laboratory animal facility with different hygienic levels. Different colors of the animal rooms (2) represent different sanitary levels from green (highest level) via yellow to red (lowest level). *Credit: Helmut Fuchs.*

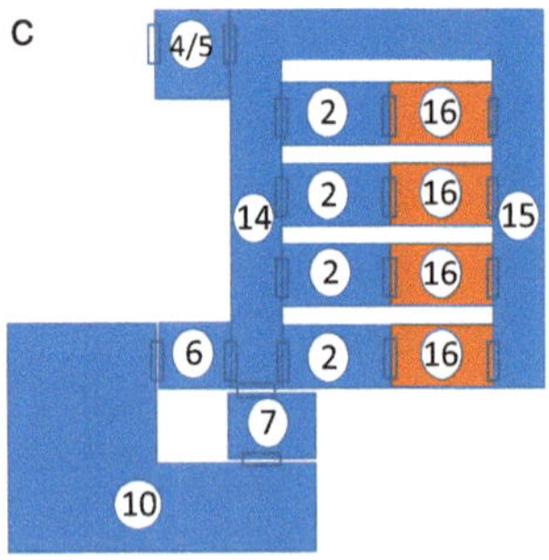

FIGURE 4.2C 1: clean corridor, 2: animal room, 3: dirty corridor, 4: entry for stuff, 5: exit for staff, 6: entry for material, 7: sluice to discharge material, 8: storage for clean material, 9: storage for used material, 10: cleaning area (see Annex III, 11.4b of European Directive), 11: storage (see Annex III, 1.4a of European Directive), 12: quarantine (see Annex III, 1.3b of European Directive), 13: cadaver (see Annex III, 1.4c of European Directive), 14 corridor for animal care, 15: corridor to access procedure rooms, 16 procedure rooms. *Credit: Helmut Fuchs.*

care activities and procedures with animals have to meet different sanitary demands, or where the use of procedure rooms does not always need the same protective equipment (e.g., masks, gloves, etc.) and this can be avoided during certain procedures for the sake of health and safety measures (e.g., permanent wearing gloves might promote allergies).

In many cases, the design of a laboratory animal facility cannot be matched with a perfect design, as the details of the design of the facility always have to fit to the local circumstances. If there is an existing building that needs to be connected, or the area where the facility will be situated has a specific scene, compromises have to be found (Figs. 4.3 and 4.4).

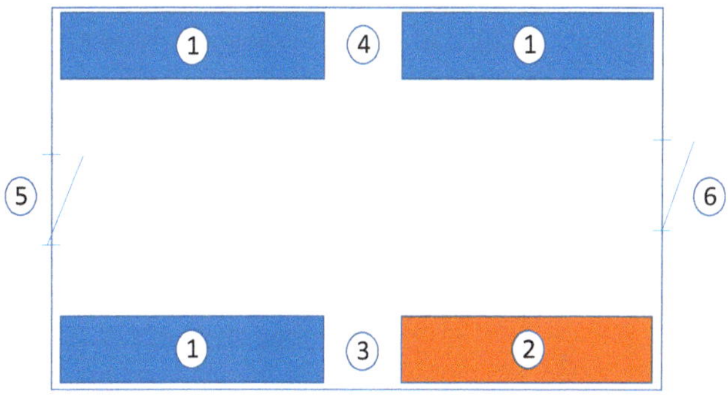

FIGURE 4.3 1: Example for an animal room. 1: mouse rack, 2: cage changing station, 3: space for computer, 4: space for material/documentation, 5: corridor, 6: procedure room. *Credit: Helmut Fuchs.*

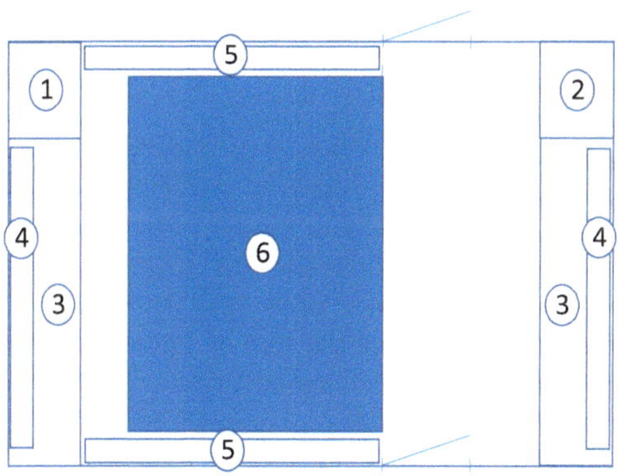

FIGURE 4.4 1: Example schematic view for a big procedure room. 1: fumigator (e.g., for anaesthesia), 2: lavatory and cleaning area, 3: desk, 4: installations for gas, power, LAN, 5: installations for gas, power, LAN at the wall, 6: space for flexible use, e.g., big machines, freezers, tables, etc., 7: installations for power and LAN coming down from ceiling. *Credit: Helmut Fuchs.*

Architecture and design of animal rooms

Size and capacity

The first aspect to consider when planning an animal facility is the size and capacity. In this respect, it is important to realize that the planned maximum capacity usage might not exceed about 85%. This is due to circumstances where unexpected space demands might come up. Thus, one has to keep space empty for special situations for sick or injured animals (see Annex III, 1.3c of the EU Directive) to house them separately, or just in case breeding resulted in bigger litters as expected.

When planning a new animal facility, a decision has to be taken for the size of the animal rooms. In general, the size should fit to fill the room with a maximum of cages for the planned species without to create unusable areas. But still there is the option for rather big rooms or smaller rooms to house the animals. From the point of view of construction costs, for a given number of cages, small rooms will be more expensive than bigger rooms, and big rooms will be more flexible for cases where the usage of the room may change over time. Smaller rooms are calmer, which will have certain advantages for animals that underwent surgeries or for behavior experiments. In big rooms, it might be easier to share the room when people for animal care and experimentation have to work in parallel. In order to save space, big animal rooms for rodents can be designed using a so-called library system where the rack containing the cages are movable similar to the shelves in a library. In many cases, one is best advised with a good mixture of small and big rooms to serve the different purposes.

Lightening

An important aspect for the design of animal rooms that is regulated in the Annex of the EU Directive is the lightening. In topic 2.2 of Annex III, an appropriate light/dark cycle is stipulated. Of importance are the biological requirements as well as a satisfactory working environment. The light has to be adequate for husbandry and suitable for the proper inspection of the animals. Illumination has to fit to the specific demands of the housed species. Special care has to be taken for albino animals as their eyes are sensitive to light. For the generation of artificial light/dark periods, there are different technical realizations on the market. They span from simple local controller units to complex integrated systems that can be operated via the internet from apart. One feature might be the inclusion of sunrise and sunset periods to carefully adapt the animals to the lights-on and lights-off situation. In some cases, the perfect light situation for the animals might not be compatible with safety at work conditions for the employees. For example, mice prefer a darker atmosphere with around 400 lux daylight, while this is too dark for safe labor conditions. As the expenditure of time for animal care will only take a small fraction of the daylight time, a possible solution is the implementation of a switch to change the light intensity in a higher level when humans have to carry out some work in the room. After termination, the light can be switched back to the lower level, or even this can be automated after a given period of time, or by a motion sensor.

Special experiments might need special settings for the light regime (e.g., for metabolic analysis or behavioral assessment on certain conditions). Or a complete inversion of the light cycle might be helpful if experiments are designed for the dark phase, which is the active phase of the animals. This will facilitate the analysis during normal working hours, instead of coming during night time to the facility. But this change requires careful consideration, as the inversion of the light cycle may be a burden for the animals, and they will need a long period to adapt to the new light regime. From the point of view of the design of an experimental animal facility, it is important that the light controller system is able to realize all the needed settings without hours of complicated programming. Even increments or decrements of the light period over a given time should be feasible. In most cases, these kinds of experiments with special light/dark settings are restricted to a very low number of experiments and animals. Thus it is important that the controlling system is able to run different light regimes in different rooms.

For the design of an experimental animal facility, additional features might be helpful. For the surveillance or experiments during light off periods in mouse facilities, the supply with red lights can be helpful. As the mice won't see the red light, the experimenter can observe the mice without to disturb them. In the best case, the red light is also available in the corridors, which facilitates the entry to and exit of the animal rooms without other light coming into the room.

Washing and cleaning facility

Kollmus et al. (2012) visited a series of mouse facilities, infrastructures, and commercial breeders and concluded that the washing and cleaning facility is an essential element of the support infrastructure. It is usually equipped with tunnel washers, rack washers, and bottle washers. In many cases, there are automated washing systems with robots that feed the cages in and out of the washer, and the bedding is automatically dispensed in the cages. Attention has to be taken that the favored bedding material is compatible with the automated system. For example, there might be problems arising from a too sticky bedding that is not able to trickle into the cages.

Architecture and design of laboratory and procedure rooms

The EU Directive demands the availability of procedure rooms (Annex III, 1.3a) for simple diagnostics, postmortem examinations, or sample collection. In addition, procedure rooms are required for procedures or observations that are undesirable to carry out in the animal room. But also for other reasons, the installation of special laboratory rooms close to the animal rooms is of advantage. This applies particularly for phenotyping facilities, where specialized equipment with requirements for space, installations for supplies or safety precautions have to be met, just to mention a magnet resonance tomography system as one possible example. Similar as for animal rooms, the size of the rooms requires attention during the planning of the facility. Smaller rooms (around 10 square-meters) will be the best choice for operations that need a calm environment, with one or maximum two experimenters in the room. This applies, for example, to experiments with behavioral observations of the animals, and for running machines where the operator has to take special safety precautions, or might disturb other labor in the same room. Bigger room sizes are in many cases much more flexible to use, and might be cheaper to construct. In order to be flexible, the lab furniture might be best organized with a mixture of fixed and moveable installations. The more mobile elements are used, the easier will be a change of applications inside the room. Consider that in science technological progress and innovations are major drivers, and the implementation of ground-breaking new systems will be crucial for the success of research.

An essential component of the procedure rooms compartment is the availability of sufficient storage space: not all phenotyping equipment will be needed all the time, and might be disassembled and stored to safe lab space. But this option might be considered with caution as assembling and disassembling of sensitive lab equipment can risk damages or at least a recalibration of the system. Consider a sufficient number of electric power supplies. A few thoughts might be worth to spend on the desk height. Depending on the kind of work that will be operated, a sitting or standing position might be more advantageous.

For diagnostics and phenotyping, imaging systems such as X-ray, (micro-) CT, but also higher sophisticated systems such as PET, SPECT, or MRT are of special interest. These rooms have special demands for control of entry, safety, shielding, installations, supplies, and even tonnage of the ground floor. An MRT system for mice, for example, requires with around 5 tons a special tonnage of the ground floor and a shielding from environmental magnetic waves. The positioning of the system close to moving metal (as for example elevators) should be avoided. Laboratory rooms might need special safety regimes and different biosafety levels might apply. The biosafety level will have consequences on the air pressure regulation of the room. Normally, there should be a positive pressure with highest levels in the mouse room and with a gradient via the lab rooms and corridors to the outside world. For laboratory rooms in higher biosafety levels than 1, a negative pressure is needed to avoid a flow of dangerous agents outside the laboratory. For many experiments in the lab, but also for euthanasia of animals, gas provision is needed. A centralized supply via a tube system that is charged with bottles that are located outside the facility will help to keep the sanitary concept. Thus, no bottles that might imply a certain risk for the sanitary status of the facility have to be entered.

Hybrids of procedure rooms and animal rooms: for long-term experiments where animals stay for 24 h or longer. A day night cycle regime has to be installed, and climatization has to be similar to a mouse room. Installations for experimental settings might be needed inside the room. A special variant might be a climate chamber, where mice can be housed under certain temperature conditions, or even temperature regimes, while physiological data can be retrieved.

An important component of a laboratory animal facility is the computer and IT infrastructure. Equipment of procedure rooms and especially imaging devices might produce immense amounts of data that should be stored, if possible, on a central data storage system that is located and organized outside the facility. Data handling and analysis should be possible in offices outside the animal facility. For this purpose, a connection of the equipment via LAN as well as WLAN connections should be foreseen. Especially for devices that produce big amounts of data, a fast transmission is desirable. A concept for a LIMS system, electronic lab books, electronic breeding books, or animal databases for the documentation of animal breeding operations will be helpful. A good solution for an animal database might directly support the collection and storage of all data that is required to fulfill the documentation of animals according to legislations (see EU Directive article 30). It might be convenient to connect the centralized database system to a documentation system for humidity and temperature (Annex III, 2.1b). There is a selection of commercial systems for the documentation of experimental animals available. In special cases, tailor-made systems as, for example, MouseNET (Pargent et al., 2000) for ENU-projects have been programmed. There are also free systems available (e.g., MausDB (Maier et al., 2008) that is able to run the documentation according to German laws).

If tablets, monitors, or handheld systems can be implemented and connected to a central database system, there is the possibility to directly transfer all kinds of actions or observations in front of an animal cage in real time to the database system. As well, all information about a single animal can be retrieved directly in front of the animal (Table 4.9).

Management in vivo research infrastructures and laboratory animal facilities (Olga Boruc)

Running of animal facilities has become more complex in recent years creating more challenges for the managers. Changes in the legislation led to adjustments in training, competency assessment, housing conditions, and local procedures. This required modification in staffing structures and operations. Introduction of EU Directive 63/2010 started the process of establishing a

TABLE 4.9 Checklist for the planning of a laboratory animal facility.

Area	Topic	Option (s)
General design Data connections	Water supply	Bottled versus automated system
	Diet provision	Irradiated versus autoclaved
	Sanitary concept	Are all animals same sanitary level?
	Storage	Is sufficient space available? Inside and outside the barrier?
	Are there machines with special demands for data transfer?	Install powerful LAN system
	Online animal documentations and databases	Make sure that all areas can be connected, is WLAN everywhere accessible? Sufficient capacities?
Animal room	Purpose of the rooms	Decide for adequate room size
	Lightening	Special measures for the care of the animals needed?
	Lightening	Special conditions required?
Procedure rooms	Purpose of the rooms	Decide for adequate room size
	Special machines	Any special installations for supplies needed?
	Special machines	Any special safety precautions needed?
	Special machines	Control of entry, shielding, tonnage of the ground floor
	Special machines	Shielding, tonnage of the ground floor to consider?
	Biosafety level	Which level(s) apply?
	Installations	Special gas supply required?
	Storage	Is sufficient space available?
	Desk height	Sedentary versus upright

common framework for EU members in promoting 3Rs and is improving transparency to the general public on the use of laboratory animals. Impact of the Directive still varies among the Member States, mostly due to differences in national legislation so the structures of the animal facilities within different EU countries still vary.

However, some aspects are developing well like promotion of a culture of care, introduction of animal welfare bodies, or appointment of named personnel. Those are contributing positively to local 3Rs applications and animal welfare.

There are many different types of animal facilities and there is no one management plan that is suitable for all. It will depend on multiple factors like the size of the facility, location, species housed, available resources, and type of research being performed. Laboratory animals are widely used in colleges, universities, contract research organizations, and pharmaceutical companies, each of them heaving their management challenges. Type of the facility will impact on staff structure and competence, record keeping, and daily operational requirements.

Staff

Continues advancement in the area of animal welfare impacted on the type of personnel needed to fulfill all the tasks in daily activities (Table 4.10). More training and education were introduced and new job posts created. The staff

TABLE 4.10 Job specifications in animal facilities.

Job title	Job description
Director	Advanced degree in a related field and experience in managing animal facilities. Ensures overall compliance and is familiar with current legislation. Responsible for general policy, planning, and administration.
Facility manager	Usually requires a college degree in a related field, animal care qualifications, and extensive practical experience. Provides direct supervision of the personnel. Depending on the size of the facility may manage personnel directly or through the supervisory staff. Ensures high standards of operations and compliance with local and national legislation and is responsible for recruitment, appraisals, and disciplinary procedures.
Supervisor	Usually requires a college degree in a related field or animal care qualification. Provides direct supervision of animal facility staff and prepares work schedules. Responsible for training of the personnel and duty assignments.
Veterinarian	Fully licensed with experience in the animal research environment. Provides advisory duties concerning the well-being and treatment of the animals. Reviews projects, deliver surgical and procedural support, and safeguards the welfare of the animals kept in the establishment.

Continued

TABLE 4.10 Job specifications in animal facilities.—cont'd

Job title	Job description
Senior animal technician/ senior caretaker	Animal care qualification with a good understanding of local animal welfare legislation. Experienced in daily animal husbandry duties, including keeping detailed records, handling sexing and identification of animals. Carries out regular observation, health checks, and assists research staff with regulated and nonregulated procedures. Responsible for training other members of personnel and scientific staff in animal handling and procedures.
Animal technician/ caretaker	Animal care qualification and practical experience caring for animals. Responsible for caring out daily welfare checks and husbandry duties and assisting research staff with nonregulated procedures. Able to keep records and provide help to junior staff.
Junior animal technician/junior caretaker	Basic animal care qualification or willingness to acquire those and basic practical experience with laboratory animals. Carries out daily welfare checks and provides animal husbandry duties and sanitation in the facility. Carries out regular observations including environmental monitoring and health.
Service technician	Usually, no qualifications and experience are needed. Training is provided on-site with the relevant equipment. Responsibilities include an operation of sanitizing/sterilizing equipment, preparation and use of disinfecting and sanitizing solutions, preventive maintenance and service record keeping.
Named animal care and welfare officer (NACWO)	Responsible for overseeing the day-to-day husbandry, care and welfare of the protected animals held at their establishment. They should be a source of independent advice on welfare and care to ensure high standards of care. Generally, this post is filled by experienced animal technicians who attend additional training. NACWO should have appropriate personal authority to promote high standards of care. Good communication skills will also be essential to work well with animal care personnel and researchers.
Named information officer	Usually holds a degree in a relevant subject and extensive experience in an animal research environment. Familiar with the species used and the types of research performed. Responsible for ensuring that those dealing with animals have access to information about the species and the procedures they are performing.
Named training and competency officer	Responsible for ensuring that all those dealing with animals are adequately educated, trained, and supervised until they are competent. NTCO also ensures that they continue to undertake further training to maintain their expertise.
Study director/ research technicians	Generally, required to have a college degree in a related scientific field. Assists with research projects by performing specific procedures or surgeries. Collects research data and maintain accurate research records.

structure, types of positions and number of personnel depend on the size of the facility, species housed, and type of research performed. Some positions like animal technicians, managers will be present in all the facilities and others like directors, supervisors and study directors will be present only in larger institutions.

The bigger establishment will have higher job specialization where employees spend most of the time on specific tasks gaining more efficient skills. However, a more complicated staff structure can be challenging to manage. Clear task division, close supervision, and written work schedules are essential to ensure successful operations. In the larger team, there is usually also higher staff rotation. Well-structured training programs and supervision will ensure that new members have a good understanding of animal welfare requirements in their facilities.

The staff in smaller institutions will generally have a wider range of techniques but be less efficient with individual procedures. Generally, their work will be located in one particular area or room. Small teams can become problematic in case of annual leave/sickness. Cross-training within different areas and projects can ensure that covering staff can provide the best possible care to the animals.

Animal care staff has a substantial influence on animal welfare therefore, each position within the animal facility should be carefully considered. Not only experience should be taken into account but also the attitude toward animals and consideration for their well-being.

Regardless of the staff structure and number of personnel, the focus should always be on good communication within a team and a clear division of responsibilities. This ensures a high level of care and prevents any potential mistakes which could impact the welfare of the animals.

Records

A good record system is essential for efficient operations in the animal facilities and also a requirement under EU Directive2010/63/EU. Facilities that work to GLP will generally have a higher level of record keeping. The intent of all GLP guidance is to ensure that the study design and experimental methods as well as the acquisition, analysis, interpretation, and documentation are always done in a reliable and reproducible manner. Infrastructure for GLP compliance is labor-intensive and time-consuming, but GLP standards contribute to the quality of the data from each animal. This is critical for the results interpretation and reduction of animals used. Even is in smaller facilities, not GLP approved the aim should always be to record in a way that ensures reliable data collection and reduction of the animals used.

Animal records

The well-being of research animals depends upon careful and frequent monitoring and appropriate recording. Every experimental plan should include observation and monitoring of animals for signs of pain or distress. The frequency of observation will depend on the possible effect of the procedures on animal health. All observations, treatments, and procedures must be documented in the animal procedural records (Fig. 4.5). Rabbits, guinea pigs, and hamsters will require individual records while rats and mice can be recorded as a group. Records of animals that died are also important when it comes to animal welfare. Where possible cause of death should be established, especially following experimental procedures. Steps should be taken to prevent further deaths within that experiment. If severity limits have been exceeded and pain and suffering cannot be alleviated, the experiment should be terminated.

Training records

Training records should reflect the level of training and competence to allow skills transfer across the EU and different facilities. Currently, there is high variability in types of records being kept by the facility and individuals. However appropriate record-keeping and compliance history should be one of the priorities of any facility. Accurate records are an essential part of all training schemes and should incorporate professional education and relevant competencies acquired before present employment.

Policies: Supporting Documents **Record of Procedures**

Name: **Research Group:**

Licence No:

Date	Cage ID	Number of animals	Procedure/Protocol No	Recovery Y/N?	Date of Termination

No.:	Version:1	Effective date:	Site:	Page:

FIGURE 4.5 Example of procedure record. *Credit: Olga Boruc.*

Records should detail type of the procedure, level of skills, and be species-specific. They should identify the level of supervision required and competence of each member of personnel working with animals.

For smaller facilities, simple files or spreadsheets might be sufficient (Fig. 4.6) while for larger facilities more complex training database systems may be needed. Regardless of the system used, all need regular updating and monitoring to provide up to date information. It should be easily accessible for those responsible for training and for the staff/students. The institutes with a larger number of various research projects and staff will require extra attention. Training and competency persons should be employed to follow up with the education and skills assessments. Named Training and Competency Officer should also be able to identify experienced trainers for the procedure within the establishment or if not available, within another institutes.

Operations

When setting up operations in the animal facility, the goal should always be to promote the humane care of animals used in education and biomedical research. Essential in this process are:

1. Standard operating procedures (SOPs)

Clear guidelines for all staff and researchers in the form of SOPs. SOPs are written procedures or a set of practical, specific instruction on how to

Training: Supporting Documents **Training Record Form**

Name: **Research group:**

Procedure	Species	Trainee under supervision			Competence Achieved			Trainer Status Achieved			Review date	Review date	Review date
		Date	Trainee (initials)	Trainer (initials)	Date	Trainee (initials)	Trainer (initials)	Date	Trainee (initials)	Trainer (initials)			

No.:	Version:	Effective date:	Site:	Page:

FIGURE 4.6 Example of basic training record form. *Credit: Olga Boruc.*

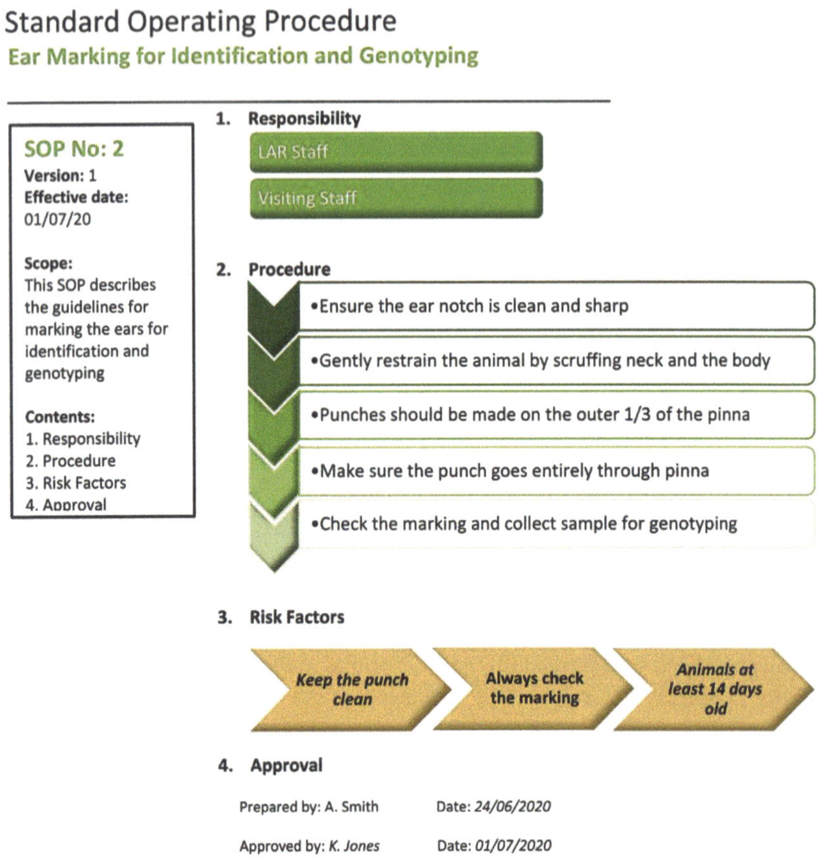

FIGURE 4.7 Example of standard operating procedures form. *Credit: Olga Boruc.*

implement a policy or establish standardized processes (Fig. 4.7). SOPs are important for animal care and husbandry in any animal care program. They cover all areas of facility operations like husbandry, handling, housing conditions, as well as analgesia, anaesthesia, and euthanasia. The SOP should be written clearly and serve more as addition to practical training rather that the only source of information. The written procedures should be stored in the location accessible to those involved in the care and use of animals and reviewed before starting any in vivo activity. Policies and procedures must be updated regularly to reflect changing conditions and regulations.

2. Daily checks

Performing a daily check of animal health and environment are important aspects of daily operations. Cages should be checked first thing every day to note the condition of the animals. A staff member should focus on visible signs

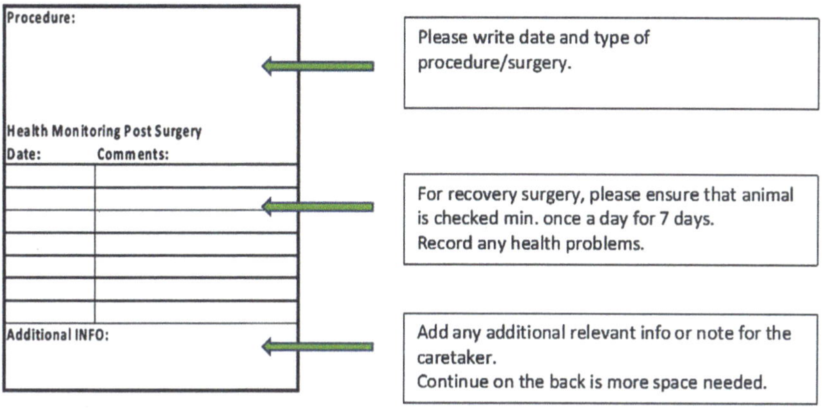

FIGURE 4.8 Example of health monitoring card. *Credit: Olga Boruc.*

of pain or distress, decrease in food or water consumption, blood in cage, wounds, swelling, secretions around the eyes, nose and genital area, respiratory distress, diarrhoea, piloerection, or loss of hair. All concerns should be reported to the supervisor or veterinarian. Care staff should also ensure that animals have enough food available and a clean cage. Special attention should be given to animals recovering from procedures or surgery. Health monitoring card (Fig. 4.8) could be added to help staff identify those cages quickly. Additionally, the caretaker must ensure that each cage has an identification card with the following information: protocol number, investigator's name, strain, sex, date of birth, and the number of animals per cage. Environmental conditions and equipment should also be checked daily to ensure correct operation. A good practice is to set up a room monitoring checklist in the room to ensure all tasks are completed as required (Fig. 4.9).

3. Animal Welfare Body (AWB)

EU Directive 63/2010 requires that institutes using animals for experimental or other scientific/educational purposes set up an AWB. AWB aims to promote awareness of the 3Rs in the facility. It provides a forum for discussion on all matters related to animal welfare, care and use and helps to promote a "culture of care" within an establishment. Specifically, the task of AWB is to present technical, scientific, and ethical opinions on research projects and any further amendments to the project. The AWB is made up of members responsible for the welfare of the animals, veterinarians, experts from the scientific community with a good understanding of animal experimentation and lay members. Lay members may be drawn from various backgrounds and fields of work that are unrelated to animal use. They can bring a different perspective and ask questions that more deeply involved in the research might

Daily Observation: Supporting Documents **Room Monitoring Checklist**

MM/YY: **Area/Room:**

Date	Room T°	IVC T°	Room RH	IVC RH	Water Top-up	Diet Top-up	Cages changed	Mop & Sweep	Racks Wiped	Remarks (dead/sick animals etc.)	Sign

No.:	Version:	Effective date:	Site:	Page:

FIGURE 4.9 Example of room monitoring check list. *Credit: Olga Boruc.*

not identify or ask. The AWB should meet regularly throughout the year to actively monitor issues and raise awareness to ensure the effective use of any animals and maintain principles of 3Rs.

References

Andrè, V., Gau, C., Scheideler, A., Aguilar-Pimentel, J.A., Amarie, O.V., Becker, L., Gar-rett, L., Hans, W., Hölter, S.M., Janik, D., Moreth, K., Neff, F., Östereicher, M., Racz, I., Rathkolb, B., Rozman, J., Bekeredjian, R., Graw, J., Klingenspor, M., Klopstock, T., Ollert, M., Schmidt-Weber, C., Wolf, E., Wurst, W., Gailus-Durner, V., Brielmeier, M., Fuchs, H., Hrabé de Angelis, M., 2018. Laboratory mouse housing conditions can be improved using common environmental enrichment without compromising. Data PLOS Biol. 16 (4), e2005019.

Bailoo, J.D., Murphy, E., Boada-Sañna, M., Hintze, J.A.S., Baussière, C., Hahn, K.C., Göpfert, C., Palme, R., Voelkl, B., Würbel, H., 2018. Effect of cage enrichment on behaviour, welfare and outcome variability in female mice. Front. Behav. Neurosci. 12, 232. https://doi.org/10.3389/fnbeh.2018.00232.

Brown, S.D., Moore, M.W., 2012a. The International Mouse Phenotyping Consortium: past and future perspectives on mouse phenotyping. Mamm. Genome 23, 632–640.

Brown, S.D., Moore, M.W., 2012b. Towards an encyclopaedia of mammalian gene function: the International Mouse Phenotyping Consortium. Dis. Model. Mech. 5, 289–292.

Chen, K.K., Anderson, R.C., Steldt, F.A., Mills, C.A., 1943. Environmental temperature and drug interaction in mice. J. Pharmacol. Exp. Therapeut. 79 (2), 127–132.

Consortium, I., 2015. INFRAFRONTIER–providing mutant mouse resources as research tools for the international scientific community. Nucleic Acids Res. 43, D1171–D1175.

Dawkins, M.S., 2003. Behaviour as a tool in the assessment of animal welfare. Zoology 106 (4), 383–387.

Dawkins, M.S., 2004. Using behaviour to assess animal welfare. Anim. Welf. 13 (1), 3–7.

Directive 2010/63/EU of the European Parliament and of the Council of 22 September 2010 on the Protection of Animals Used for Scientific Purposes.

Farm Animal Welfare Council (FAWC), 1993. Second Report on Priorities for Research and Development in Farm Animal Welfare. DEFRA, London, UK.

Garner, J., 2005. Stereotypies and other abnormal repetitive behaviours: potential impact on validity, reliability and replicability of scientific outcomes. ILAR J. 46 (2), 106–117.

Gaskill, B.N., Karas, A.Z., Garner, J.P., Pritchett-Corning, K.R., 2013. Nest building as an indicator of health and welfare in laboratory mice. J. Vis. Exp. 82, 51012. https://doi.org/10.3791/51012.

Gordon, C.J., 2012. Thermal Physiology of laboratory mice: defining thermoneutrality. J. Therm. Biol. 37, 654–685.

Gouveia, K., Hurst, J.L., 2019. Improving the practicality of using non-aversive handling methods to reduce background stress and anxiety in laboratory mice. Sci. Rep. 9, 20305. https://doi.org/10.1038/s41598-019-56860-7.

Hankenson, F.C., Marx, J.O., Gordon, C.J., David, J.M., 2018. Effects of rodent thermoregulation on animal models in the research environment. Comp. Med. 68 (6), 425–438. https://doi.org/10.30802/AALAS-CM-18-000049.

Haselimashhadi, H., Mason, J.C., Munoz-Fuentes, V., Lopez-Gomez, F., Babalola, K., Acar, E.F., Kumar, V., White, J., Flenniken, A.M., King, R., et al., 2020. Soft windowing application to improve analysis of high-throughput phenotyping data. Bioinformatics 36, 1492–1500.

Hawkins, P., Littlefair, P., Golledge, H., Richardson, C., Allden, S., Boden, T., Hendrie, C., Wells, D., Osborne, N., Hutchison, M., Ryder, S., Jennings, M., Hubrecht, R., 2014. Report of the 2013 RSPCS/UFAW Rodent Welfare Group Meeting. Animal Technology and Welfare, pp. 165–180.

Hendrie, C.A., Pickles, A.R., 2011. Behavioural scientists appear not to have specialist knowledge of laboratory animal behaviour. J. Psychopharmacol. 25 (Suppl. 8), A35.

Howerton, C.L., Garner, J.P., Mench, J.A., 2008. Effects of a running wheel-igloo enrichment on aggression, hierarchy linearity, and stereotypy in group-housed male CD-1 (ICR) mice. Appl. Anim. Behav. Sci. 115 (1–2), 90–103.

Hylander, B.L., Repasky, E.A., 2016. Thermoneutrality, mice, and cancer: a heated opinion. Trends Cancer 2 (4), 166–175. https://doi.org/10.1016/j.trecan.2016.03.005.

Ioannidis, J.P., 2014. How to make more published research true. PLoS Med. 11, e1001747.

Kollmus, H., Post, R., Brielmeier, M., Fernandez, J., Fuchs, H., McKerlie, C., Montoliu, L., Otaegui, P.J., Rebelo, M., Riedesel, H., et al., 2012. Structural and functional concepts in current mouse phenotyping and archiving facilities. J. Am. Assoc. Lab. Anim. Sci. 51, 418–435.

Koscielny, G., Yaikhom, G., Iyer, V., Meehan, T.F., Morgan, H., Atienza-Herrero, J., Blake, A., Chen, C.K., Easty, R., Di Fenza, A., et al., 2014. The International Mouse Phenotyping Consortium Web Portal, a unified point of access for knockout mice and related phenotyping data. Nucleic Acids Res. 42, D802–D809.

Leach, M.C., Thornton, P.D., Main, D.C.J., 2008. Identification of appropriate measures for the assessment of laboratory mouse welfare. Anim. Welf. 17 (2), 161–170.

Maier, H., Lengger, C., Simic, B., Fuchs, H., Gailus-Durner, V., Hrabe de Angelis, M., 2008. MausDB: an open source application for phenotype data and mouse colony management in large-scale mouse phenotyping projects. BMC Bioinf. 9, 169.

Mallon, A.M., Iyer, V., Melvin, D., Morgan, H., Parkinson, H., Brown, S.D., Flicek, P., Skarnes, W.C., 2012. Accessing data from the International Mouse Phenotyping Consortium: state of the art and future plans. Mamm. Genome 23, 641–652.

Maloney, S.K., Fuller, A., Mitchell, D., Gordon, C., Overton, J.M., 2014. Translating animal model research: does it matter that our rodents are cold? Physiology 29 (6), 413–420. https://doi.org/10.1152/physiol.00029.2014.

Martin, P., Bateson, P., 2007. Measuring Behaviour: An Introductory Guide, third ed. Cambridge University Press, Cambridge.

Meehan, T.F., Conte, N., West, D.B., Jacobsen, J.O., Mason, J., Warren, J., Chen, C.K., Tudose, I., Relac, M., Matthews, P., et al., 2017. Disease model discovery from 3,328 gene knockouts by the international mouse phenotyping consortium. Nat. Genet. 49, 1231–1238.

Mellor, D.J., Reid, C.S.W., 1994. Concepts of animal well-being and predicting the impact of procedures on experimental animals. In: Improving the Well-Being of Animals in the Research Environment; Australian and New Zealand Council for the Care of Animals in Research and Teaching. Glen Osmond, Australia, pp. 3–18.

Munoz-Fuentes, V., Cacheiro, P., Meehan, T.F., Aguilar-Pimentel, J.A., Brown, S.D.M., Flenniken, A.M., Flicek, P., Galli, A., Mashhadi, H.H., Hrabe de Angelis, M., et al., 2018. The International Mouse Phenotyping Consortium (IMPC): a functional catalogue of the mammalian genome that informs conservation. Conserv. Genet. 19, 995–1005.

Olsson, I.A.S., Nevison, C.M., Patterson-Kane, E.G., Sherwin, C.M., Van de Weerd, H.A., Würbel, H., 2003. Understanding behaviour: the relevance of ethological approaches in laboratory animal science. Appl. Anim. Behav. Sci. 81 (3), 245–264.

Pargent, W., Heffner, S., Schable, K.F., Soewarto, D., Fuchs, H., Hrabe de Angelis, M., 2000. MouseNet database: digital management of a large-scale mutagenesis project. Mamm. Genome 11, 590–593.

Poole, T., 1997. Happy animals make good science. Lab. Anim. 31, 116–124.

Prescott, M.J., Lidster, K., 2017. Improving quality of science through better animal welfare: the NC3Rs strategy. Nat. Am. 46 (4), 152–156.

Raess, M., de Castro, A.A., Gailus-Durner, V., Fessele, S., Hrabe de Angelis, M., Consortium, I., 2016. INFRAFRONTIER: a European resource for studying the functional basis of human disease. Mamm. Genome 27, 445–450.

Spangenberg, E.M., Keeling, L.J., 2016. Assessing the welfare of laboratory mice in their home environment using animal-based measures—a benchmarking tool. Lab. Anim. 50 (1), 30–38. https://doi.org/10.1177/0023677215577298.

Wilkinson, P., Sengerova, J., Matteoni, R., Chen, C.K., Soulat, G., Ureta-Vidal, A., Fessele, S., Hagn, M., Massimi, M., Pickford, K., et al., 2010. EMMA—mouse mutant resources for the international scientific community. Nucleic Acids Res. 38, D570–D576.

Recommended reading

Allen, E.D., Czarra, E.F., DeTolla, L., 2018. In: Weichbrod, R.H., Thompson, G.A.H., Norton, J.N. (Eds.), Management of Animal Care and Use Programs in Research, Education, and Testing, second ed.

An overview of impact of various kind of enrichment on mice strains can be found in Bayne, K., 2018. Environmental enrichment and mouse model: current perspectives. Anim. Model Exp. Med. 1, 82–90.

https://www.nc3rs.org.uk/how-to-pick-up-a-mouse.

http://storage.googleapis.com/ecourses/Rat%20Tickling%20Certification/story_html5.html.
https://www.nc3rs.org.uk/tickling-rats-improved-welfare.

Mellor, D.J., 2016. Updating animal welfare thinking: moving beyond the "Five Freedoms" towards "a life worth living". Animals (Basel) 6 (3), 21. https://doi.org/10.3390/ani6030021. Published March 14, 2016.

National Competent Authorities for the Implementation of Directive 2010/63/EU on the Protection of Animals Used for Scientific Purposes. A Working Document on the Development of a Common Education and Training Framework to Fulfil the Requirements under the Directive. Re-placing Consensus Document of 18-19 September 2013. Brussels, February 19-20, 2014.

Guide for the Care and Use of Laboratory Animals by the Institute of Laboratory Animal Resources. Commission on Life Sciences, National Research Council, 1996. National Academy Press, Washington.

Resources available on the NC3Rs website: https://www.nc3rs.org.uk/3rs-resources.

A new relevant biologically environmental enrichment for mice can be found in Vogt, M.A., Mertens, S., Serba, S., Palme, R.,Chourbaji, S., 2020. The "Cage Climber" – a new enrichment for use in large-dimensioned mouse facilities. Appl. Anim. Behav. Sci. https://doi.org/10.1016/j.applanim.2020.105078.

Webster, J., 2016. Animal welfare: freedoms, dominions and "A life worth living". Animals (Basel) 6 (6), 35. https://doi.org/10.3390/ani6060035. Published May 24, 2016.

Weichbrod, R.H., Thompson, G.A., Norton, J.N., 2018. Management of Animal Care and Use Programs in Research, Education, and Testing, second ed. CRC Press/Taylor&Francis.

An example of enrichment safety assessment can be found in Winnicker, C., Gaskill, B., Garner, K., Pritchett-Corning, K., 2016. A Guide to the Behaviour and Enrichment of Laboratory Rodents, Charles River Laboratories, Boston.

Yeates, J.W., Main, D.C., 2008. Assessment of positive welfare: a review. Vet. J. 175 (3), 293–300.

Chapter 5

Methods of handling and procedures

Chiara Attanasio[1], Livia D'Angelo[1], Luisa Corsi[2]
[1]*Department of Veterinary Medicine and Animal Production, University of Naples Federico II, Naples, Italy;* [2]*Laboratory Animal Veterinarian Consultant, Parma, Italy*

Introduction

Handling and substance administration are basic procedures performed on laboratory animals, both during husbandry and experimental practices. In addition to be carried out by experienced and adequately trained staff, handling, tagging, and substance administration should be carried out to minimize animal stress. In this chapter, we report common and well-established approaches for handling, tagging, and substance administration in the most used rodents species, rabbits, and zebrafish. We also discuss refined approaches of each method for all species considered.

Rodent and rabbit handling methods (Luisa Corsi, Chiara Attanasio)

Handling is the most common procedure experienced by laboratory animals over their entire lifespan; therefore, it should be considered a mild stressful but unavoidable practice.

Handling and restraining techniques are different for each species depending on specific anatomical features and experimental needs.

In this chapter, the most common techniques for rodents and rabbits will be presented.

When approaching animals, any fast or erratic movement should be avoided; therefore, handler movements should be slow and calm. Sharp noises, loud voices, or scents of other animals, particularly predators, can evocate fear responses (Arakawa et al., 2008).

Even mild manipulations, such as moving the animal from one room to another, may increase cortisol levels resulting in temporary but significant changes in endocrine markers (Gartner et al., 1980; Rubio and Huang, 1992).

Practical Handbook on the 3Rs in the Context of the Directive 2010/63/EU
https://doi.org/10.1016/B978-0-12-821180-9.00001-5

Acute stress can be also induced by transport, which is able to transiently alter psychophysiological parameters, thus impairing research outcomes (Obernier and Baldwin, 2006).

Any modification in animal's everyday life, indeed, can become a stressful event especially when the animal hasn't any possibility to control the situation (Arts et al., 2012; Obernier and Baldwin, 2006).

In view of this, animals should be allowed to recover after transport, or any other stressful procedure, through an adequate acclimation period.

Laboratory rodents and rabbits, similarly to what observed in their natural environment, exhibit three common responses to stressful events: freeze behavior, fleeing behavior and, in rare occasions, fighting or aggressive behavior (Brown, 2012).

Freezing is a common reaction observed in prey animals with the purpose of not being seen by a predator. Fleeing is usually the second choice, but it is not applicable in animals living in cages; for this reason, combative or aggressive behaviors can occur in captivity. Animals could sequentially exhibit all the described behaviors, selecting the most effective for each situation.

Whenever an aggressive behavior removes an adverse stimulus, the behavior is reinforced. Aggressive behavior can usually be prevented by avoiding the animal to use this strategy (Brown, 2012).

All the species commonly used in scientific research are extremely sensitive to interaction with humans, particularly if associated with negative experiences, resulting in anxiety, and acute or chronic stress response.

Stress and anxiety in laboratory animals can lead to the generation of unreliable experimental data by inducing behavioral modifications as well as changes in physiological parameters related to immune response, autonomic nervous and neuroendocrine systems (Hölscher, 1999). Minimizing these responses may significantly reduce variability (Nakamura et al., 2018) thus improving the consistency of experimental results (Gouveia and Hurst, 2017, 2019; Clarkson et al., 2018; Mertens et al., 2019).

A further source of distress in laboratory animals is inadequate handling, especially in case of restraining by personnel who is not familiar with the procedures or without a proper training.

In order to reduce stress and anxiety responses, animals should gradually learn that the contact with humans itself is not harmful (Pekow, 2005; Wilson et al., 1998).

Staff training is crucial to reduce the different sources of stress associated to animal handling.

Furthermore, training the animals is generally a very helpful tool for laboratory animal management as required by EU Directive 2010/63/EU.

Training and habituation serve as a form of refinement, help to prepare animals for experiments, and contribute to sustain and enhance animal welfare (Balcombe et al., 2004).

One of the most successful method of training behaviors is positive reinforcement, a conditioning process in which a reward is offered every time a desired action or behavior is performed by the subject. Positive reinforcement training can be applied to several animal species including laboratory rodents and rabbits. The reward can be food, sweet juice, or tasty solution, as well as the access to a specific area (Carp et al., 2006).

In some circumstances, physical restraint by using appropriate devices can be applied to perform sample collection, drug administration, and different kinds of manipulations as well.

Devices such as Plexiglas restraining cylinders allowing access to mouse or rat tail are available and can be used for intravenous injections or blood collection. Additional devices commonly used to perform injections are perforated plastic sleeves in which the animal can be placed. For rabbits, specific restraining devices allow to contain the entire animal body giving access to ears (Truitt et al., 2008).

The use of these restraining devices should be limited to the time needed to perform the procedure, and they should never be used for routine animal handling. A restraint lasting longer than 15–30 min is considered prolonged and alternative practices to reduce distress need to be considered. Training the animal with positive reinforcement and habituation to the device can be beneficial especially for prolonged restraint. Animals refusing the device and failing to accept the training should be excluded from the procedure. The device should be big enough to avoid any lesions to animal limbs or testis and has to be cleaned after each use to minimize the potential stress caused by scents deriving from other subjects.

Refined handling and restraining methods have a positive impact on animal welfare and can be applied also for some husbandry procedures such as routine cage changing or basic practices like weaning or sexing. Sexing in rodent species is performed by checking the distance between the anus and genital papilla which is greater in the male than in the female. In rabbits, a gentle pressure to allow penis extrusion is additionally needed. In general, in animal husbandry, the least invasive and aversive manipulation techniques should be considered to avoid unnecessary stress.

Mouse handling

The most common way to lift a mouse up is to hold it firmly at the tail base. However, recent studies demonstrated that mice picked up by the base of the tail are very unwilling to interact voluntarily with the handler and show high anxiety in behavioral tests, even when they are familiar with this handling method (Nakamura and Suzuki, 2018; Gouveia and Hurst, 2019).

In mouse behavioral phenotyping studies, handling stress has been identified as one of the most likely causes of failure in replicating phenotypes within and between experiments (Mandillo et al., 2008; Crabbe et al., 1999; Schellinck et al., 2010).

The use of nonaversive handling could significantly reduce the stress related to manipulations.

Numerous studies confirmed that handling by nonaversive methods like tunnel or cupping handling can induce: a reduction in plasma corticosterone and blood glucose levels as well as an improvement in glucose tolerance (Ghosal et al., 2015), a reduction of glomerular lesions in a glomerulonephritis model (Ono et al., 2016), an improved reliability in behavioral tests requiring exploration of test stimuli (Gouveia and Hurst, 2017), an improved responsiveness to sucrose reward (Clarkson et al., 2018).

Mice handled by using a tunnel show a less fearful behavior compared to tail-handled mice. In addition, the mice handled with the tunnel continue to interact readily with their handler even after having experienced procedures such as oral gavage, intraperitoneal injection, tattooing, or ear-tagging (Nakamura and Suzuki, 2018; Gouveia and Hurst, 2017).

Just brief experience of tunnel handling such as for a few seconds during cage cleaning for 10 days or less is sufficient to ensure lack of aversion and low anxiety during handling (Hurst and West, 2010).

Handling by cup consists in scooping up the mouse in open hands without a direct restraint as showed in Fig. 5.1.

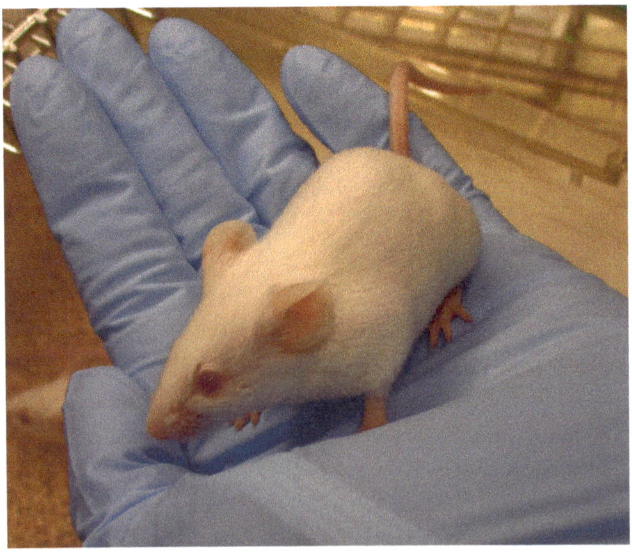

FIGURE 5.1 **Handling by cup.** Handling by cup consists in scooping up the mouse in open hands without a direct restraint.

This nonaversive handling method may be unsuitable for jumpy strains or young mice, when tunnel handling is preferable.

Handling with a tunnel consists of using a smooth plastic tunnel, where the mouse is encouraged to enter without direct contact with the operator's hands (Fig. 5.2).

Tunnel handling may offer many benefits, allowing minimal contact with the animal while allowing a full inspection in case of transparent plastic tunnel. Mice tunnel handling can be of help in performing basic procedures like sexing at weaning age.

Despite the evidence that tail handling can impair welfare, this method remains the most commonly used to handle mice. Other more refined methods, such as tunnel or cup handling, should be implemented across research institutions, using tail handling only when different systems are unapplicable.

In the case restriction is needed, the animal should be placed on a rough surface (e.g., the cage lid) and the skin of the neck should be quickly and gently pinched between the thumb and the index finger of the nondominant hand of the operator. The mouse should be then lifted, being the tail held between the ring finger and the palm of the hand as showed in Fig. 5.3.

If it is necessary to lift a mouse by the tail, it is imperative to grasp the tail at the base near the rump of the animal to avoid tail lesions.

The mouse restraining techniques described above allow to perform most of basic procedures.

Rat handling

Rats do not like to be lifted by the tail. If this method cannot be avoided, it is important to grasp the base of the tail because the impact of the body weight on tail skin can induce painful lesions. For this reason, it is strongly recommended to lift rats gently by the body for routine handling, placing one hand under the belly or around the chest, with the other hand supporting the hind legs.

Tunnel handling already extensively described for mice can be used also for rats.

In rats, tickling, also defined "playful handling," mimicking aspects of the playful rough-and-tumble behavior can be used to mitigate stressful effects of some experimental procedures.

Numerous studies support the conclusion that tickling is a beneficial form of social contact with humans, especially for young rats (LaFollette et al., 2018, 2019; Cloutier et al., 2018). Tickling triggers the playful social behavior of these animals turning handling in a positive experience. This concept was originally developed as a model for investigating the neurobiological basis of positive affective states evoked by social play behavior (Burgdorf and Panksepp, 2001).

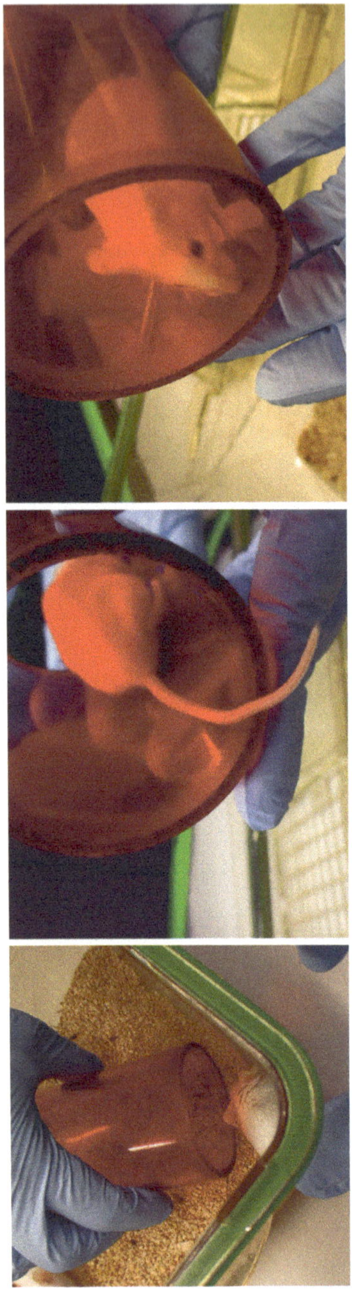

FIGURE 5.2 **Tunnel handling.** For tunnel handling, the tunnel is brought toward the animal and held resting on the cage floor while the animal is guided toward it. Tunnels are transparent or opaque plastic cylinders, measuring approximately 50 mm in width 23 and 150 mm in length.

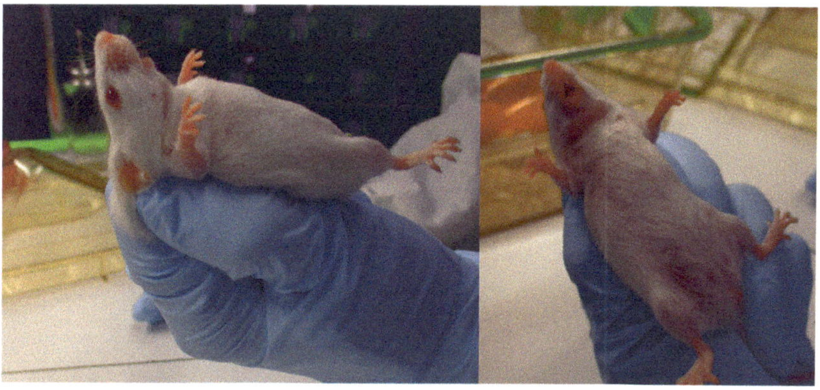

FIGURE 5.3 **Mouse restraining technique.** The mouse is placed on a rough surface (e.g., the cage lid) and the skin of the neck is quickly and gently pinched between the thumb and the index fingers of the nondominant hand of the operator. The mouse should be then lifted, holding the tail between the ring finger and the palm of the hand.

Human handlers can mimic rat playful behavior by quick movements of the fingers.

It has been also shown that social play generates resilience to anxiety and depression (Popik et al., 2012).

Small rats can be restrained by grasping and holding the skin at the nape, a practice generally defined as "scruffing."

In rats, one or two handed restrain methods can be applied.

The first one consists of placing the rat on a surface, holding the base of the tail with one hand while the other hand is brought up and over the back of the rat with the index and middle finger split, and then placed over the shoulder so that the index finger is put on one side of the head and the middle finger on the other side. The body of the rat is then encircled behind the forelegs with the third, the fourth finger, and the thumb, at this point the animal is lifted, and the tail is stabilized by grasping it and the hindquarters (Fig. 5.4).

This method works well with larger rats and is commonly defined as "the claw," "the baseball method," or "T-rex grip" (Frohlich, 2020).

It is crucial to limit the pressure on the chest of the animal and to verify that it can breathe normally, by checking the buccal mucous membranes that appear blue when the breathing is impaired.

The second method is applied by placing the rat on a grippy surface (e.g., the cage lid). The tail is gently grasped at the base with the dominant hand applying a light backward traction. By doing so, the rat body is gently stretched. The other hand grasps the rat behind the shoulders putting the finger on one side of the body and the thumb on the other side pressing gently the forelegs so that they will cross under the chin as showed in Fig. 5.5.

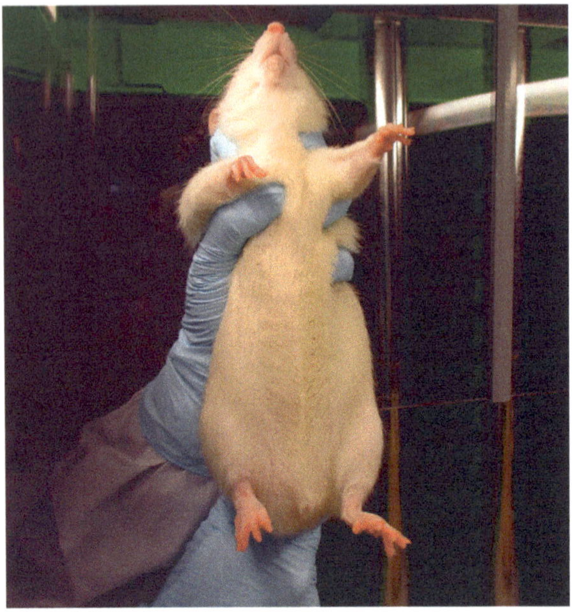

FIGURE 5.4 **Rat T-rex grip.** The rat is placed on a rough surface, holding the base of the tail with one hand while the other hand is placed over the back of the animal. The index and the middle fingers are placed over the shoulder of the rat that is then lifted while the tail is stabilized with the other hand.

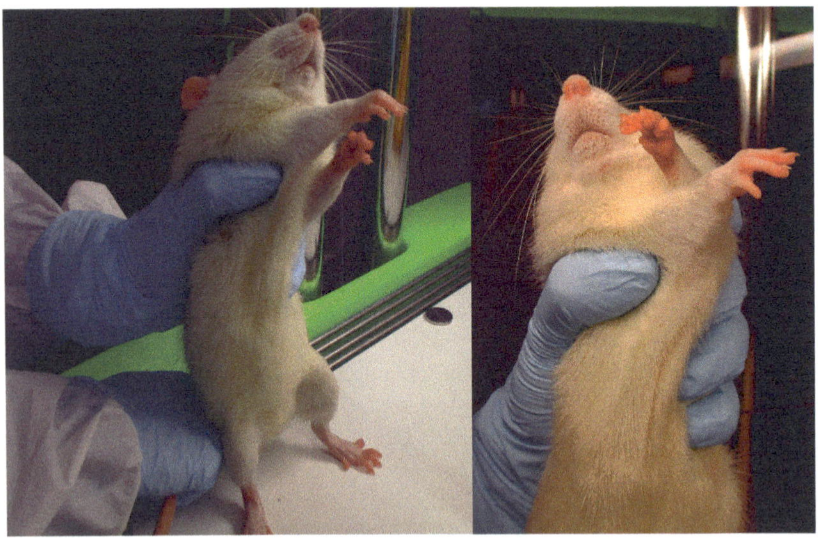

FIGURE 5.5 **Rat restraining technique.** The rat is placed on a rough surface and its tail is gently grasped at the base. The rat is then grasped behind the shoulders.

This position will create a physical barrier preventing the rat from moving or biting the handler.

During this procedure, it is always suggested to check that the breathing is not impaired by excessive pressure on the chest.

The described methods allow to perform the most common procedures.

Conventional approaches for substances administration include restraining the animal by using a firm grip around neck and shoulders or by using the scruffing technique.

To this purpose, most guidelines recommended for rats report that the procedure should be done by two operators, one restraining the animal and the other one performing the injection.

Additionally, for intraperitoneal injection, a modified method of handling, which involves minimal animal restraint and minimizes the stress has been described and can be implemented to improve animal welfare (Stuart and Robinson, 2015). This modified procedure implies that the rat is gently kept with only one hand surrounding the chest.

Guinea pig handling

Guinea pigs are docile and easy-to-handle animals that rarely scratch or bite the operator. When first approached, these animals appear immobilized while they frequently run afterward before being grasped.

In addition, guinea pigs often squeal when being grasped and this can discourage the handler. Furthermore, to reduce the risk of injuries, it is important to approach a guinea pig as slowly and quietly as possible.

For this reason, guinea pigs should be always grasped firmly and carefully with both hands around the chest, while young guinea pigs can be grasped with just one hand.

The best method to restrain a guinea pig is to hold the anterior portion of the animal up by the chest fixing the forelegs and containing the pelvis and hindlegs in the other hand (Fig. 5.6).

For simple procedures, it is also possible to restrain the animal by holding its anterior portion while allowing the rear legs to remain on the surface (Fig. 5.7).

Guinea pigs can also be held firmly against the operator body or in his elbow crease.

Aversive stimuli such as restraint, inversion, and presence of predators can induce the tonic immobility response often described as "hypnosis" or "the freeze response."

Tonic immobility is a profound somatic and autonomic motor inhibition state mediated by the periaqueductal gray matter, the limbic forebrain, and the spinal areas. Therefore, tonic immobility should not be considered as a suitable restraint method (Vieira et al., 2011).

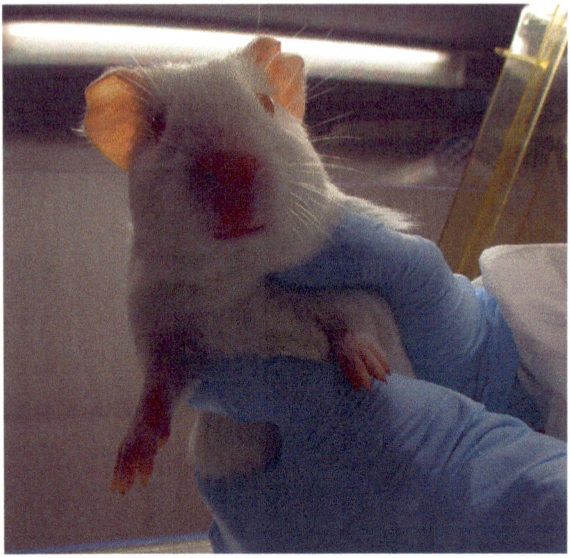

FIGURE 5.6 **Guinea pig restraining technique.** A Guinea pig is grasped with one hand placed between the neck and the chest while it is stabilized with the other hand placed around the pelvis and the hindlegs.

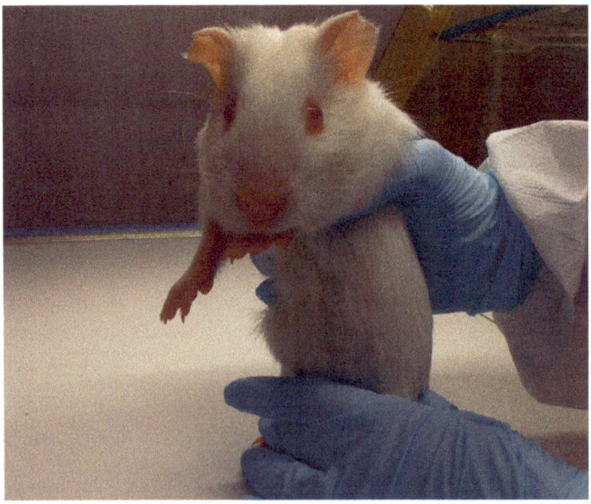

FIGURE 5.7 **An alternative guinea pig restraining technique.** For simple procedures, the animal can be by lifted by placing one hand between the neck and the chest while keeping the hindlegs on a rough surface.

Rabbit handling

Many techniques can be applied to handle rabbits, in any case a gentle and firm approach is preferable. Rabbits should never be picked up by their ears as this is painful and lead to struggle to escape which can result in severe injuries. Spinal muscles of rabbits are strong enough to cause spinal fractures when they contract forcibly, for example, when trying to kick out with hind legs. In view of this, during rabbit handling, it has to be considered that isometric contraction of the spinal muscles can induce significant spinal damage (Harcourt-Brown, 2008).

The best method to pick a rabbit up is scruffing the skin of the neck supporting with the other hand the rest of the body (Fig. 5.8).

For nervous subjects, an effective technique is to hold them with the face hidden in the elbow crease of the handler (Fig. 5.9).

Dorsal recumbency can evoke the immobility response often described as "hypnosis" or the freeze response; however, it can be a source of stress leading to kicking. The immobility response is exhibited in prey species in stressful situations when terrified subjects lie on their backs expecting predators' strike. This phenomenon is characterized by lack of spontaneous movements and failure to respond to external stimuli for several minutes.

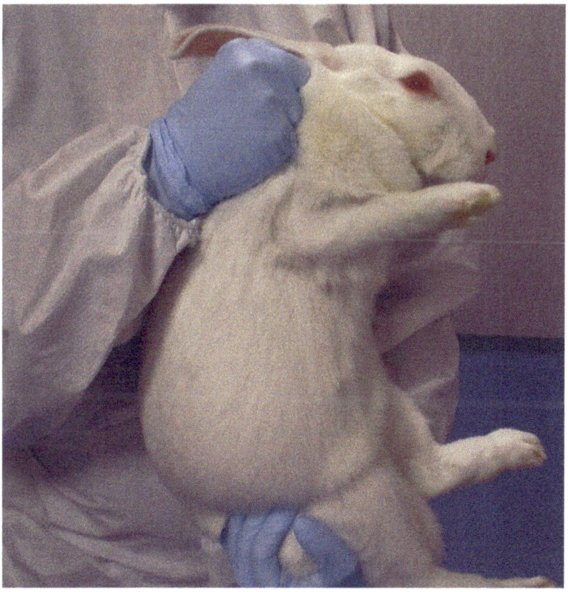

FIGURE 5.8 **A method to pick a rabbit up.** The best method is scruffing the skin of the neck while the rest of the animal body is supported by the other hand.

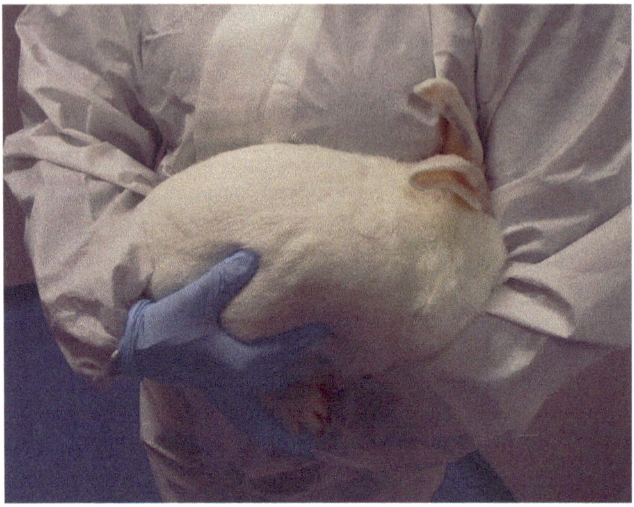

FIGURE 5.9 **A method for rabbit handling.** An effective technique is to hold the rabbit keeping its face hidden in the elbow crease of the operator.

The immobility response cannot be evoked in 25% of rabbits as there is a considerable interindividual susceptibility (Danneman et al., 1988). This response cannot be considered a good restraining technique due to the fear and anxiety that it induces on the animals.

Marking methods of rodents and rabbits (Luisa Corsi, Chiara Attanasio)

It is widely known that laboratory animals need social housing, therefore a method to identify each animal in a specific cage is required.

Individual identification is critical for breeding records and also to link essential data to single animals (genotype, study parameters, and results) in long-term studies. Loss of identification jeopardizes the experimental results making them useless; therefore, a proper and accurate identification system contributes to the reduction principle by avoiding animal waste.

The main international guidelines concerning marking methods suggest selecting a method that causes only a temporary discomfort or pain compatibly with the research purposes (Dahlborn et al., 2013).

Identification methods can be temporary or permanent.

Temporary marking methods include shaving fur, writing marks or numbers on hairless parts of the body, e.g., ears and tail as well as on the skin by felt-tip markers or alcohol-based pens.

When a temporary marker system is selected, it is important to use nontoxic inks or food colorants to avoid any unexpected reaction.

A felt-tip marker used on rodent tail or on rodent and rabbit ears usually disappear on 1−2 days due to the normal grooming activity of these animals.

Food coloring or henna, usually lasting 1 or two weeks, can be used on albino or light-colored rodents and rabbits. Shaved areas can last for a maximum of four weeks depending on the stage of hair cycle (Chase et al., 1951; Orasan et al., 2016).

The methods for permanently identifying rodents and rabbits include ear-tagging, tattooing, ear or toe clipping (only for mice and rats), and subcutaneous microchip implantation.

The most common system used for rodents is the ear clipping, also known as ear punching, even though in some facilities, it is preferred the use of individual ear tags. In the case of the ear punch system, rodents can be identified by making small holes in specific areas of the ear to create a numerical code for individual identification. It is important to clean the puncher after each use, particularly when the collected tissue is used for genotyping to avoid any DNA cross contamination.

This method of identification cannot be performed on mice or rats younger than 14 days of age due to the small size of the ears.

An additional system used to identify and collect samples for rodent genotyping is toe clipping, a refined procedure of distal phalanx removal. In the past, a similar nonrefined method based on a total toe amputation without considering the age of the animal, has been extensively used. Due to its mutilating effect, toe amputation is now discouraged or even prohibited in several European countries and institutions.

The toe clipping method refers to the distal phalanx removal used as a source of DNA as described in the FELASA working group reports (Dahlborn et al., 2013).

The effects of ear or toe clipping on animal welfare has been extensively investigated (Castelhano-Carlos et al., 2010; Cinelli et al., 2007; Miller and Leach, 2015; Paluch et al., 2014; Rasid et al., 2012; Schaefer et al., 2010; Williams et al., 2008; Wever et al., 2017).

Unfortunately, the outputs of the several studied performed on this topic did not allow to identify the gold standard method. The international guidelines currently available report that toe clipping should not be performed after seven days of age, since pups become more and more active with age which increases the risk of performing incorrect clipping and requires a stronger restraining method (Norecopa, 2008; Dahlborn et al., 2013).

On the other hand, toe clipping should not be performed before the three days of age when toes are still partially fused together and are too small to allow the accurate clipping of the distal phalanx, resulting in the removal of an excessively extended toe segment (Schaefer et al., 2010). All these considerations drive to the conclusion that the age of 5−7 days (being day 0 the day of birth) is a preferable time frame to remove phalanges for mice identification and genotyping (Dahlborn et al., 2013).

The selection of the most adequate identification method is also influenced by the required quantity of tissue that needs to be collected, by the kind of technique to be applied for quantitative or qualitative genotyping, by the duration of the experimental procedure as well as by its invasiveness affecting animal welfare.

Alternative less common methods include tattooing or microchipping for identification, and hair biopsies or rectal swabs for DNA sampling (Bonaparte et al., 2013).

However, most of these techniques cannot be used in new born or very young animals; therefore, they cannot be considered a valid alternative to toe clipping (Wever et al., 2017).

In addition, some of these techniques may cause higher discomfort than either toe or ear clipping (Castelhano-Carlos et al., 2010; Wever et al., 2017; Kasanen et al., 2011).

Ear tags with identification codes can be applied to the base of the ear, approximately 3 mm from the edge of the pinna. These tags are relatively heavy and can drag on the ear causing it to fold over or the tag to become a site of chronic infection and irritation (Kitagaki and Hirota, 2007).

Tail or ear tattooing can be performed with a microtattoo or an electric tattoo system; according to the FELASA guidelines, the use of the microtattoo system on the tail is recommended only in young animals before ossification occurs (between two and three weeks of age) since the system has to be completely inserted into the tail. The microtattoo system can also be used to mark toes or footpad by inserting the needle through the skin so that only the pads are marked (Dahlborn et al., 2013).

An important advantage of this method is its suitability for the identification of animals of all ages, as toe tattooing can be performed even on newborn subjects. Both toe and footpad tattoo methods are potentially painful, however, the possibility to adapt the needle used to the size of the animal offers a refinement of the procedure making it acceptable.

The electric tattoo machine allows to perform tail or skin tattooing both in rodents and rabbits.

The tattoo performed with this machine can induce temporary pain and discomfort due to the restraint but also to the machine noise and vibrations.

Each selected method has both advantages and disadvantages which are summarized in Table 5.1. In view of this, it would be preferable to select a method which enables the removed tissue fragment use for genotyping to reduce the number of procedures on the animals. However, there are some cases in which this strategy cannot be applied.

Moreover, some of the mentioned techniques can be used only within a specific range of age and the best one would represent a compromise between the experimental needs and the stress induced on animals.

TABLE 5.1 Permanent marking systems: advantages and disadvantages.

Marking method	Species	Advantages	Disadvantages
Tattoo/ microtattoo	Mouse, rat, guinea pig, rabbit	No anaesthesia required, fast procedure, fast readable, no/small manipulation for reading. Applicable at all ages.	Possible inflammation of the site, possible misreading; second technique needed for tissue collection (genotyping purpose). Special equipment needed.
Ear clipping	Mouse, rat, guinea pig	No anaesthesia required, fast procedure. Collection of the removed tissue for genotyping.	Manipulation required for reading, possible misreading, training required on the numbering schema for the operators. Not applicable in rodents before 14 days of age. Still unclear effects on animal welfare.
Ear tag	Mouse, rat, guinea pig,	No anaesthesia required, fast procedure, fast readable, no/small manipulation for reading.	Possible inflammation of the site, possible misreading; second technique needed for tissue collection. If metallic tags are used, it impacts on some procedures (e.g., RMI, impaired response).
Toe clipping	Mouse, rat	No anaesthesia required. Training needed to remove only the distal phalanx. Collection of the removed tissue for genotyping.	Possible bleeding, possible misreading, preferable for rodents of four to seven days of age. Effects on animal welfare still unclear.
Microchip transponder	Mouse, rat, guinea pig, rabbit	Anaesthesia required, aseptic procedures for implantation. No risks of misreading.	Possible pain, restraining needed for reading. Second technique needed for genotyping. Training required.

Basic procedures, substance administration and sampling in rodents and rabbits (Luisa Corsi, Chiara Attanasio)

Substance administration can be performed on laboratory animals by different routes to achieve either local or systemic exposure. Systemic routes include oral (through the digestive tract) or parenteral/intravenous administration. The choice of the best route is based on several factors, including the chemical features of the molecule, the purpose of the study, pharmacokinetics, and pharmacodynamics as well as the related regulatory requirements. An adequate knowledge of the substance pharmacological features is crucial (Brunton et al., 2007; Rowland and Tozer, 2010; Loftsson, 2015). The administration frequency is another critical factor to consider in choosing the administration route.

Refinement should also be carefully considered to minimize the negative impact of the procedure on animal welfare. All the methods commonly used for substance administration require animal handling and restraining. In addition to what has already been reported in the handling paragraph, before substance administration animals should be trained to make them accustomed to the procedure, thus minimizing the stress during the effective procedure.

In order to reduce pain and distress as well as to avoid tissue damage, key features of the drug such as pH, solubility, biocompatibility, viscosity, sterility, and quality of the substance, as well as temperature and the correct needle size need to be checked before administration (Morton et al. 2001). The pain and distress related to the needle insertion can be effectively reduced by applying a topical anaesthetic (Flecknell et al., 1990).

The subcutaneous route implies a slower absorption rate compared with other parenteral routes, providing a sustained effect of the therapeutic compound. The exact mechanism of absorption is thought to be due to the uptake of macromolecules within the subcutis through the small capillaries underlying the skin, with minimal lymphatic absorption (Kagan et al., 2007).

Substances delivered subcutaneously can be aqueous or oily fluids, agents for slow absorption, solid pellets, or substances injected via osmotic minipumps or other implantable devices surgically inserted into a subcutaneous pocket. Subcutaneous space is a virtual space; therefore, it can be an excellent site to deliver a large volume of fluid in small or dehydrated animals, avoiding the technical difficulties related to intravenous administration. Subcutaneous administration is an easier method compared to intravenous injection; however, a good training is always needed to avoid the accidental injection of the compound into the venous system and to ensure that the substances are adequately delivered. The appropriate size of the needle, and humane and aseptic techniques to be performed before and during the injection procedure should be carefully considered (Turner et al. 2011).

The needle needs to be inserted at a shallow angle to minimize any potential damage to the underlying tissues. It is essential to ensure the needle tip integrity, therefore, the needle must be changed at each injection. If asepsis is not respected, subcutaneous injections can result in abscess formation. The loose skin over the scruff or over the chest and the flanks are the usual subcutaneous injection sites in rodents and rabbits.

TABLE 5.2 Maximum injectable volume per species in relation to the different administration routes.

Species	Subcutaneous injection	Intramuscular injection	Oral administration	Intraperitoneal administration
Mouse	10 mL/kg	0.05 mL/kg	20 mL/kg	20 mL/kg
Rat	10 mL/kg	0.1 mL/kg	20 mL/kg	10 mL/kg
Guinea pig	10 mL/kg	0.05 mL/kg	10 mL/kg	10 mL/kg
Rabbit	10 mL/kg	0.1 mL/kg	10 mL/kg	5 mL/kg

The maximum injectable volume for each species is reported in the table below. Splitting the total amount for both subcutaneous and intramuscular injections into multiple sites is also recommended (Table 5.2).

The intramuscular administration technique requires higher skills compared to subcutaneous injection; therefore, it should be made only by trained personnel. In general, this method is less used in small mammals due to their reduced muscle mass which allows the administration of only small volumes making it necessary to split large volumes into smaller ones to be administered in multiple sites.

The quadriceps in the hind limb is the preferred site, however, in general, the caudal muscle mass can be used by identifying the semimembranous, the semitendinous, and the biceps femoris muscles, and avoiding the sciatic nerve that can be identified through palpation. The sciatic nerve is situated deeply in the posterior (caudal) femoral muscle mass and along the line of the bone; therefore, it is preferable to perform the injection into the muscle mass on the anterior of the thigh without inserting the needle too deeply. The animal leg should be firmly held, and the needle must be carefully inserted. The syringe plunger should be withdrawn slightly prior to injecting, to ensure that the needle hasn't passed a blood vessel wall, then, the substance can be slowly injected and at the end of the procedure, the site should be delicately massaged (Morton et al., 2001).

Alternative injection sites are triceps and dorsal lumbar muscles. Self-mutilation of the foot has been reported in rabbits as result of nerve damage during intramuscular injection into the caudal muscle mass (Beyers et al., 1991).

Intramuscular injection of irritating substances or accidental injection into the nerves may result in paresis, paralysis, muscle necrosis, and localized muscle sloughing (Rasmussen, 1978) while repeated injections may result in muscle inflammation and necrosis (Diehl et al., 2001).

The intramuscular route should only be used if a less painful alternative is impossible, or if it is required to be consistent with, e.g., a clinical trial design (Morton et al., 2001).

The most used method of administration in small rodents is intraperitoneal injection that is extensively performed in small-sized species in which it can be used to safely administer large volumes of fluid in contrast to intravenous access that can be challenging. The time required for the absorption of fluids delivered intraperitoneally is much slower than that required for fluids administered by intravenous injection. The pharmacokinetics of substances after intraperitoneal injection is more similar to that following oral administration, being mesenteric vessels the primary route of absorption by draining into the portal vein and passing through the liver. Therefore, substances administered intraperitoneally may undergo hepatic metabolism before reaching the systemic circulation (Lukas et al., 1971).

Even though intraperitoneal administration of pharmacological agents results in faster and more complete absorption compared to oral, intramuscular, and subcutaneous administration routes, this procedure holds some limitations. Fluids injected intraperitoneally should be isotonic, nonirritating, and sterile. The intraperitoneal route is used mainly in rodents but can be applied also on rabbit pups.

The injection is performed by holding the rodent in supine position with the head tilted lower than the posterior part of the body and by inserting the needle in the lower quadrant of the abdomen (at ~30 degrees angle) and by avoiding the accidental penetration into the viscera.

This procedure correctly applied enables the administration of large volumes of solution (up to 10 mL/kg) (Morton et al. 2001); however, training is required to avoid the unintentional puncture of the abdominal organs: with this technique, plunger withdrawal is not usually helpful being gut content too viscous to be drawn into the needle (Turner et al., 2011).

In case of an intraperitoneal injection wrongly performed, adverse consequences may occur, including painful lesions of the internal organs and peritonitis which may lead to death if bacteria from the intestinal tract enter the circulatory system causing bacteriemia and septic shock (Coria-Avila et al., 2007).

Studies performed on rats demonstrated that 19.6% of intraperitoneal injections can result in fluid injection into the gastrointestinal tract, subcutaneously, retroperitoneally, or even into the urinary bladder (Lewis et al., 1966). Potential complications related to this route include infection, pain, local irritation, chemical peritonitis, formation of fibrous tissue with adhesions within the abdominal cavity, perforation of an abdominal organ, haemorrhages, or even respiratory distress in case of administration of very large volumes.

Moreover, repeated administration can result in a cumulative irritant effect and needle-induced damage (Gotloib et al., 2005).

The maximum frequency allowed for intraperitoneal injections is once per day. In some circumstances, depending on specific experimental settings, brief anaesthesia can make the technique more reliable and less stressful; however, animal training and refinement in animal handling can be of help in reducing distress (Stuart and Robinson, 2015).

The oral route is commonly used when systemic exposure is required provided good absorption from the gastrointestinal tract and little or no first-pass hepatic metabolism are known; the administration route can be applied by oral gavage or by including the substances to be administered in food or water (Turner et al., 2011). This latter solution can reduce the palatability of some substances resulting in insufficient intake. This is the least stressful and easiest oral administration method. An alternative option is the gavage, consisting in a feeding tube passed into oesophagus and stomach. The tube is positioned by premeasuring the distance from the oral cavity to the xiphoid process, which is the caudal point of the sternum, situated at the bottom of the rib cage.

There are metal or flexible plastic feeding needles of various sizes. The use of too long feeding needles may result in oesophagus perforation whereas the use of too short needles may cause the animal to aspirate the substance.

In case of resistance during the insertion or if signs of distress are observed, the tube should be immediately withdrawn. Incorrect insertion, excessive force applied, or animal movements during the procedure may result in trachea penetration, oesophagus or stomach perforation with substance injection into the thoracic or peritoneal cavity.

Several factors can influence the success of gavage, including the level of experience of the technician, the size and the type of probe used, the volume administered, the repetitive dosing as well as the vehicle used (Vandenberg et al., 2014; Okva et al., 2006).

Due to the relatively high frequency of gavage-related complications and the operator high skill level required, alternative dosing methods should be used whenever possible. In many circumstances, oral gavage is not well tolerated by the animal and some evidence suggest that this method is associated with a significant amount of stress induction.

The use of a soft Teflon probe as feeding tube and the respect of maximum volumes of 6–8 mL/kg are refinement parameters associated to this procedure (Okva et al., 2006).

Due to limitations related to oral gavage, particularly for long-terms studies, and given its impact on animal welfare, considerable efforts have been devoted to developing novel methods of preclinical oral drug delivery (Brown et al., 2000).

Current alternative approaches include incorporating the test compound into peanut butter mixtures, wafer crackers, chocolate pellets, or drinking water (Walker et al., 2012; Zhang, 2011; Goldkuhl et al., 2008; Marx et al., 2014; Ferguson and Boctor, 2009).

Different authors investigated the effects of novel methods to refine the gavage procedure. Precoating needles with sucrose reduces the time to passage, decreases detectable stress-related reactions, and maintains baseline plasma corticosterone levels (Hoggat et al., 2010).

Furthermore, brief inhalant anaesthesia has been proposed as a potential refinement option to reduce adverse consequences of oral gavage. The comparison of gavage performed in anesthetized or awake animals showed that anaesthesia resulted more effective in mice and rats when serial oral gavage had to be performed (Jones et al., 2016; Murphy et al., 2001).

Other successful alternative techniques were syringe feeding method and orally dissolving strips that represent a safe, convenient, and humane alternative (Atcha et al., 2010; Tillmann et al., 2018).

The intravenous route is often used to ensure that maximum plasma exposure can be rapidly achieved. In most cases, substances are infused by a single dose into a suitable peripheral vein. The respect of sterility is mandatory to avoid potential adverse effects such as phlebitis, sepsis, or infection at the site of needle entry.

Many of the common intravascular administration sites, summarized in the table below (Table 5.3), are the same used for blood collection (Aasland et al., 2010; Frohlich et al., 2018; Golde et al., 2005; Heimann et al., 2010; Oliveira et al., 2009).

During blood collection, especially from small rodents, a limiting factor is the maximum volume that can be collected at one time without impairing animal health and welfare. This amount depends on the total blood volume of the animal, which is a species-specific parameter that can vary with age, sex, bodyweight, and blood withdrawn frequency.

The mean values for species and body weight are summarized in Table 5.4. Different values may be adopted for different strains. According to the main international guidelines, the maximum volume of blood that can be collected at one time from a healthy animal should not exceed the 10% of the total blood volume. A second collection of the same volume can be performed after two weeks. If a weekly frequency is required, the suggested limit is 7.5%, in any case without exceeding the value of 15% in 28 days. For daily collection, a maximum of 1% of the total volume every 24 h is recommended.

When applicable, warm saline administered subcutaneously or intraperitoneally after the collection is helpful in preventing hypovolemia. Any sign of anaemia such as pale mucous membranes, restlessness, hyperventilation, and hypothermia should be accurately monitored.

In some studies, the collection of animal urine and faeces might be necessary for analyzing different metabolites and toxic products or to gain information about the clinical conditions of the animal. In rodents, the collection requires the use of metabolic cages with grid floor, which enable to accurately control food and water intake as well as urine and faeces excretion by collecting the material in an external tube (Kurien et al., 2004).

TABLE 5.3 Common sites for blood collection/intravascular administration in different species, max volumes and impact on animal welfare.

Species	Common site for blood collection	Max volume withdrawable from the site	Common site for intravascular administration	Max volume injectable for site	Impact on animal welfare
Rabbit	Marginal ear vein	>20 mL	Marginal ear vein	5 mL/kg	+
	Jugular vein	>20 mL	Jugular vein	5 mL/kg	++
	Cephalic vein	Up to 10 mL	Cephalic vein	5 mL/kg	+
	Lateral saphenous vein	Up to 10 mL	Lateral saphenous vein	5 mL/kg	+
Guinea pig	Marginal ear vein	Up to 0.3 mL	Marginal ear vein	Up to 5 mL	+
	Jugular vein	Up to 2 mL	Jugular vein	>5 mL	++ (anaesthesia required)
	Cephalic vein	0.1–0.3 mL	Cephalic vein	Up to 5 mL	+
	Lateral saphenous vein	0.1–0.3 mL	Lateral saphenous vein	Up to 5 mL	+
	Gingival vein	0.2 mL	Gingival vein	Not applicable	+ (anaesthesia required)

Continued

TABLE 5.3 **Common sites for blood collection/intravascular administration in different species, max volumes and impact on animal welfare.—cont'd**

Species	Common site for blood collection	Max volume withdrawable from the site	Common site for intravascular administration	Max volume injectable for site	Impact on animal welfare
Rat/ mouse	Pedal vein puncturing + heparinized capillary tube	Rat: up to 0.05 mL	Not applicable site for IV injection	Not applicable	+
		Mouse up to 0.01 mL			
	Saphenous vein	Rat: up to 0.2 mL	Saphenous vein	Up to 0.1 mL (rat)	++ (anaesthesia recommended for injection)
		Mouse: up to 0.15 mL			
	Jugular vein (Rat only)	Up to 2 mL	Not applicable site for IV injection surgical cannulation needed	Not applicable	++
	Tail vein	Rat: up to 2 mL	Tail vein	Up to 0.5 mL	+
		Mouse: up to 0.2 mL		Up to 0.2 mL	
	Gingival vein	Rat: up to 0.8 mL	Gingival vein	Up to 1 mL	+ (anaesthesia required)
		Mouse: up to 0.1		Up to 0.15 mL	
	Sublingual collection	Rat: up to 1 mL	Not applicable site for IV injection	Not applicable	+ (anaesthesia required)
		Mouse: up to 0.3 mL			
	Facial vein (mouse only)	Up to 0.3 mL	Not applicable site for IV injection	Not applicable site for IV injection	++

+, low impact (not painful, minimally invasive and quick); ++, Medium impact (painful, stressful restraint or possible damage to the surrounding tissues).

TABLE 5.4 **Circulating blood in different species and volume which can be safely collected.**

Species	Total blood volume (mL/kg) (Average considering sex differences)	Body weight (grams)	Total blood volume	Safe volume for collection
Mouse	58.5 mL/kg	18–40 g	1.25–2.4 mL	Single collection 0.12–0.24 mL Weekly collection 0.09–0.18 Daily collection 0.012–0024 mL
Rat	64 mL/kg	250–500 g	16–32 mL	Single collection 1.6–3.2 mL Weekly collection 1.2–2.4 Daily collection 0.16–0.32 mL
Guinea pig	72 mL/kg	700–1200 g	50.4–86.4 mL	Single collection 5–8 mL Weekly collection 3.75–6.48 Daily collection 0.5–0.8 mL
Rabbit	61 mL/kg	1000–6000 g	61–366 mL	Single collection 6.1–36 mL Weekly collection 4.5–27 Daily collection 0.6–3.66 mL

However, metabolic cages entail the animal isolation and deprivation of nesting material or any kind of environmental enrichment. Rodents live in social groups ruled by a complex dominance system (Gil et al., 1999) so that the confinement in an unfamiliar cage is undoubtedly a stressful condition. Habituation to metabolic cages and auditory, visual, and olfactory contact with conspecific animals should, therefore, be provided.

Recently, the use of hydrophobic sand has been proposed with successful results as a refined method for urine collection. Rats and mice involved in these studies were more relaxed and comfortable in the sand environment than in metabolic cages and urine proprieties were perfectly preserved. Therefore, hydrophobic sand could be a suitable method for collecting urine with a less stressful method in compliance with the refinement principle (Hoffman et al., 2018).

Beyond the difference in sampling characteristics, also handling, processing, and storage of samples are important aspects since these procedures can significantly impact the results of the analysis.

Methods of zebrafish handling (Livia D'Angelo, Chiara Attanasio)

Source and transport

Exchange of zebrafish among research laboratories or shipment from stock centers (United States: https://zebrafish.org/home/guide.php; Europe: https://www.ezrc.kit.edu/; China: CZRC http://en.zfish.cn/) is the most appropriate source of zebrafish used for scientific purposes.

Zebrafish can be shipped as eggs and/or adult, although it is highly preferable to ship eggs.

The shipment (either eggs or adult fish) must be accompanied by a certificate of the health status provided by the delivering laboratory including also a brief description of husbandry conditions and parameters.

Shipment of eggs: Surface disinfected eggs (using bleach or iodine) should be shipped (50% embryo medium and 50% sterile water or water system). At reception, embryos/larvae should be visually inspected, and live embryos should be transferred to a small tank or Petri dish.

Shipment of adult fish: Animals should be starved for at least 24 h prior shipping. Low fish density (two adult fish/0.5 L) and a 1:1 or higher ratio of air or oxygen to water volume in each container is recommended (Varga, 2016). For shipment longer than a day, it is suggested to add ammonia binder to the shipping water, to limit health risks to the fish (Harper and Lawrence, 2011). On the arrival in the facility, new fish should be handled as little as possible, and care should be taken to prevent thermal shock (less than 2–3°C), visually inspected and introduced in the quarantine system and should undergo general health screening, some of which may be influenced by the aims of the research. For sanitary measures, only bleached embryos deriving of quarantined adults can be introduced in the main water system.

Housing systems

Zebrafish are generally maintained in recirculating water systems, fully integrated with filters, germicidal irradiation, and light and temperature control units (Aleström et al., 2020). Housing tanks should be transparent to allow visual inspection of personnel and visual contact among fish housed in close tanks.

Water supply and quality: City water is a common source, although it should be dechlorinated through a reverse osmosis system, as chlorine is toxic for fish also at humane safe levels. Water chemistry needs to be adjusted to ensure:

- a range of pH between 6.5–8
- total water hardness ($CaCO_3$) stable at 75–200 mg/L
- water conductivity of 300–1.500 μS/cm.

Zebrafish should be maintained at the temperature of 26–29°C, and in a standard lighting regime of either 14 h light:10 h dark or 12:12 cycle. Light influences almost all physiological and behavioral processes in fish, including growth, development, and reproduction. A gradual brightening and dimming period of 20–30 min in the morning and evening can be incorporated to avoid startling the fish, rather than switching lights abruptly on and off. Abrupt activation of lights during dark hours negatively impacts zebrafish welfare. In the facility environment, loud noises or novel vibrations should be avoided for the welfare safeguard. For more details, see Aleström et al. (2020).

Differently from terrestrial vertebrates, fish survival strictly depends on aquatic environment and thus variations of water parameters cause suffering and stress to animals.

Acceptable ranges of water parameters for adult zebrafish maintained under recirculating water systems are reported in Table 5.5.

TABLE 5.5 Optimal range of water parameters for keeping adult zebrafish in recirculating water systems (Alestrom et al., 2020).

Water	Optimal range
Water hardness (mg/L $CaCO_3$)	4–5
Conductivity	above 150 and up to 1700 μS/cm
pH	7–8
Ammonia	<0.1 mg/L
Nitrite	<0.3 mg/L
Nitrate	<25 mg/L
Temperature	24–29°C

Handling of zebrafish

Handling is one of the most stressful husbandry tasks for all fishes. Zebrafish are frequently handled during husbandry and experimental procedures. Even minimal handling causes a profound stress response in fish influencing the animal response to experimental manipulations, weakening the endocrine and immune systems, predisposing fish to infection, ultimately causing disease or even death. Increased levels of cortisol persisted up to one hour postnet handling in zebrafish (Ramsay et al., 2009). Furthermore, handling may damage the mucus layer covering and protecting the external surface of zebrafish, thus exposing animals to secondary infections. Therefore, proper handling of zebrafish is essential, both to prevent injury and to reduce stress.

Zebrafish bred under laboratory conditions generally tolerate better and recovered rapidly to the net stress compared to wild and or captive animals. This is likely due to the extensive domestication which has selected traits favorable for survival in a laboratory environment, including the ability to endure net handling and air exposure. Despite the ability of domesticated zebrafish to recover more rapidly from handling stressors, attention should be paid to the frequency and duration of handling which may alter the dynamics of the cortisol response and necessitate longer recovery periods. Furthermore, differences in handling procedures among laboratories may increase the husbandry-associated sources of variation in experiments (Ramsay et al., 2009).

Equipment and general handling procedures

Handling equipment for adult zebrafish is represented by a soft net with a smooth inner surface and/or a container. The use of container minimizes the stress of air exposure and reduces the potential for scales to be lost due to abrasions caused by the transfer net (Ruhl et al., 2009).

Each room of the facility and or aquatic system (quarantine, main system, etc.) should have a dedicated equipment to avoid cross contamination and should be sanitized prior each use. All equipment has to be easily sanitizable and sterilizable between use with different groups of fish. It is highly recommended to avoid chemicals, whose residues are harmful for fish. If possible, fish should not be fed prior to handling to lessen the risk of water contamination from gut contents. The length of time that fish should be fasted prior to handling depends on the procedures involved: a couple of hours may be sufficient for simple breeding procedures, while more invasive procedures, such as those requiring anesthetic (e.g., intramuscular or intraperitoneal injections, elastomer tagging or fin clipping), may require up to 24 h. Fasting is not necessary when fish are only moved between tanks.

Repeated handling during experimental procedures

During experimental procedures, i.e., behavioral tests, zebrafish may be handled repeatedly with consequent increase of a period of recovery between handling procedures. Recovery from stress can be prolonged. Repeated handling may require an increased level of monitoring, and the stress may be alleviated by administration of opioids (i.e., morphine by intracoelomic and/or immersion) (Sloman et al., 2019).

Methods of identification and marking of zebrafish (Livia D'Angelo, Chiara Attanasio)

Animal identification and marking are certainly more challenging and complex compared to terrestrial vertebrates. Up to now, there is no cost-effective system to distinguish fish individuals in a tank, meaning that groups of animals must be marked and tracked.

On a large-scale level, a dedicated and scalable inventory database represents the solution to track groups of the animals and their rearrangements noninvasively. The inventory system should be capable of storing the breeding history as well as creating animal usage reports, tailored specifically to the needs of a fish facility. Minimum parameters to track are name of fish strain/line, date of fertilization, number of fish (initially entered into the system/current number calculated by subtraction of dead fish), and molecular modifications/genetic information.

Depending on the research purposes, individual identification of zebrafish may be required along the entire experimental period. Individual tagging may cause pain, therefore it must be planned and executed cautiously and using anaesthetics and analgesics. Different tagging methods are in place for zebrafish (Table 5.6). The choice of the optimal method should me made depending on the specific experimental setting (duration of use, laboratory budget, handling skills, scientific purpose, etc.).

Basic procedures performed on zebrafish (Livia D'Angelo, Chiara Attanasio)

Fish have been shown to possess nociceptors that are physiologically identical to those found in mammals, brain structures, and opioid compound receptors necessary to feel pain, and a capacity to associate specific events with noxious stimuli (Sloman et al., 2019).

Fish respond to noxious stimuli with altered behavioral, physiological and hormonal parameters, in accordance with the intensity of stimuli.

Fin clipping

Common technique used to acquire a DNA sample for genotyping adult fish is fin clipping, a basic procedure characterized by the amputation of a small part of the caudal fin using a sterile razor blade, scalpel, or surgical scissors. Only the minimum amount of tissue necessary (2–3 mm) should be taken (Fig. 5.10A),

TABLE 5.6 Overview of the most common individual tagging methods for zebrafish.

Method	Equipment/ Furniture	Developmental stage	Body location	Animal welfare impact	Anaesthesia/ Analgesia	References
Subcutaneous dye injection	Kit of marking dyes	Juvenile/adult	Skin	Not impact on social interactions	Yes	Cheung et al. (2014)
Visual elastomer implanting (VIE)	Colorant and injection tools	Juvenile/adult	Skin	Not impact on social interactions	Yes	Hohn and Petrie-Hanson (2013), Delcourt et al. (2018)
Fin clipping	Dissection kit	Adult	Fin	Caudal fin section	Yes	Schroeder and Sneddon (2017), Delcourt et al. (2018)
Radio frequency identification (RFID) microtags	Tags, antennas, readers, and computer programs	Adult (some approaches may be applied also on juvenile)	Intracoelomic	No social impact	Yes	Cousin et al. (2012), Delcourt et al. (2018)

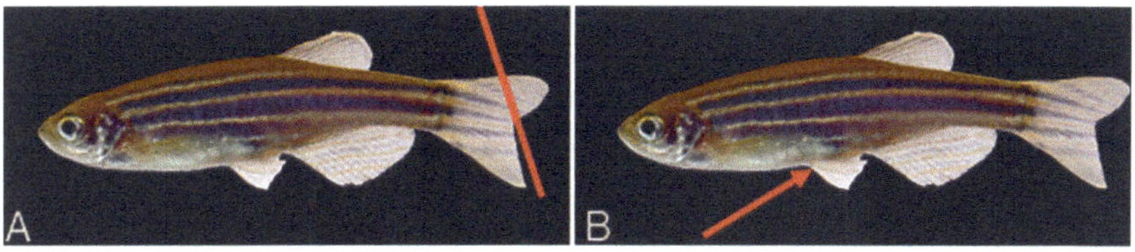

FIGURE 5.10 Basic procedures in adult zebrafish to be conducted under anaesthesia. (A) *Fin clipping in adult zebrafish*: the red line shows the section cut to collect enough amount of DNA for genotyping. Fin clipping is also used as tagging method for individual adult animals. (B) *Intra-peritoneal injection in adult zebrafish*: the *red arrow* shows the site to inject the needle.

as the caudal fin is innervated and clearly important for locomotion. This procedure is usually carried out under anaesthesia but can potentially cause pain, and an additional use of analgesic has been recommended (Schroeder and Sneddon, 2017). Fin clipped zebrafish are individually housed. To minimize the stress caused by single housing, zebrafish tanks should be placed in order to allow visual contact among animals (Stevens et al., 2021).

A recently validated method as a noninvasive alternative to obtain DNA in zebrafish is represented by skin swabbing (Breacker et al., 2017). This technique involves swabbing the skin to collect mucus and epithelial cells. Although swabbing does not require anaesthetic and appears to be less invasive than fin clipping, it still requires fish to be netted, held in air and handled, all procedures that can cause stress. However, smaller change in cortisol release and behavior on the first day of analysis has been reported in swabbed compared to fin clipped fish (Tilley et al., 2020). However, care should be taken to minimize contamination and cross contamination of the DNA sample when using a swab.

Administration and removal of substances

Branchial absorption

The most frequent route of exposure of fish for chemical agents is to dissolve them in water medium. The substance will be absorbed by gills which have a large surface area thanks to a series of lamellae protruding from the surface, and an epithelium extremely thin and designed to facilitate the diffusion of respiratory gases. Diffusion or uptake efficiency of chemicals by the gills depends primarily on their hydrophobicity and molecular size.

Oral administration

To deliver precise quantities of compounds and/or pathogens for experimental reasons, a gavage procedure can be put in place (Collymore et al., 2013). This procedure can be developed in adult fish, which are kept under anaesthesia (150 mg/L of buffered MS-222) and the volume dose rate should not exceed 1% body weight (100 μL/100 mg). A flexible catheter implantation tubing attached to a cut 22-G needle tip can be used. The flexible tubing needs to be directed lower into the oral cavity of the fish until the tip of the tubing extended past the gills (approximately 1 cm). The solution is then injected slowly into the intestinal tract. A protocol of microgavage of zebrafish larvae is also available (Cocchiaro and Rawls, 2013).

Injection

The most prevalent route for injection in fish is intracoelomic (intraperitoneal). Intraperitoneal injections should be carried out into the abdominal cavity, posterior to the pelvic girdle (Fig. 5.10B), avoiding penetrating abdominal viscera, as substances that cause inflammation may lead to adhesion formation. The entire procedure is conducted with animal under anaesthesia and precautions are required to minimize stress throughout the procedure (Kinkel et al., 2010). A small injection volume and a 35G needle need to be used and chemicals, depending on the composition, can be dissolved directly in sterile physiological saline or in very small quantities of cosolvent (e.g., ethanol, methanol, or dimethyl sulfoxide) prior to dilution in saline. An alternative route of injection (either compounds or cells) is represented by retro-orbital injections (Pugach et al., 2009).

When the drug must cross the blood–brain barrier, which limits the extent of drug action to the CNS, the only route is represented by the intracerebroventricular injection (Imperatore et al., 2018; Gutzman and Sive, 2009).

Blood sampling

Blood collection should be carried out only under anaesthesia and is generally followed by euthanasia. The amount of blood sample depends on the size of animal. More frequent methods for blood sampling are:

- caudal severance (Sloman et al., 2019)
- cardiac puncture (Sloman et al., 2019)
- retro-orbital bleeding (Vliegenthart et al., 2014).

Repeated blood collection protocol is available (Zang et al., 2015). The site for blood collection is along the body axis and posterior to the anus in the region of the dorsal aorta. just ventral to the spine. The volume for repeated blood sampling at intervals should be $\leq$0.4% of body weight every week or $\leq$1% every two weeks, evaluated by measurements of blood haemoglobin.

References

Aasland, K.E., Skjerve, E., Smith, A.J., 2010. Quality of blood samples from the saphenous vein compared with the tail vein during multiple blood sampling of mice. Lab. Anim. 44, 25–29. https://doi.org/10.1258/la.2009.009017.

Aleström, P., D'Angelo, L., Midtlyng, P.J., Schorderet, D.F., Schulte-Merker, S., Sohm, F., Warner, S., 2020. Zebrafish: housing and husbandry recommendations. Lab. Anim. 54 (3), 213–224.

Arakawa, H., Blanchard, D.C., Arakawa, K., Dunlap, C., Blanchard, R.J., 2008. Scent marking behavior as an odorant communication in mice. Neurosci. Biobehav. Rev. 32 (7), 1236–1248. https://doi.org/10.1016/j.neubiorev.2008.05.012. Epub 2008 May 15. PMID: 18565582; PMCID: PMC2577770.

Arts, J.W., Kramer, K., Arndt, S.S., Ohl, F., 2012. The impact of transportation on physiological and behavioral parameters in Wistar rats: implications for acclimatization periods. ILAR J. 53 (1), E82–E98. https://doi.org/10.1093/ilar.53.1.82. PMID: 23382273.

Atcha, Z., Rourke, C., Neo, A.H., Goh, C.W., Lim, J.S., Aw, C.C., Browne, E.R., Pemberton, D.J., 2010. Alternative method of oral dosing for rats. J. Am. Assoc. Lab. Anim. Sci. 49 (3), 335–343. PMID: 20587166; PMCID: PMC2877307.

Balcombe, J.P., Barnard, N.D., Sandusky, C., 2004. Laboratory routines cause animal stress. Contemp. Top. Lab. Anim. Sci. 43 (6), 42–51.

Beyers, T.M., Richardson, J.A., Prince, M.D., 1991. Axonal degeneration and self mutilation as a complication of the intramuscular use of Ketamine and xylazine in rabbits. Lab. Anim. Sci. 41, 519–520.

Bonaparte, D., Cinelli, P., Douni, E., et al., 2013. FELASA guidelines for the refinement of methods for genotyping genetically-modified rodents: a report of the Federation of European Laboratory Animal Science Associations Working Group. Lab. Anim. 47, 134–145.

Breacker, C., Barber, I., Norton, W.H.J., McDearmid, J.R., Tilley, C.A., 2017. A low-cost method of skin swabbing for the collection of DNA samples from small laboratory fish. Zebrafish 14 (1), 35–41.

Brown, S.A., 2012. Small mammal training in the veterinary practice. Vet. Clin. Exot. Anim. Pract. 15, 469–485.

Brown, A.P., Dinger, N., Levine, B.S., 2000. Stress produced by gavage administration in the rat. Contemp. Top. Lab. Anim. Sci. 39 (1), 17–21.

Brunton, L., Blumenthal, D., Buxton, I., Parker, K., 2007. Goodman and Gilman's Manual of Pharmacology and Therapeutics, eleventh ed. McGraw-Hill, New York (NY).

Burgdorf, J., Panksepp, J., 2001. Tickling induces reward in adolescent rats. Physiol. Behav. 72, 167–173.

Carp, J., Tennissen, A.M., Xiang Yang Chen, J., 2006. Wolpaw H-reflex operant conditioning in mice Psychology. Med. J. Neurophysiol.

Castelhano-Carlos, M.J., Sousa, N., Ohl, F., Baumans, V., 2010. Identification methods in newborn C57BL/6 mice: a developmental and behavioural evaluation. Lab. Anim. 44, 88–103.

Chase, H.B., Rauch, R., Smith, V.W., 1951. Critical stages of hair development and pigmentation in the mouse. Physiol. Zool. 24, 1–8.

Cheung, E., Chatterjee, D., Gerlai, R., 2014. Subcutaneous dye injection for marking and identification of individual adult zebrafish (*Danio rerio*) in behavioral studies. Behav. Res. Methods 46 (3), 619–624.

Cinelli, P., Rettich, A., Seifert, B., Burki, K., Arras, M., 2007. Comparative analysis and physiological impact of different tissue biopsy methodologies used for the genotyping of laboratory mice. Lab. Anim. 41, 174–184.

Clarkson, J.M., Dwyer, D.M., Flecknell, P.A., Leach, M.C., Rowe, C., 2018. Handling method alters the hedonic value of reward in laboratory mice. Sci. Rep. UK 8, 2448. https://doi.org/10.1038/s41598-018-20716-3.

Cloutier, S., LaFollette, M.R., Gaskill, B.N., Panksepp, J., Newberry, R.C., 2018. Tickling, a technique for inducing positive affect when handling rats. JoVE e57190.

Cocchiaro, J.L., Rawls, J.F., 2013. Microgavage of zebrafish larvae. JoVE 72, e4434.

Collymore, C., Rasmussen, S., Tolwani, R.J., 2013. Gavaging adult zebrafish. JoVE 78, 50691.

Coria-Avila, G.A., Gavrila, A.M., Ménard, S., Ismail, N., Pfaus, J.G., 2007. Cecum location in rats and the implications for intraperitoneal injections. Lab. Anim. 36 (7).

Cousin, X., Daouk, T., Péan, S., Lyphout, L., Schwartz, M.E., Bégout, M.L., 2012. Electronic individual identification of zebrafish using radio frequency identification (RFID) microtags. J. Exp. Biol. 215, 2729–2734.

Crabbe, J.C., Wahlsten, D., Dudek, B.C., 1999. Genetics of mouse behavior: interactions with laboratory environment. Science 284, 1670–1672.

Dahlborn, K., Bugnon, P., Nevalainen, T., Raspa, M., Verbost, P., Spangenberg, E., 2013. Report of the Federation of European laboratory animal science Associations working group on animal identification. Lab. Anim. 47, 2–11.

Danneman, P.J., White, W.J., Marshall, W.K., Lang, C.M., 1988. An evaluation of analgesia associated with the immobility response in laboratory rabbits. Lab. Anim. Sci. 38, 51–57.

Delcourt, J., Ovidio, M., Denoël, M., Muller, M., Pendeville, H., Deneubourg, J.L., Poncin, P., 2018. Individual identification and marking techniques for zebrafish. Rev. Fish Biol. Fish. 28, 839.

Diehl, K.H., Hull, R., Morton, D., Pfister, R., Rabemampianina, Y., Smith, D., VidalJM, van de Vorstenbosch, C., European Federation of Pharmaceutical Industries Association and European Centre for the Validation of Alternative Methods, 2001. A good practice guide to the administration of substances and removal of blood, including routes and volumes. J. Appl. Toxicol. 21, 15–23.

Ferguson, S.A., Boctor, S.Y., 2009. Use of food wafers for multiple daily oral treatments in young rats. J. Am. Assoc. Lab. Anim. Sci. 48 (3), 292–295.

Flecknell, P.A., Liles, J.H., Williamson, H.A., 1990. The use of lignocaine-prilocaine local anaesthetic cream for pain-free venepuncture in laboratory animals. Lab. Anim 24, 142–6.

Frohlich, J., 2020. Ferrets, Rabbits, and Rodents, pp. 345–367.

Frohlich, J.R., Alarcón, C.N., Toarmino, C.R., Sunseri, A.K., Hockman, T.M., 2018. Comparison of serial blood collection by facial vein and retrobulbar methods in C57BL/6 mice. JAALAS 57 (4), 382–391. https://doi.org/10.30802/AALAS-JAALAS-17-000134.

Gartner, K., Buttner, D., Dohler, K., Friedel, R., Lindena, J., Trautschold, I., 1980. Stress response of rats to handling and experimental procedures. Lab. Anim 14 (3), 267–274.

Ghosal, S., et al., 2015. Mouse handling limits the impact of stress on metabolic endpoints. Physiol. Behav. 150, 31–37. https://doi.org/10.1016/j.physbeh.2015.06.021.

Gil, M.C., Aguirre, J.A., Lemoine, A.P., Segura, E.T., Barontini, M., Armando, I., 1999. Influence of age on stress responses to metabolic cage housing in rats. Cell. Mol. Neurobiol. 19 (5), 625–633.

Golde, W.T., Gollobin, P., Rodriguez, L.L., 2005. A rapid, simple, and humane method for submandibular bleeding of mice using a lancet. Lab. Anim. (NY) 34, 39–43.

Goldkuhl, R., Carlsson, H.E., Hau, J., Abelson, K.S.P., 2008. Effect of subcutaneous injection and oral voluntary ingestion of buprenorphine on post-operative serum corticosterone levels in male rats. Eur. Surg. Res. 41 (3), 272–278.

Gotloib, L., Wajsbrot, V., Shostak, A., 2005. A short review of experimental peritoneal sclerosis: from mice to men. Int. J. Artif. Organs 28, 97–104.

Gouveia, K., Hurst, J.L., 2017. Optimising reliability of mouse performance in behavioural testing: the major role of non-aversive handling. Sci. Rep. 7, 44999. https://doi.org/10.1038/srep44999.

Gouveia, K., Hurst, J.L., 2019. Improving the practicality of using non-aversive handling methods to reduce background stress and anxiety in laboratory mice. Sci. Rep. 9, 20305. https://doi.org/10.1038/s41598-019-56860-7.

Gutzman, J.H., Sive, H., 2009. Zebrafish brain ventricle injection. JoVE 26, 1218.

Harcourt-Brown Textbook of Rabbit Medicine, 2008.

Harper, C., Lawrence, C., 2011. The Laboratory Zebrafish. CRC Press, Boca Raton, FL.

Heimann, M., Roth, D.R., Ledieu, D., Pfister, R., Classen, W., 2010. Sublingual and submandibular blood collection in mice: a comparison of effects on body weight, food consumption and tissue damage. Lab. Anim 44 (4), 352–358.

Hoffman, J.F., Fan, A.X., Neuendorf, E.H., Vergara, V.B., Kalinich, J.F., 2018. Hydrophobic sand versus metabolic cages: a comparison of urine collection methods for rats (*Rattus norvegicus*). JAALAS 57 (1), 51–57.

Hoggatt, A.F., Hoggatt, J., Honerlaw, M., Pelus, L.M., 2010. A spoonful of sugar helps the medicine go down: a novel technique to improve oral gavage in mice. J. Am. Assoc. Lab. Anim. Sci. 49, 329–334.

Hohn, Petrie-Hanson, 2013. Evaluation of visible implant elastomer tags in zebrafish (*Danio rerio*). Biol. Open 2, 1397–1401.

Hölscher, C., 1999. Stress impairs performance in spatial water maze learning tasks. Behav. Brain Res. 100 (1–2).

Hurst, J., West, R., 2010. Taming anxiety in laboratory mice. Nat. Methods 7, 825–826. https://doi.org/10.1038/nmeth.1500.

Imperatore, R., D'Angelo, L., Safari, O., Motlagh, H.A., Piscitelli, F., de Girolamo, P., Cristino, L., Varricchio, E., di Marzo, V., Paolucci, M., 2018. Overlapping distribution of orexin and endocannabinoid receptors and their functional interaction in the brain of adult zebrafish. Front. Neuroanat. 12, 62.

Jones, C.P., Boyd, K.L., Wallace, J.M., 2016. Evaluation of mice undergoing serial oral gavage while awake or anesthetized. J. Am. Assoc. Lab. Anim. Sci. 55 (6), 805–810. PMID: 27931321; PMCID: PMC5113884.

Kagan, L., Gershkovich, P., Mendelman, A., Amsili, S., Ezov, N., Hoffman, A., 2007. The role of the lymphatic system in subcutaneous absorption of macromolecules in the rat model. Eur. J. Pharm. Biopharm. 67, 759–765.

Kasanen, I.H., Voipio, H.M., Leskinen, H., Luodonpaa, M., Nevalainen, T.O., 2011. Comparison of ear tattoo, ear notching and microtattoo in rats undergoing cardiovascular telemetry. Lab. Anim. 45, 154–159.

Kinkel, M.D., Eames, S.C., Philipson, L.H., Prince, V.E., 2010. Intraperitoneal injection into adult zebrafish. JoVE 42, 2126.

Kitagaki, M., Hirota, M., 2007. Auricular chondritis caused by metal ear tagging in C57BL/6 mice. Vet. Pathol. 44, 458–466.

Kurien, B.T., Everds, N.E., Scofield, R.H., 2004. Experimental animal urine collection: a review. Lab. Anim 38, 333–361.

LaFollette, M.R., O'Haire, M.E., Cloutier, S., Gaskill, B.N., 2018. Practical rat tickling: determining an efficient and effective dosage of heterospecific play. Appl. Anim. Behav. Sci. 208, 82–91.

LaFollette, M.R., Cloutier, S., Brady, C., Gaskill, B.N., O'Haire, M.E., 2019. Laboratory animal welfare and human attitudes: a cross-sectional survey on heterospecific play or "rat tickling". PLoS One 14, e0220580.

Lewis, R.E., Kunz, A.L., Bell, R.E., 1966. Error of intraperitoneal injections in rats. Lab. Anim. Care 16, 505–509.

Loftsson, T., 2015. Essential Pharmacokinetics, first ed. Elsevier.

Lukas, G., Brindle, S.D., Greengard, P., 1971. The route of absorption of intraperitoneally administered compounds. J. Pharmacol. Exp. Therapeut. 178, 562–566.

Mandillo, S., et al., 2008. Reliability, robustness, and reproducibility in mouse behavioral phenotyping: a cross-laboratory study. Physiol. Genom. 34, 243–255. https://doi.org/10.1152/physiolgenomics.90207.2008.

Marx, J.O., Vudathala, D., Murphy, L., Rankin, S., Hankenson, F.C., 2014. Antibiotic administration in the drinking water of mice. J. Am. Assoc. Lab. Anim. Sci. 53 (3), 301–306.

Mertens, S., et al., 2019. Effect of three different forms of handling on the variation of aggression-associated parameters in individually and group-housed male C57BL/6NCrl mice. PLoS One 14.

Miller, A.L., Leach, M.C., 2015. Using the mouse grimace scale to assess pain associated with routine ear notching and the effect of analgesia in laboratory mice. Lab. Anim. 49, 117–120.

Morton, D.B., Jennings, M., Buckwell, A., Ewbank, R., Godfrey, C., Holgate, B., Inglis, I., James, R., Page, C., Sharman, I., Verschoyle, R., Westall, L., Wilson, A.B., 2001. Refining procedures for the administration of substance report of the BVAAWF/FRAME/RSPCA/UFAW Joint working group on refinement laboratory animals Ltd. Lab. Anim 35, 1–41.

Murphy, S.J., Smith, P., Shaivitz, A.B., Rossberg, M.I., Hurn, P.D., 2001. The effect of brief halothane anesthesia during daily gavage on complications and body weight in rats. Contemp. Top. Lab. Anim. Sci. 40 (2), 9–12. PMID: 11300680.

Nakamura, Y., Suzuki, K., 2018. Tunnel use facilitates handling of ICR mice and decreases experimental variation. J. Vet. Med. Sci. 80, 886–892. https://doi.org/10.1292/jvms.18-0044.

Norwegian Consensus Platform for Replacement, Reduction and Refinement of Animal Experiments (Norecopa). Toe-Clipping in Mice: An Evaluation of the Method and Alternatives, 2008. Norecopa, Oslo.

Obernier, J.A., Baldwin, R.L., 2006. Establishing an appropriate period of acclimatization following transportation of laboratory animals. ILAR J. 47 (4), 364–369. https://doi.org/10.1093/ilar.47.4.364.

Okva, K., Tamoseviciute, E., Ciziute, A., Pokk, P., Ruksenas, O., Nevalainen, T., 2006. Refinements for intragastric gavage in rats. Scand. J. Anim. Sci. 33 (4), 243–252.

Oliveira, D.T., Souza-Silva, E., Tonussi, C.R., 2009. Gingival vein punction: a new simple technique for drug administration or blood sampling in rats and mice. Scand. J. Lab. Anim. Sci. 36, 109–113.

Ono, M., et al., 2016. Does the routine handling affect the phenotype of disease model mice? Jpn. J. Vet. Res. 64, 265–271. https://doi.org/10.14943/jjvr.64.4.265.

Orasan, M.S., Roman II, Coneac, A., Muresan, A., Orasan, R.I., 2016. Hair loss and regeneration performed on animal models. Clujul Med. 89 (3), 327–334. https://doi.org/10.15386/cjmed-583.

Paluch, L.R., Lieggi, C.C., Dumont, M., Monette, S., Riedel, E.R., Lipman, N.S., 2014. Developmental and behavioral effects of toe clipping on neonatal and preweanling mice with and without vapocoolant anesthesia. J. Am. Assoc. Lab. Anim. Sci. 53, 132–140.

Pekow, C., 2005. Defining, measuring, and interpreting stress in laboratory animals. Contemp. Top. Lab. Anim. Sci. 44 (2), 41–45.

Popik, P., Potasiewicz, A., Pluta, H., Zieniewicz, A., 2012. High-frequency ultrasonic vocalizations in rats in response to tickling: the effects of restraint stress. Behav. Brain Res. 234, 223–227.

Pugach, E.K., Li, P., White, R., Zon, L., 2009. Retro-orbital injection in adult zebrafish. JoVE 34, 1645.

Ramsay, J.M., Feist, G.W., Varga, Z.M., Westerfield, M., Kent, M.L., Schreck, C.B., 2009. Whole-body cortisol response of zebrafish to acute net handling stress. Aquaculture 297 (1–4), 157–162.

Rasid, O., Chirita, D., Iancu, A.D., Stavaru, C., Radu, D.L., 2012. Assessment of routine procedure effect on breathing parameters in mice by using whole-body plethysmography. J. Am. Assoc. Lab. Anim. Sci. 51, 469–474.

Rowland, M., Tozer, T.N., 2010. Clinical Pharmacokinetics and Pharmacodynamics: Concepts and Applications, fourth ed. Lippincott, Williams & Wilkins, Philadelphia (PA).

Rubio, C.A., Huang, C.B., 1992. Quantification of the sulphomucin-producing cell population of the colonic mucosa during protracted stress in rats. In Vivo 6 (1), 81–84.

Ruhl, N, McRobert, S.P., Currie, W.J.S., 2009. Shoaling preferences and the effects of sex ratio on spawning and aggression in small laboratory populations of zebrafish (Danio rerio). Lab Animal Europe 9, 19–30.

Schaefer, D.C., Asner, I.N., Seifert, B., Burki, K., Cinelli, P., 2010. Analysis of physiological and behavioural parameters in mice after toe clipping as newborns. Lab. Anim. 44, 7–13.

Schellinck, H., 2010. How many ways can mouse behavioral experiments go wrong? Confounding variables in mouse models of neurodegenerative diseases and how to control them. Adv. Stud. Behav. 41, 255–366.

Schroeder, P.G., Sneddon, L.U., 2017. Exploring the efficacy of immersion analgesics in zebrafish using an integrative approach. Appl. Anim. Behav. Sci. 187, 93–102.

Sloman, K.A., Bouyoucos, I.A., Brooks, E.J., Sneddon, L.U., 2019. Ethical considerations in fish research. J. Fish. Biol. 94 (4), 556–577.

Stevens, C.H., Reed, B.T., Hawkins, P., 2021. Enrichment for laboratory zebrafish-A review of the evidence and the challenges. Animals (Basel) 11 (3), 698.

Stuart, S., Robinson, E., 2015. Reducing the stress of drug administration: implications for the 3Rs. Sci. Rep. 5, 14288.

Tilley, C.A., Carreño Gutierrez, H., Sebire, M., Obasaju, O., Reichmann, F., Katsiadaki, I., Barber, I., Norton, W.H.J., 2020. Skin swabbing is a refined technique to collect DNA from model fish species. Sci. Rep. 10 (1), 18212.

Tillmann, S., Wegener, G., 2018. Syringe-feeding as a novel delivery method for accurate individual dosing of probiotics in rats. Benef. Microbes 9 (2), 311–315. https://doi.org/10.3920/BM2017.0127. Epub 2017 Dec 21. PMID: 29264968.

Truitt, P.W., Hete, B.F., Starr, E.W., 2008. Small Animal Restraining Device with Non-traumatic Animal Loading Device. US Patent no. US20080168948. United States Patent Office.

Turner, P.V., Brabb, T., Pekow, C., Vasbinder, M.A., 2011. Administration of substances to laboratory animals: routes of administration and factors to consider. JAALAS 50, 600–613.

Vandenberg, L.N., Welshons, W.V., Vom Saal, F.S., Toutain, P., Peterson, J., 2014. Should oral gavage be abandoned in toxicity testing of endocrine disruptors? Environ. Health 13, 46.

Varga, Z.M., 2016. Aquaculture, husbandry, and shipping at the zebrafish international resource center. Methods Cell Biol. 135, 509–534.

Vieria, E.B., Menescal-de-Oliveira, L., Leite-Panissi, C.R.A., 2011. Functional mapping of the periaqueductal gray matter involved in organizing tonic immobility behavior in Guinea pigs. Behav. Brain Res. 216, 94–99.

Vliegenthart, A.D., Starkey Lewis, P., Tucker, C.S., Del Pozo, J., Rider, S., Antoine, D.J., Dubost, V., Westphal, M., Moulin, P., Bailey, M.A., Moggs, J.G., Goldring, C.E., Park, B.K., Dear, J.W., 2014. Retro-orbital blood acquisition facilitates circulating microRNA measurement in zebrafish with paracetamol hepatotoxicity. Zebrafish 11 (3), 219–226.

Walker, M.K., et al., 2012. A less stressful alternative to oral gavage for pharmacological and toxicological studies in mice. Toxicol. Appl. Pharmacol. 260 (1), 65–69.

Wever, K.E., Geessink, F.J., Brouwerl, M.A.E., 2017. Alice Tillema and Merel Ritskes-Hoitinga A systematic review of discomfort due to toe or ear clipping in laboratory rodents. Lab. Anim. 51 (6), 583–600.

Williams, W.O., Riskin, D.K., Mott, A.K., 2008. Ultrasonic sound as an indicator of acute pain in laboratory mice. J. Am. Assoc. Lab. Anim. Sci. 47, 8–10.

Wilson, L.M., Baldwin, A.L., 1998. Effects of environmental stress on the architecture and permeability of the rat mesenteric microvasculature. Microcirculation 5 (4), 299–308.

Wilson Rasmussen, F., 1978. Tissue damage at the injection site after intramuscular injection of drugs. Vet. Res. Commun. 2, 173–182.

Zang, L., Shimada, Y., Nishimura, Y., Tanaka, T., Nishimura, N., 2015. Repeated blood collection for blood tests in adult zebrafish. JoVE 102, e53272.

Zhang, L., 2011. Voluntary Oral Administration of Drugs in Mice. Protocol Exchange. Nature Publishing Group.

Recommended reading

Bailey, J., 2017. Does the stress inherent to laboratory life and experimentation on animals adversely affect research data? Atla-Altern Lab. Anim. 45, 299–301. https://doi.org/10.1177/026119291704500605.

Birck, M.M., Tveden-Nyborg, P., Lindblad, M.M., Lykkesfeldt, J., 2014. Non-terminal blood sampling techniques in Guinea pigs. JoVE e51982. https://doi.org/10.3791/51982.

Brudzynski, S.M., Ociepa, D., 1992. Ultrasonic vocalization of laboratory rats in response to handling and touch. Physiol. Behav. 52 (4), 655–660. https://doi.org/10.1016/0031-9384(92)90393-g. PMID: 1409936.

Deacon, R.M., 2006. Housing, husbandry and handling of rodents for behavioral experiments. Nat. Protoc. 1, 936–946. https://doi.org/10.1038/nprot.2006.120.

Gariepy, J.L., Rodriguiz, R.M., Jones, B.C., 2002. Handling, genetic and housing effects on the mouse stress system, dopamine functi and behavior. Pharmacol. Biochem. Behav. 73, 7–17. https://doi.org/10.1016/S0091-3057(02)00789-X.

Gonzales, C., et al., 2014. Alternative method of oral administration by peanut butter pellet formulation results in target engagement of BACE1 and attenuation of gavage-induced stress responses in mice. Pharmacol. Biochem. Behav. 126, 28–35.

Gouveia, K., Hurst, J.L., 2013. Reducing mouse anxiety during handling: effect of experience with handling tunnels. PLoS One 8, e66401. https://doi.org/10.1371/journal.pone.0066401.

Hori, M., Yamada, K., Ohnishi, J., Sakamoto, S., Furuie, H., Murakami, K., et al., 2014. Tickling during adolescence alters fear-related and cognitive behaviors in rats after prolonged isolation. Physiol. Behav. 131, 62–67.

Keino, M., Hirasawa, K., Watari, T., Kanno, M., Mikawa, M., Saito, K., et al., 2002. Minimally invasive method for collection of blood from Guinea pigs. Jpn. J. Infect. Dis. 55, 27–28.

Mazlan, N., López-Salesansky, N., Burn, C., Wells, D., 2014. Mouse identification methods and potential welfare issues: a survey of current practice in the UK. ATW 13, 1–10.

McIntyre, K., 2001. Rodent gavage technique concerns: avoiding excess mortality. Contemp. Top. Lab. Anim. Sci. 40, 7–7.

Murphy, S., 2001. Training could prevent deaths due to rodent gavage procedure. Contemp. Top. Lab. Anim. Sci. 40, 8.

Rodrigues, M.V., de Castro, S.O., de Albuquerque, C.Z., Mattaraia, V., Santoro, M.L., 2017. The gingival vein as a minimally traumatic site for multiple blood sampling in Guinea pigs and hamsters. PloS One 12 (5), e0177967. https://doi.org/10.1371/journal.pone.0177967.

Roughan, J.V., Sevenoaks, T., 2019. Welfare and scientifc considerations of tattooing and ear tagging for mouse identifcation. J. Am. Assoc. Lab. Anim. 58, 142–153. https://doi.org/10.30802/Aalas-Jaalas-18-000057.

The European union European Parliament and the council of the European union directive 2010/63/EU of the European Parliament and of the council of 22 September 2010 on the protection of animals used for scientific purposes. Off. J. Eur. Union L 276/33-79, 2010.

Vachon, P., 1998. Anatomical and histological observations of fore- and hind limb toes in adult mice after amputations performed at the age of two weeks. Can. J. Vet. Res. 62, 311–313.

Zeller, W., Weber, H., Panoussis, B., Bürge, T., Bergmann, R., 1998. Refinement of blood sampling from the sublingual vein of rats. Lab. Anim. 32 (4), 369–376. https://doi.org/10.1258/002367798780599910. PMID: 9807750.

Chapter 6

Animal health management and hygiene

Ferdinando Scavizzi[1], Viola Galligioni[2], Valentina Vasina[3], Marcello Raspa[1]
[1]*Institute of Biochemistry and Cell Biology, Italian National Research Council (CNR), Rome, Italy;* [2]*Comparative Medicine Unit, Trinity College Dublin, Dublin, Ireland;* [3]*University of Bologna, Bologna, Italy*

Introduction (Ferdinando Scavizzi and Marcello Raspa)

In recent decades, the microbiological status of animal colonies, especially mice and rats, has been highly improved due to the amelioration of animal facilities, house caging systems, experimental procedures, and health monitoring programs. The microbiological status of experimental animals can dramatically influence the animal welfare, the robustness, and the reproducibility of research results. The Federation of European Laboratory Animal Science Associations (FELASA) has published several updated editions of recommendations for rodent health monitoring (1994, 1996, 2002, and 2014). These recommendations are intended to harmonize health monitoring programs and to standardize the microbiological quality of the animals to ensure reproducible results. Microorganisms can influence experimental results even if symptoms are not detected since many pathogens cause subclinical diseases. Subclinical disease may perturb biological parameters (behavior, growth rate, organ weight, immune response), and they may generate interferences with research. Symptom manifestations depend on many variables such as bacterial or viral infectious strains, the host immune status, the simultaneous presence of other pathogenic or opportunistic microorganisms, and the genetic background/mutation. Some pathogen of laboratory animals (LAs) can also infect humans (zoonoses). For all these reasons, all institutions should establish a health monitoring program to guarantee the quality of the animals and the safety of personnel. It is important that a qualified person be responsible for health management (HM) of the facility since the decision whether a microorganism is acceptable or not in a given animal colony is a complex issue.

All the HM results of a microbiological unit (room, rack, isolator) are reported in the HM report.

Practical Handbook on the 3Rs in the Context of the Directive 2010/63/EU
https://doi.org/10.1016/B978-0-12-821180-9.00007-6

Classification of animals and basic structures to their microbiological characteristics (Ferdinando Scavizzi and Marcello Raspa)

Classification of animals and structure are summarized in Figs. 6.1 and 6.2.

Cleaning, sanitizing, and sterilization procedures (Ferdinando Scavizzi and Marcello Raspa)

The choice of methods and frequency of cleaning and disinfection of the premises and equipment of an enclosure depend on many factors, such as the microbiological state of the animals that it is intended to maintain (germ-free, specific pathogen-free [SPF], and conventional), the physiological and behavioral characteristics of the species hosted, the type of equipment, the type of sawdust, the ambient temperature and humidity, the efficiency of the air-conditioning system, etc.

In this context, general indications will be given on some aspects that are dealt with daily by staff working in an enclosure, such as cage changing and equipment cleaning.

Sawdust must be changed with a suitable frequency to keep the cages clean and dry. There is no minimum or ideal cage change frequency, as it depends on the number and size of the animals housed, the type of sawdust, the type and size of cages, the conditions in which the animals are bred. The frequency, therefore, can be variable, ranging from weekly change to a change every two or three weeks. It is advisable, however, to avoid changing the litter in the days immediately before and after the birth and during the mating period, to contain animal stress.

Conventional	Specific Pathogen Free (SPF) and (SOPF) Specific and Opportunistic Pathogen Free	Germ-free (GF) and gnotobiotic
Animals housed in conventional enclosure and subject to experimental procedure	Animals are barrier maintained only for breeding	Germ-free: animals devoid of any living micro-organism. Gnotobiotic: germ-free animals inoculated with few and well-defined bacteria
Introduction of animals from outside often necessary	Animals enter only after embryo derivation process	Enclosures must meet management criteria and equipped with diversified isolators

FIGURE 6.1 Laboratory animals, according to their microbiological quality, are generally classified into three main categories.

Conventional	Barriered	Germ-free and gnotobiotic
Animals do not need to be negative to all microbiological agents tested, but it should always be considered that microorganisms could affect animal welfare and experimental outcome. These type of enclosure are the most common facilities.	Only SPF and SOPF animals are kept. These facilities are ideal for immunocompromised and immunodeficient animals.	Germ-free and gnotobiotic animals are kept in isolators with positive pressure for bioexclusion. Note: nowadays there are systems that are isolators at the cage level. High cost and complex management, hence numbers of animals are kept at minimum
Both for breeding and experimental animals	Mainly for breeding purposes	Both for breeding and experimental animals
Not strict access rules nor working procedures	Strict access rules and working procedures to keep at minimum risk of microbiological contamination	High level containment and working procedures
Users can freely access and use animals in this area to carry out procedures	Preferrable if only specialized staff access the area and researchers get in touch with animal care staff. Ability to adhere to this depends on organization of the facility. When experimental units are inside the barriered area, researchers must strictly follow the access procedures like (air or wet) shower and wear sterile clothes	Enclosures must meet management criteria and equipped with diversified isolators.
Hygiene: - Materials (including food and water) do not undergo particular sanitization process; - good cleaning of equipment and strict working procedures are enough	Hygiene: - All material, including food and water, go through disinfection/sterilization process; - rigorous cleaning and working procedures need to be in place	All materials coming into contact with animals are sterile through the use of autoclaves and / or sterilization through physical and / or chemical means

FIGURE 6.2 Animal holding areas can be classified according to various parameters.

There is also no general rule for washing cages and equipment, such as shelves, watering bottles, etc., but the procedures will depend on the characteristics of the enclosure. In general, the cages must be replaced, washed, and sterilized with each litter change, along with the watering bottles. Some types of cages and shelves may require a lower frequency of washing and disinfection, for example, when there are very large cages with a small number of animals, or animals housed individually in ventilated cages. Conversely, a higher frequency may be needed in cases where cages are densely populated, or animals are treated with experimental procedures that increase excrement.

A large variety of machines currently exist on the market for washing equipment. Most use wash cycles where an alternation of alkaline and acid products is provided. The former is effective in removing organic residues, the others are necessary for the neutralization of alkaline products which, if used alone, could ruin the equipment in the long run. The temperatures must be around 55−60°C for washing and higher, around 80°C, for rinsing. In order to reduce or eliminate the bacterial load, it can be indicated to combine these washing cycles with disinfection and/or sterilization procedures. The sterilization of all equipment and materials to be introduced into the enclosure can take place through the use of autoclaves (autoclave cycles should be at a T of 121°C and for a duration not exceeding 20 min), gas (hydrogen peroxide [H_2O_2], ethylene oxide, and formaldehyde), radiation (2.5 Mrad gamma irradiation), and dry (T of 160°C for 120 min) depending on the size and nature of the materials to be treated. The large material and, where possible, even an entire housing room are sterilized by fumigation using a H_2O_2 nebulizer and leaving the products to act in the sealed room for 24 h before turning the ventilation back on. Smaller materials can be sterilized using an autoclave; the material that cannot be autoclaved (e.g., plastic material) and must be introduced into a barrier animal is passed inside a spray box (sealed chamber for product sterilization), where a nebulizer sprays a peracetic acid solution at 4% (or 2.5% H_2O_2) for about 10 min or irradiated by specialized companies. Finally, the dip tank represents a further valid aid to quickly sterilize and pass biological material such as embryos and newborns in barrier areas. The dip tank is a closed tank filled with a sterilizing liquid (e.g., 300ppm chlorinated solution) changed regularly. Tests for the cultivation of the sanitizing liquid are carried out weekly to verify the sterilizing power.

All sterilization processes can be controlled by using tubes containing spores of *Bacillus subtilis* and *Bacillus stearothermophilus*, placed inside the room, equipment, or other material. Subsequently, the contents of the tubes must be plated on culture media to verify the absence of bacterial growth.

All enclosure environments such as housing, warehouses, washrooms, corridors, etc., must be cleaned and disinfected regularly using appropriate products. A periodic rotation of products must be foreseen in order to avoid that the microorganisms develop a resistance to the active principles contained in them. Products based on quaternary ammonium salts, potassium sodium monopersulfate, or products with multiactive formulation (containing, for example, surfactants, organic acids, oxidizing agents, etc.) are recommended.

Animal drinking water can be a vehicle of pathogens and therefore must be subject to regular microbiological treatments and controls. Generally, filtration (up to 0,2μ) and chlorination or acidification are used. Culture of drinking water every time bottles are filled is crucial. Testing water for presence of metals and oligominerals is also important to maintain proper quality and avoid animal welfare issues.

The effectiveness of sterilization should be periodically verified, for example, by performing microbiological analysis of the materials subjected to this process.

It should be kept in mind that most of the products used in cleaning and disinfection can be toxic, flammable, corrosive, and, therefore, adequate use and storage procedures must be provided.

Risks (Viola Galligioni and Ferdinando Scavizzi)

Strictly considering the microbiological status of the rodent colony, the risk of introducing microbiological agents (potential threat to the unit) comes mainly from three different areas:

- introduction of animals
- introduction of material: flow of material into the unit
- people: flow of animal unit personnel, researchers, visitors, contractors, and maintenance workers.

Risks are summarized in Fig. 6.3.

Health and environmental monitoring in animal unit (Ferdinando Scavizzi and Viola Galligioni)

As mentioned earlier in this chapter, the health of LAs influences their experimental use (Nicklas, 1999, 2002) therefore, an adequate health monitoring program assumes a fundamental role for every modern research facility. The enclosure is conventionally defined as a controlled environment with limited access (for personnel, materials, animals, and their products), in which the infrastructure, the procedures adopted, and the surveillance exercised, guarantee the category to which they belong (barrier or conventional). The main health control activity is carried out on animals and their products and is a critical procedure necessary for assessing the health status of the animals themselves; however, the indirect controls of the hygienic-sanitary conditions of the environment, the implementation of the procedures, the continuous training of personnel and related activities are not to be

Risk factors	Explanation	Mitigation strategies
Introduction of animals	Movement of live animals across institutes is quite common in research, because of collaborations and sharing of unique genetically modified strains. It is important to consider that this is also one of the best way to introduce microbiological agents in the animal unit.	- The Health Monitoring Report should always be provided when moving animals and interpreted correctly (Mahler Convenor et al., 2014). - Not all pathogens are vertically transmitted, therefore moving embryos rather than live animals can be considered safer. Of course, this means technical know-how and proper equipment should be available in the recipient institute. We don't have to forget that the transportation of embryos rather than live animals is also a way to provide better animal welfare - In case there is no possibility to rederive animals and the Health Report is not clear, organize a quarantine or a small isolator, where the animals coming from another institute can be placed. They can be tested for the agents causing more concern. If they result negative, they can go into the unit. In case they are positive, a thorough discussion is needed between DV and researcher about the intended use of these animals and for how long.
Introduction of material	Cages, diet, bedding, nesting material, water, instruments and consumables for experiments, biological material for injection are just some examples of what we introduce daily in the animal unit. How much we want a particular material to be disinfected depends on the level status of the animal unit we are aiming.	- For higher level status, everything needs to be disinfected/sterilized. - In case of biological material (cells, Antibodies), it needs to be tested before being introduced into the the animal (e.g. IMPACT testing, Mahler Convenor et al., 2014). Considering immunocompromised animal models and the use of cell-line derived and patient derived xenograft models (respectively CDX and PDX) particular attention should be given to microbiological agents that are a risk for animals and humans. - With regard to biological material, collaboration between the animal unit staff, the DV and the researcher is very important from the time of planning the experiment, to understand what and which volume should be tested. This will help to avoid surprises of not having enough material for the testing and for the experiment itself.
Flow of polpe	Animal facility staff and researchers are not the only people who would need access to the unit. Measures should be in place so that routine equipment maintenance can be undertaken, as well as allowing access to contractors/personnel looking after electrical parts, ventilation of the unit, gas detectors (CO_2 and O_2 sensors) and plumbers.	- Engineers maintaining the Air Handling Units of IVC or autoclaves, can be asked to avoid entering any other animal unit in the 48 hrs beforehand, but of course this is dependent on the severity and urgency of the problem. - If the animal facility consists of different units, then plans should be in place for having every unit maintained, starting from the one with the higher health status. - In case of units with different health status, one option is to have a main tools box (screwdrivers, drills and spanners) in every unit, so that contractors do not have to wait for disinfection of their tools. - Personnel working in the enclosures must be trained and provided with suitable personal protective equipment. These precautions are essential to avoid any zoonoses and to protect animals as staff can function as a carrier and transmit pathogens from one animal to another. - Personnel entering the barriers must take a complete shower (air or wet) and wear sterile and protective clothing. It is advisable to regularly carry out microbiological checks on PPE and showers used by staff. - It is not recommended that the same personnel work in more than one animal facility/room with different health levels. If necessary, personnel should follow all the procedures to decrease the risk of transmitting the infections (work from clean to dirty, shower, change clothes, etc).

FIGURE 6.3 Risks affecting microbiological status of animal units and related mitigation strategies.

underestimated. The health program must be designed to detect the presence of pathogens. Beyond the standard programs of health surveillance, the essays chosen, the samplings and the frequency of the tests, the daily observations on animals by the researchers are always very important, and sometime crucial.

The FELASA recommendations (Mahler Convenor et al., 2014) are aligned with those in use on other continents, including the type of tests and the frequency of tests. Furthermore, a new joint working group "AALAS—FELASA working group on health monitoring of rodents" was established, whose aim was to harmonize international health programs and thus facilitate the exchange of animals and their products (Pritchett-Corning, 2014). Although these guidelines have no normative value, they are now commonly recognized and used as a reference by most national and international institutions that use animals for scientific research programs (Nicklas, 2010). From International estimates (Silverman, 2002), it can be calculated that about 3% of the entire budget of the animal welfare should be allocated to health surveillance programs.

Health monitoring program (Ferdinando Scavizzi and Marcello Raspa)

To develop an effective health monitoring plan, it is necessary to divide the enclosure into microbiological units (Mahler Convenor et al., 2014). A microbiological unit is intended to mean a separate environment, whose microbiological characteristics are homogeneous and where the potential exposure of animals to pathogens is practically the same (e.g., barrier enclosure, housing room, individually ventilated cage [IVC] rack, isolator, etc.). The microbiological unit determines the type, frequency, and number of sampling. The definitions of Area SPF and Specific and Opportunistic Pathogen Free do not refer only to structural and procedural aspects but are closely associated with a specific exclusion list.

A health control program involves the periodic execution of diagnostic tests on animals (Percy and Barthold, 2007). The choice of animals to be subjected to analysis is carried out using epidemiological criteria. To perform health controls on animal colonies, both resident animals and so-called sentinel animals should be used. The former is represented by animals belonging to experimental or breeding colonies already present in the enclosure. Sentinel animals, on the other hand, can be external, purchased for this purpose by an authorized supplier, or internal, animals already present in the enclosure, but dedicated only to this specific function. In any case, regardless of the type of animals selected for health monitoring, they must have mature immune systems (if serology has to be performed) and have been adequately exposed to all possible pathogens present in the colony.

Regarding the use of external sentinel animals, the choice of these makes sampling more flexible, both for the possible selection of the most suitable strain for analysis, and for the choice of their age. The sentinel cages must be positioned in the lower part of the shelf and stabilized in the same conditions as the other animals; they must also be regularly exposed to pathogens through contact with exhausted sawdust, food, and water taken in rotation from the other cages according to a predefined schedule. To avoid "the dilution effect," clean sawdust should not be added to the cages of sentinels. To use sentinels for diagnostic purposes, it is necessary to wait at least six weeks from the start of exposure to allow the development of the antibody response. A longer exposure (10–12 weeks) is recommended to favor the infection and/or the seroconversion of certain agents such as *Mycoplasma pulmonis* and *Pasteurella pneumotropica*.

It is good practice to run a complete screening (parasitology, bacteriology, serology, and polymerase chain reaction [PCR]) only on a part of the sentinels available, replacing them with new animals, and leaving them in contact with the remaining subjects. These precautions allow to increase the effectiveness of the exposure, but also to have back up animals in case of retesting.

For these reasons, different strategies are associated with the use and contemporary analysis of both sentinel and resident animals. In the case where the functional unit is represented by IVC cages or by isolators, it is good to also use external contact sentinels, such animals introduced directly into the cages should be checked.

With regard to the frequency of sampling, the FELASA recommendations suggest a quarterly monitoring program for prevalent and annual pathogens for nonprevalent and rare pathogens. This program can be changed in the event of suspected contamination, eradication plans, and the control of established pathogens, but also during embryonic or cesarean rederivation programs, for batches of animals that from quarantine must pass in barrier or clean-conventional enclosures.

The sentinel animals must come from a safe and controlled breeding (SPF-FELASA category), to prevent them from being a source of contamination. They must also be representative of the population to be controlled (e.g., mutant mice with background B6 are associated with sentinels B6). The health screening of animals housed in IVC racks (Compton, 2004) may be less representative of the conditions of the colony. In fact, every single IVC cage can represent a microbiological subunit of the entire rack and therefore more cages should be controlled to have a reasonable estimate of the health status of the entire colony. The application of the Institute of Laboratory Animal Research formula is also not appropriate to define the size of the sample to be monitored. There is no perfect strain to be chosen for animal sentinels; the best solution would be to choose the same strain or the more genetically related to the animals housed in that microbiological unit.

Even though the use of sentinels (animals bred and sacrificed only for pathogen surveillance) is under discussion and their use is decreasing as new methods arise and develop (see Diagnostic tests), they are still largely used in many animal facilities.

Not all agents are transmitted through sawdust (e.g., lymphocytic choriomeningitis virus, Sendai virus, *P. pneumotropica*) or air (e.g., Helicobacter spp, mouse rotavirus, mouse parvovirus) and therefore in some cases, contact sentinels or colony animals are the best choice. In particular, colony animals are extremely useful when genetically modified animals are maintained because they may have compromised immunocompetent status and different sensitivity to specific pathogens especially if they are subjected to experimental procedures (e.g., irradiation or surgery), therefore exposed to differential microbiological threats. These animals are generally precious and cannot always be available for health monitoring, but it is crucial they are often analyzed because they represent more than the sentinels the health status of the experimental colony. Sampling these animals is possible (and advisable), simply collecting fecal pellets, oral and fur swabs. These samples may be processed for bacteriology and molecular methods for all pathogens without any harm to the animals. Particular attention must be paid to immunodeficient strains nude or NOD scid gamma mouse (NSG), testing them more frequently and accurately with respect to the opportunistic microorganisms (like *Pseudomonas aeruginosa* or *Staphylococcus aureus*) because these strains may develop infectious symptoms.

Particular attention must be paid to avoid contamination of founder colonies, expansion colonies, and production colonies, especially if the animals are kept in barrier or protected areas.

It should be kept in mind that sensitivity to infection may decrease with age (e.g., Mouse Parvo Virus (MPV), Mouse Hepatitis Virus (MHV)) and that most pathogens generate asymptomatic infections.

It is advisable to keep separate and under specific control areas of the enclosure, the different colonies with regard to procedures, animals, materials, and personnel. The sample size (random sample) depends on the infectivity of the pathogen and the probability of error of the result according to the formula:

$$S = \frac{\log p}{\log N}$$

where S is the sample size, p the probability of the result error, and N the percentage of uncontaminated animals. The formula applies to animal populations of at least 100 individuals.

The sampling width is usually between 5 and 25 individuals. For example, in the presence of an infectivity of approximately 30%, a sample of 10 subjects (or sera) per functional unit is sufficient to identify a positive with a 95% confidence interval (Nicklas, 2002; Dubin, 1991).

Diagnostic tests (Ferdinando Scavizzi and Viola Galligioni)

Indirect diagnostic tests (serological analysis) or direct (molecular or cultural analysis) can be used to highlight the presence of pathogens. Obviously, only the critical combination of the various analysis systems, together with the anamnesis, even historical, and the clinical symptomatology can give an overall and more precise picture of the health status of animal colonies.

Serology (indirect methods) is generally performed by enzyme-linked immunosorbent assay (ELISA), hemagglutination inhibition test, immunofluorescent assay, or multiplexed fluorometric immunoassay. These methods are advisable for routine preliminary screening as they are practical, allow an immediate overview of the presence or absence of different pathogens, have a high sensitivity and contain costs; on the other hand, false positives can occur due to cross reactions because of the presence of antibodies that react with antigens that are not always absolutely specific.

The direct methods are represented by culture, parasitological, necroscopic, and histopathological examinations and by molecular analysis such as PCR. Culture tests or analysis of nucleic acids by PCR are very specific and can identify an ongoing infection. PCR, in particular, directly detects the DNA/RNA of the infectious agent and is particularly useful for identifying viruses and all persistent infectious agents in tissues and also those for which there is no adequate serological or cultural technique. Direct methods are also methods of selection for the detection of diseases in immunosuppressed or immunodeficient animals, in which the antibody response is absent or very poor. In addition, they can be used in cases where it is not possible to detect immune responses or wait for the time necessary for them to develop (seroconversion minimum 14 days), such as in the case of:

- animals in which the immune system is not yet fully developed;
- animals of new arrival (quarantine);
- animals exposed to possible contamination (even experimental);
- suspect animals.

The necropsy and histopathological examinations, although they do not represent fundamental elements for diagnostic purposes, still make it possible to acquire supplementary and confirmatory information and for this reason they are usually performed.

Parasitological methods play a fundamental role in the search for endo- and ectoparasites. With regard to endoparasites, and especially to protozoa, these can be sought in the small and large intestines by microscopic analysis of the mucous membranes or intestinal contents. Organisms such as giardia and spironucleus, larvae and adults of helminths, and also flagellate or entamoeba can be directly identified under the microscope. The ectoparasites (e.g., fleas, mites, and lice), on the other hand, can be directly detected at the stereomicroscope through observation and analysis of the hair of the cranial, facial, neck, and back of the animals. In doubtful cases, both endo- and ectoparasites can also be detected directly by molecular analysis (e.g., PCR) after a cutaneous swab.

The direct search for bacteria can be performed by culture using, when possible, selective media. Culture examinations allow a precise identification. However, they require the time necessary for the development of the colonies and the respect of the specific conditions (in particular temperature or oxygen concentration) for the development of the searched bacteria. It is therefore not always a suitable method for providing quick answers. Identification at species level, if necessary, is carried out by means of classical microbiological identification (gram staining, microscopical observation, and biochemical tests) but also by systems that are easily available on the market (e.g., Microbial Identification kit from Biomerieux (API) or Microbial Identification kit from Becton Dickinson (BBL)).

A new direct approach for comprehensive health monitoring is emerging, based on whole genome sequencing. It has been shown (Scavizzi et al., 2021) that shotgun sequencing of nucleic acids extracted from fecal pellets (from single animal or from a pool of colony animals) can provide, in a single test, complete information both on microbiota and on pathogenic microorganisms. This information includes DNA and RNA viruses, fungi, bacteria, and eukaryotic endo- and ectoparasites. The use of Next Generation Sequencing for health monitoring of laboratory mice is able to broaden the concept of health monitoring from "pathogen surveillance" to a more exhaustive and useful "microbiota surveillance."

Exhaust air dust filters (Viola Galligioni and Ferdinando Scavizzi)

Some IVC manufacturers have developed fast, cheap, and effective systems for the control and detection of pathogenic microorganisms. These systems use exhaust air and particulate from ventilated shelves which, through PCR, lead to the detection of any pathogens present inside the cage and the environment where the animals live without the need to directly test any animal. This technique clearly reduces or replaces the use of animals for health monitoring purposes. Even though studies on exhaust air dust (EAD) are rapidly progressing, it has not yet been demonstrated if all pathogens can be detected on filters and what is the minimum load or time of exposure to ensure detection. The EAD approach should ideally be used in combination with classical bedding sentinel and direct colony animal testing.

Environmental monitoring (Ferdinando Scavizzi and Marcello Raspa)

To guarantee environmental conditions that minimize the possibility of transmission and spread of pathogens among animals, it is necessary, in the first instance, to guarantee the microclimate conditions (temperature, pressure, humidity, light, etc.). The continuity of these conditions requires periodic controls and maintenance. In particular, a critical element is the condition of the air filters. The air conditioning filters must be regularly checked and replaced; they themselves serve as a vehicle of contamination. It is also advisable to periodically check the microbiological conditions of all the enclosure environments, including those equipped with high-efficiency particulate air filters at the air conditioning system inlet. These tests are carried out using sterile swabs, they are smeared on the surfaces of the various rooms and the equipment of the enclosure. The swabs are then plated on culture media to check for bacterial growth or assayed by PCR. Where there are cases of contamination, more accurate checks and subsequent cleaning and sanitation can be carried out. Alternatively, or in combination, it is also possible to use surface air system system, which aspirates a determined volume of air in the room (generally 500–1000 L) on an agar plate, to measure the quantity and quality of bacteria present, microbiologically monitoring the facility structure, environment, procedures, and air.

Interventions in case of pathogen contamination (facility outbreak) (Marcello Raspa and Viola Galligioni)

The health programs of the enclosures and the related periodic reports (Health Monitoring Report) have for years been fundamental tools for the exchange and export of animals to other institutes. The results of the health check must always be reviewed and checked by the person in charge of the enclosure who must intervene, in agreement with the veterinarian designated (Directive 63/2010/EU) in the event of contamination or infection conditions, planning emergency management, and implementing an intervention plan for eradication.

In the case of positivity, prompt intervention is needed:

- a confirmation test using the same method.
- an examination on the same sample or other samples from the same animal using an alternative method if available (e.g., PCR if the first analysis was performed with ELISA).
- test more animals of the same cage and colony.
- an examination on samples sent to an analytical laboratory other than the first one.

Once the positivity is confirmed, all the procedures of the area concerned should be blocked, in particular the flows of animals, personnel, and materials. Moreover, additional tests should be carried out on the same colony and on the neighboring colonies of different rooms, with numerous sampling concentrating the attention on the pathogen under examination.

The next phase is that of examining the damage and identifying the source of the contamination. Often procedural errors can be detected, technical deficiencies of the facility or equipment that once fixed may prevent further contamination. Once the agents have been identified, it is possible to decide whether to accept the new situation by varying the exclusion list of pathogens, whether to eliminate the pathogen with a specific therapeutic treatment if there are any effective ones (they exist only for parasites and for some bacteria), or whether to eradicate the pathogen, by completely eliminating contaminated and exposed animals.

In the event that complete suppression is not possible for experimental reasons and/or due to the unique characteristics of the strain, it is advisable to proceed at least to the reduction of the animal population concerned to minimize transmission. In these cases, the animals not suppressed are placed in quarantine, if possible inside isolators, in negative pressure. For strains of particular scientific importance, one can proceed to derivation (cesarean or embryonic derivation) and/or to the freezing of the embryos and/or gametes pending their reimplantation. The cross-fostering procedure can be considered as a valid alternative to the embryo or cesarean derivation procedure in some cases (Buxbaum, 2011).

After eliminating colonies and consumables in use (e.g., feed, sawdust, gloves, etc.), it is always advisable to empty the rooms, dismantle all the equipment, and sanitize the components. In particular, the shelves, cages, and accessories must be disassembled, washed, and disinfected or, where possible, sterilized. It is important that the filters of the cages, the conditioning system, and the hoods are also replaced with new filters.

Therefore, thorough cleaning, sanitization, and sterilization of the areas involved are carried out.

At the end of this phase, the rooms must not be used for at least 7–10 days (absolute absence of animals and personnel) applying the so-called sanitary vacuum.

Then the premises are cleaned and sanitized again.

To verify the effectiveness of the procedures adopted, before housing the new animal colony, it is good practice to introduce sentinel animals to be screened specifically after an appropriate period of time. In the case of parasites and bacteria, microbiological environmental analysis should also be performed. If all the tests are negative, you can resume the usual activities, unblocking all the procedures and repopulating the environments with new animals, of safe and guaranteed origin.

The definitive disappearance of the contaminating pathogen can be established when all the health monitoring carried out in the following six months in the contaminated and neighboring areas is negative.

It is very important to pay attention to the correct procedures of collection, treatment, storage, and shipment of biological samples on which to carry out health analysis. The exclusion list refers to the list of pathogenic and opportunistic organisms that are considered not compatible with the health status of the enclosure (Pritchett-Corning et al., 2010) by the veterinarian and responsible for HM on the basis of risk assessment and the likelihood of occurrence and potential impact. Cesarean derivation (ysterectomy Derivation) means the removal, under sterile conditions, of the uterus of an animal at the end of the pregnancy coming from the contaminated colony, its subsequent opening in sterile conditions, and the entrusting of newborns to a noncontaminated foster mother. Embryonic derivation (Embryo Derivation) means the sampling in conditions of sterility of the embryos, at any stage, by one or more mothers of the contaminated colony. After appropriate washing, the embryos are implanted in the oviduct or uterus of a barrier-housed female.

Cross-fostering means the custody within 24 h of birth and through the adoption of procedures in sterility, of infants to a clean adoptive mother.

For viral contaminations, contaminated areas should be sterilized by fumigation with H_2O_2. How to tackle a pathogen contamination is represented in Fig. 6.4.

Health status and well-being (Marcello Raspa and Ferdinando Scavizzi)

Health status and well-being can be, first of all, maintained by knowing and monitoring normal behavior and activities of our animals, related to strain, sex, immunological status, and age. Normal behaviors are the ones allowing the animal to control its environment. Abnormal behaviors might be expression of abnormal physiology because the animal is unable to adjust to the environment. Many information on behavior of mice and rats and what they mean can be found in these useful websites: mousebehavior.org and ratbehavior.org.

It is important to be aware of strain-specific characteristics and abnormalities, since they can affect future behavioral studies. Information on strain characteristics can be found in the *Guidebook on mouse and rat colony management* (Pritchett-Corning et al., 2011) or on the Mouse Phenome Database on the Jackson Lab website (phenome.jax.org).

It is important to be mindful that defects and health issues can be congenital (e.g., micro/anophthalmia), related to husbandry (e.g., fighting males, overgrooming) or related to experimental procedures (e.g., injection). Burkholder and colleagues (Burkholder et al., 2012) summarized main clinical conditions correlated with good quality pictures. Some considerations when starting to work with an animal strain are summarized in Fig. 6.5.

Links to website

mousebehavior.org
ratbehavior.org
phenome.jax.org.

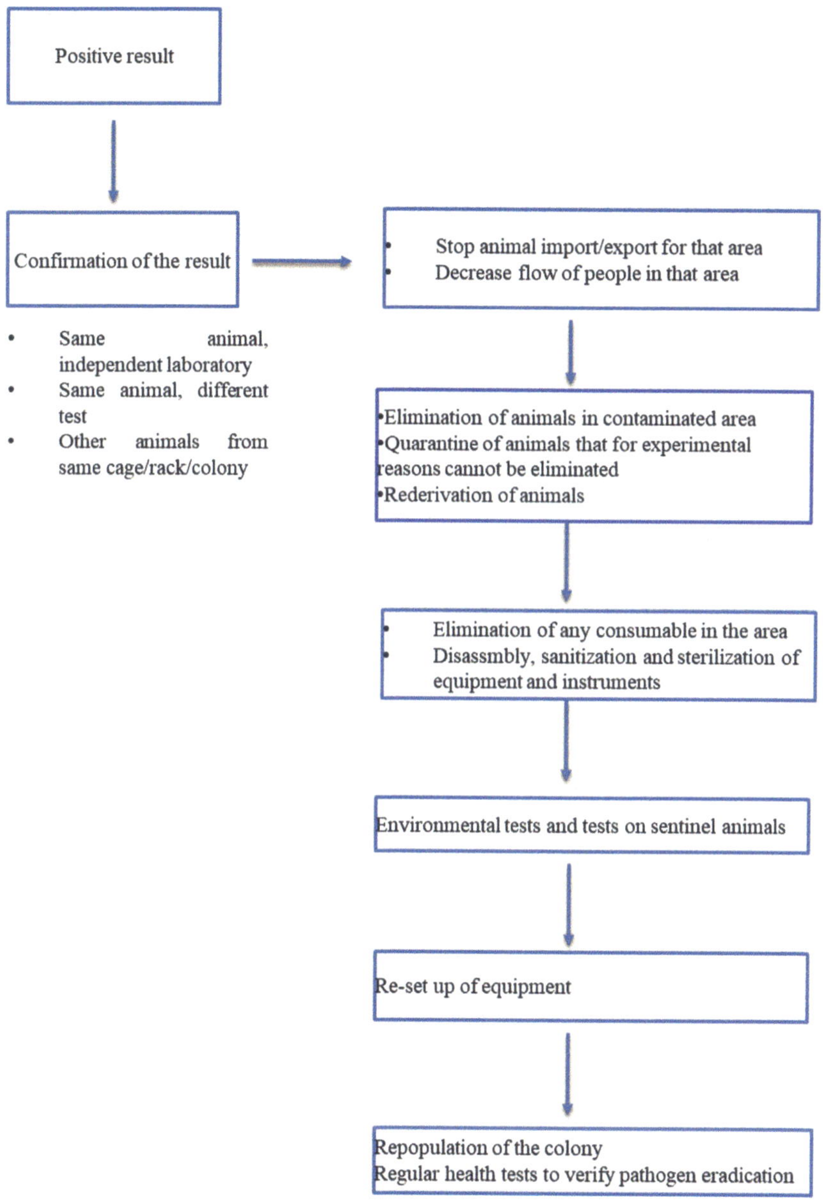

FIGURE 6.4 Outline of the master plan to tackle a pathogen contamination (Outbreak).

Traceability (Viola Galligioni)

Traceability of animals from when they are born to their use/end of life is important to identify welfare issues, health monitoring, and outbreak of

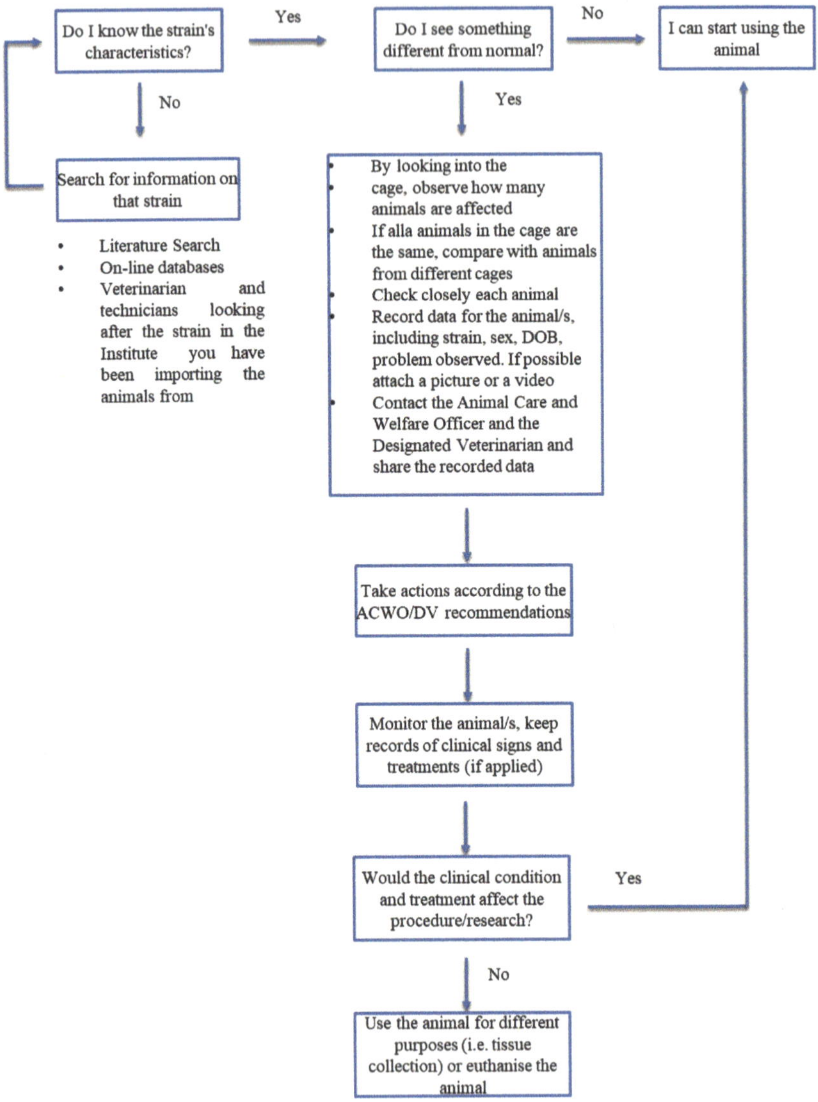

FIGURE 6.5 Steps to consider when starting using one animal strain. *Credit: Viola Galligioni.*

diseases. There are several databases on the market, but there are also examples of institutes that decided to build their own through the help of IT services. These databases help also to maintain proper inbreeding strategy and to detect genetic drift. An easy and very simple way to create a database is to use an Excel file on which to report all animals, as in Fig. 6.6A–C.

(A)

Breeding pair information							Litters											
Pair Ref	Gen	M DOB	F DOB	date M+F	End Date	Renewal date	DOB	DOW	# Pups	#Male	#Female	Wean: Born	Notes	Days to 1st litter	Interval litters	Average days between litters	Room	Rack
BP N/Year	F	DOB Parental BP	DOB Parental BP	date mating		date mating + 180 days							*found dead etc.*	Date first litter - DOB F	Date new litter - date previous litter	Average interval litters		

(B)

Males										
Age (weeks)	Cage	DOB	DOW	BP	Generation	Room	Rack	Status	Request	Notes
		11-09-2019	02-10-2019	343	F8	B3	B3_1		User XYZ	
		11-09-2019	02-10-2019	343	F8	B3	B3_1		User XYZ	
54.6		11-09-2019	02-10-2019	343	F8	B3	B3_1		User XYZ	08-10-2019
		12-09-2019	03-10-2019	361	F9	B3	B3_1		User XYZ	
		12-09-2019	03-10-2019	361	F9	B3	B3_1		User XYZ	
		13-09-2019	04-10-2019	308	F8	B3	B3_1		User XXW	
		13-09-2019	04-10-2019	308	F8	B3	B3_1		User XXW	
54.3		13-09-2019	04-10-2019	308	F8	B3	B3_1		User XXW	08-10-2019
		15-09-2019	06-10-2019	359	F9	B3	B3_1		User XXW	
		15-09-2019	06-10-2019	359	F9	B3	B3_1		User XXW	

(C)

Females										
Age (weeks)	Cage	DOB	DOW	BP	Generation	Room	Rack	Status	Request	Notes
		09-09-2019	30-09-2019	382	F9	B3	B3_1			
		10-09-2019	01-10-2019	318	F8	B3	B3_1			
54.9		10-09-2019	01-10-2019	318	F8	B3	B3_1			
		11-09-2019	02-10-2019	343	F8	B3	B3_1			
		11-09-2019	02-10-2019	343	F8	B3	B3_1			
		11-09-2019	02-10-2019	343	F8	B3	B3_1			
54.6		12-09-2019	03-10-2019	361	F9	B3	B3_1			
		12-09-2019	03-10-2019	361	F9	B3	B3 1			

FIGURE 6.6 (A): Basic file to collect data for breeding. (B, C): basic file to collect data of males (B) and females (C) in stock. *BP*, breeding pair the animal is generating from; *DOB*, date of birth; *DOW*, date of weaning; status: if animal found dead or used in any BP; request: to be moved into experimental area or to be used on a specific date (notes). *Credit: Viola Galligioni.*

Local arrangements for use of animals (Viola Galligioni)

Before housing animals for research purposes, specific licenses must be in place:

- Establishment license
- Project authorization license
- Individual authorization license (could differ according to the Country).

It is important to define all needs with facility staff already at the stage of grant proposal stage. The Animal Welfare Body is composed of: the facility manager, the designated veterinarian, the responsible for HM, and other expert facility members; discussion of the project should be held with this group: is there space to accommodate that species and those numbers of animals? Does the project need specific requirements in terms of equipment (surgical theater, phenotyping, or imaging equipment) or facility (i.e., biosafety level II, germ-free isolators)? Is there any specific need for disposal of hazards (toxic agents, radiations, infected material)?

There are several guidelines to perform animal experiments, but indeed the updated ARRIVE (Percie du Sert, 2020) and PREPARE guidelines (Smith et al., 2018) address all steps and parties that need to be involved from the beginning.

In many institutions, there is one person responsible for the import/export of animals (Animal Transport Coordinator), who helps in the purchase of animals and makes sure all the admin processes are properly followed.

Considering that not all animal facilities have the same microbiological status, when importing animals, the DV needs to check the health report of the facility from where the animals are coming. To avoid introduction of new pathogens, only animals with same or higher health status can be introduced in the animal facility. This policy not only avoids outbreak of diseases in a *naïve* population but also avoid the introduction of microbiological agents that can alter breeding or research currently done in the animal facility. For example, introducing animals coming from a facility positive for *Helicobacter* spp. will negatively impact on breeding colony efficiency (Bracken et al., 2017) and, indeed, can affect studies on gastrointestinal microbiota.

Local policies should be put in place so that welfare and health issues are addressed immediately. In fact the cooperation between staff and scientists will help on one side to reduce pain, distress, and harms to the animals and, on the other side, to avoid variables that can affect the study.

Animal transport (Viola Galligioni)

At present, with the available technologies, the transfer of live animals between facilities has decreased in favor of embryos, sperm, or eggs (for aquatic species) transport. This is considered a refinement, since transport is stressful for animals and could impact their welfare. Nevertheless, transport of live animals is still important, especially considering collaborations between institutions.

We must make sure that animals travel safely, healthily, and in humane conditions (Fig. 6.7).

The veterinarian and the responsible for HM of the facility should always check the animals and sign a health certificate recording the status of the animals on the documentations accompanying the animals during transport.

Regardless of species and shipping route (truck or flight), some general aspects should be considered to ensure safe and humane animals transport (Fig. 6.8).

In scheduling transportation of animals, consideration should be given to the following steps:

- Transport planning: all parties in the departure facility, courier, and arrival facility should be involved
- Animal preparation: veterinarian and animal care takers ensure animals are fine to be shipped. Crates are prepared and labeled properly
- Animal pickup at origin, documentation control, and transfer to airport
- Holding facilities at airport
- Plane loading (animals might wait long hours prior to loading)
- Flight
- Unloading
- Documentation control and management of transport to destination site (including through the facility)

Animals should be healthy and without sign of distress	Stressful situations, such as transport, may induce recrudescence of subclinical infections harbored by an animal
Surgically altered animals (i.e. jugular vein catheters)	They should recover completely before shipping
Pregnant females	Shipping near-term pregnant females should be avoided
Germ free animals	Risk of contamination is extremely high; proper boxing and crates should be available
Obese animals	Number of animals per container may need to be reduced in warm weather
Animals treated with hazardous agents, including toxic, carcinogenic substances or infected with microbiological agents of human or animal importance	Specific permits and licenses may be needed (Directive 63/2010/EU and national transpositions). Appropriate labels should be placed on crates. Extreme caution in boxing and unboxing of animals should be used to avoid cross-contamination and infection of humans. Transport material should be disposed of accordingly

FIGURE 6.7 Aspects to consider when preparing animals for shipping.

- Destination arrival and unpacking
- Clinical examination and acclimation/quarantine.

During each of these phases, animal health and welfare should be managed properly with general and species-specific conditions.

There are several factors that could harm welfare of animals from shipping institution to the destination facility. Potential risks affecting health and welfare of animals during transport are reported in Fig. 6.9.

Transportation will expose animals to environmental changes that can affect light cycle and food consumption. Animals can lose up to 10% of body weight because they do not drink much during first 12–15 h of transit. Water provision is prescribed by regulations, but animals still need to be monitored when they arrive at the final facility (National Academies of Sciences, Engineering, and Medicine, 2017. Transportation of Laboratory Animals: Proceedings of a Workshop, Chapter 4). Moreover, physiological changes correlated to transport and related stress-response could affect animals for some time after arrival at destination. Therefore, it is important to allow for an acclimation period for the animals once they arrive at final destination, before starting experiments or even collection of physiological data (i.e., blood sampling) (Fig. 6.10).

Crates and containers should be appropriate for the species and clean	Containers should comply with International Air Transport Association (IATA) construction standards and should be large enough to allow animals to move freely. They should be made of resistant materials, to avoid escape of animals. Interior materials should be smooth and moisture resistant. Containers should be clean to avoid contaminations. In case of shipping of SPF, germ-free and surgically modified animals, it isgood practice to disinfect shipping containers.
Appropriate ventilation	Containers should have adequate airspace over the highest part of their body to allow air movement and to prevent injury from contact with top of the container
Appropriate provision of water and food	'no-spill' water source (gel water packages) and diet to support nutritional needs for at least one extra day in case of problems with transport. Remember that crates cannot be opened till they reach final destination.
Appropriate provision of bedding and nesting	In case of low temperature animals can keep their bodies warm

FIGURE 6.8 Aspects to consider to ensure safe and humane animal transport.

Biosafety and biocontainment (Valentina Vasina)

Research with animals involves different personnel working with LAs: technicians, animal caretakers, veterinarians, and researchers, each with different exposure to risks depending on their role. We can distinguish different hazards within the animal facility related to animal staff and researchers' activities: physical (i.e., musculoskeletal injury, fall due to slippery surfaces, injury due to high or low temperature, bites, scratches, etc.), biological and chemical hazards; their origin can be environmental, animal-related, or research-related. In Fig. 6.11, the main potential occupational hazards and risks that can be encountered in a LA facility are listed.

In this handbook, we will mainly focus on biological hazard (or biohazard); we define a biological agent any microorganism, virus, and toxin of biological origin, even if genetically modified, cell culture or human endoparasites which could potentially cause infection, allergy, or intoxication in exposed workers

Risk factors	Explanations	Mitigation
Temperatures	Temperatures extreme should be avoided. Due to time differences and delays, animals might spend long time at airports for custom clearance and checks. Holding areas in most airports are open warehouses. In case of weather changes once they have started their journey, animals could wait longer in uncontrolled conditions (not immediately delivered to aircraft), re-route to other airports or wait for connection flights.	Climate controlled vehicles, extra insulation (cold weather), larger ventilated boxes and reduced density (hot weather). In case of intercontinental flights, weather forecasts should be checked regularly and in advance, to avoid extra time spent at airports in unsuitable weather conditions. Proper planning of the whole journey should be scheduled, as well as disaster planning.
Technical issues during transport	Delays or cancellation of flights, technical issues of aircrafts, road accidents, equipment malfunctions (i.e. ventilation system break-down)	Ensure there is always extra food and water that will keep animals comfortable for at least 1 extra day. Redundancy of technological systems to monitor and control ventilation and temperatures, as well as alarm systems that will be activated in case of problems.
Loss of documentation	Documents can get lost by mistake. When the journey includes crossing international borders, if documents are not available, animals may not be released.	Ensure there is always extra food and water that will keep animals comfortable for at least 1 extra day. Multiple copies should be made of all documents, considering each border inspection point during the journey. Spare copies should also be available and a copy could be sent in advance to the recipient.
Infections	Risk of animals to get infected during the transportation and to introduce infection into arrival facility.	Ensure that shipping crates are intact, made of durable material, and clean. Do not store shipping crates in dirty area, if necessary disinfect them before use.

FIGURE 6.9 Potential risks during transport.

Species	Physiological change	References
Mice	Change in gut microbiota	(Montonye et al., 2018)
	Decreased body weight	(Lee et al., 2012)
	Suppressed immune response	(Obernier and Baldwin, 2006)
Rats	Decrease in social interactions and locomotor activity	(Arts et al., 2012)
	Decreased bodyweight	(Arts et al., 2012, Lee et al., 2012, National Academies of Sciences, 2017a, National Academies of Sciences, 2017b)
	Decreased blood pressure	(Arts et al., 2012, Karicheti, 2018),
	Increased corticosterone	(Arts et al., 2012, Shim et al., 2009)

FIGURE 6.10 Physiological changes related to transport.

(EC Directive 2000/54/EC). By biohazard, we mean the risk to contract disease from a biological agent which can penetrate, reproduce, and pose a threat to the health of humans and other living organisms. As stated above, biological hazards may be related to animals (e.g., allergy and zoonosis) or to research protocols (e.g., animal infection with pathogens or genetically modified mouse models).

We can classify biological agents into four risk groups (groups 1, 2, 3, and 4), depending on the effect of the biological agent on healthy workers, that is the likelihood for an agent to be pathogenic to humans, how likely the disease could spread within the community, and the availability of any prophylaxis or effective treatment, as reported in Fig. 6.12.

Main exposure routes and pathway to biological agents are:

- inhalation of droplets, contaminated dust, and bioaerosol;
- direct contact with animal tissue, biological fluids, and faeces;
- accidental contact of mucosal surfaces (eyes, nose, mouth) with squirts and contaminated droplets or accidental ingestion;
- direct contact with dirty surfaces and devices (cages, bench, etc.);

Environmental Hazards	Animals related Hazards	Research related Hazards
Generic: fall due to slippery floor/surfaces, musculoskeletal apparatus injury	**Physical Injuries:** Bites, Scratches due to incorrect manipulation	**Use of biological agents potentially pathogen to humans:** occupational exposure to pathogens used in research protocols.
High and low Temperature: burns by using heat or steam to sanitize and sterilize equipment and instruments (cage washers, autoclaves); use of cryogens (e.g., liquid nitrogen and dry ice) can cause cold burn and lead to asphyxiation.	**Allergy:** sensitization to allergens from saliva, hair, urine and dander allergens.	**Use of genetically modified microorganism**
Sharps: inappropriate use of surgical instruments, syringes, broken glass, needles, scalpels and other sharp equipment may cause tissue injury.	**Infection:** personnel exposure to bacteria, fungi, parasites, rickettsia, viruses and bloodborne pathogens.	**Use of Drugs or chemicals**
Chemical: anaesthetic gases, cleaning agents and disinfectants.	**Zoonosis:** Human diseases acquired from animals (e.g. pasteurellosis, salmonellosis, etc.).	**Use of radioactive material**
Environmental Hazards	**Animals related Hazards**	**Research related Hazards**
Ergonomic: musculoskeletal disorders, pain, discomfort, or injuries due to repetitive motion (bending, lifting, etc).		
Machinery and Equipment: moving parts hazardous to operators.		
Electrical and electromagnetic: laser and UV light eye and skin injuries; electromagnetic hazards due to magnetic resonance imaging (MRI) (e.g life-threatening for people with implanted pacemakers or stents).		
Noises: noise-generating equipment (e.g., cage washers, pressure washers, etc).		

FIGURE 6.11 Main potential occupational hazards and risks in animal facility.

Risk Group	Pathogenicity	Ability to spread	Prophylactic and therapeutic measure
1	Low	--	--
2	Yes	Low	available
3	Yes, serious	Yes	Usually available
4	Yes, serious	Yes, high	missing

FIGURE 6.12 Classification of biological agents.

- external vectors (insect bite), traumatic lesions (bites, scratches, cuts, abrasions), and needle puncture;
- discharge of infectious biological samples due to wrong transportation procedures;
- carcass and waste disposal.

Allergy (Valentina Vasina)

Sensitization to LAs has a high prevalence among laboratory workers; historical data suggest that 15% of LA workers develop immunoglobulin (IgE) sensitization and 10% develop symptoms of LA allergy, including skin reactions, rhinitis, conjunctivitis, and occupational asthma, a more serious respiratory disorder that can result in life-threatening episodes (Feary et al., 2019). The main route of exposure to animal allergens is by inhalation, followed by eye contact, direct skin contact, and animal bites or scratches; allergens can be found in the environment or deposited on laboratory materials and equipment, and levels of airborne allergens tend to rise considerably with certain activities such as changing or cleaning animal cages (Acton and McCauley, 2007). Moreover, it has been shown that LA allergens can sensitize also subjects outside the laboratory environment, thus exposing not only workers to allergy hazard but also children whose parents were occupationally exposed to LA (Krakowiak et al., 1999). Due to the large use of these animals in biomedical research, allergy to mice and rats are the most common and result from repeated exposure to animal allergens which are proteins from urine, saliva, serum, epithelium, faeces, and hair; it has been shown that the urine of male rat and mouse represents the most allergenic matrix (Elliot et al., 2005).

Symptoms usually appear within the first 2–3 years after the first exposure, and the disease is likely to get worse with time; allergy onset includes three main phases: exposure, where allergens enter the organism; preallergy, characterized by production of specific IgE; finally, allergy with clinical symptoms manifestation due to mast cell degranulation and production of inflammatory mediators.

There are three risk factors contributing to allergy: contact frequency between animal and human, contact duration and allergen quantity; to prevent allergy risk, we can act only on this last factor by reducing allergen quantity with the use of personal protective equipment (PPE) and engineering equipment (Ferraz et al., 2013).

Zoonosis (Valentina Vasina)

A zoonosis is an infectious disease caused by a pathogen (such as virus, bacterium, parasite, or prion) that has been transmitted from a nonhuman animal to human. Zoonotic diseases are currently very rare in rodent and rabbit animal facilities, due to animal supplying only by commercial vendors that have successfully eliminated all or most zoonotic agents from their colonies, limiting policies for the transportation and importation of LA, and the use of sentinel monitoring programs. However, zoonotic threats from rodents could come from the use of pathogens or from wild caught species for research purposes, the use of contaminated rodent products or, more rarely, from feral rodents gaining access to animal facilities. Fig. 6.13 summarizes some relevant infectious agents transmissible from rodent and rabbit species to humans that can be of concern. The main routes of transmission are faecal-oral

	Infectious agent	Disease
Bacteria	Leptospira sp.	Leptospirosis
	Streptobacillus moniliformis	Rat bite fever
	Pasteurella multocida	Pasteurellosis
	Salmonella enterica	Salmonellosis
	Shigella	Shigellosis
	Mycobacterium tubercolosis	Tubercolosis
Virus	Rabies virus	Rabies
	LCM virus	Lymphocytic Choriomeningitis
	encephalomyocarditis virus	encephalomyocarditis
	Hantavirus	Hantavirus pulmonary syndrome (HPS)
Parasites	Sarcoptes scabiei	Scabies
Protozoans	Toxoplasma gondii	Toxoplasmosis
Mycosis	Trichophyton mentagrophytes	Dermatomycois (ringworm)

FIGURE 6.13 Main zoonotic diseases from rodents.

(e.g., Salmonellosis), respiratory (by inhalation of aerosol particles, e.g., tuberculosis, lymphocytic choriomeningitis, Hantavirus pulmonary syndrome), oculo-conjunctival (e.g., leptospirosis), transcutaneous (whole skin, e.g., scabies), through bites and scratches (e.g., rat bite fever, rabies, pasteurellosis, etc).

Biohazard can also be research-related, e.g., experiments using biological agents potentially pathogen to humans or research protocols involving recombinant or synthetic nucleic acid molecules, including the creation and use of organisms and viruses containing recombinant or synthetic nucleic acid molecules, which require specific risk assessment and guidelines (NIH Guidelines for Research Involving Recombinant or Synthetic Nucleic Acid Molecules).

Risk prevention and hazard mitigation (Valentina Vasina)

The development of hazard mitigation strategies arises from a combination of several elements:

- Hazard identification (risk analysis);
- Prevention: procedures and behavioral action, information and training;
- Engineering;
- PPE and containment equipment;
- Health monitoring program.

Risk assessment in animal facilities must include both animal husbandry and research protocols-related risks and it represents an ongoing and daily process. As shown in Fig. 6.14, hazard mitigation can be achieved by an

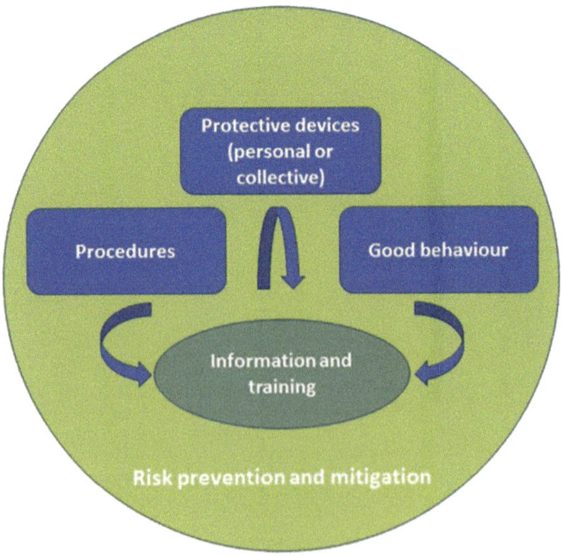

FIGURE 6.14 Risk prevention and mitigation.

integrated strategy which includes the respect of "good behavioral rules" (e.g., not to eat or drink in animal facility, use the correct clothing sequence, avoiding noises and high tone of voice, etc.) and standard operating procedure when using washing machinery, any equipment, and devices, performing animal procedure or experiments. Moreover, appropriate animal handling preventing animal injury (bites, scratches), proper carcass and waste disposal, accurate cleaning, and disinfection procedure reduce biohazard.

PPE (including goggles, gloves, respiratory protection, etc.), containment equipment, and engineering controls (ventilated animal cage racks, biological safety cabinets, cage changing stations, etc.) are essential to prevent allergy and zoonosis.

Moreover, information and training are crucial for the safety of animal facility staff and researchers; the training should meet the needs of persons working with animals, depending on their role and educational level: e.g., researchers will be trained on the general health and safety, care and use of LA and specific protocols-related hazards, while animal care staff should be trained for additional chemical and physical hazard they could encounter in their work.

References

Acton, D., McCauley, L., 2007. Laboratory animal allergy: an occupational hazard. AAOHN J. 55 (6), 241–244. https://doi.org/10.1177/216507990705500604. PMID: 17601065.

Bracken, T.C., Cooper, C.A., Ali, Z., Truong, H., Moore, J.M., 2017. Infection significantly alters pregnancy success in laboratory mice. J. Am. Assoc. Lab. Anim. Sci. 56 (3), 322–329. PMID: 28535868; PMCID: PMC5438927.

Burkholder, T., Foltz, C., Karlsson, E., Linton, C.G., Smith, J.M., 2012. Health evaluation of experimental laboratory mice. Curr. Protoc. Mouse Biol. 2, 145–165. https://doi.org/10.1002/9780470942390.mo110217.

Directive 2000/54/EC - Biological Agents at Work - Safety and Health at Work - EU-OSHA (europa.eu).

Buxbaum, LU, et al., 2011. Eliminating murine norovirus by cross-fostering. J Am Assoc Lab Anim Sci 50, 495–499.

Compton, SR, et al., 2004. Microbiological monitoring in individually ventilated cage systems. Lab Anim 33, 36–41.

Dubin, ZS, 1991. Sample size for animal health surveillance. Lab Anim 20, 719–745.

Elliott, L., Heederik, D., Marshall, S., Peden, D., Loomis, D., 2005. Incidence of allergy and allergy symptoms among workers exposed to laboratory animals. Occup. Environ. Med. 62 (11), 766–771. https://doi.org/10.1136/oem.2004.018739. PMID: 16234402; PMCID: PMC1740921.

Feary, J.R., Schofield, S.J., Canizales, J., Fitzgerald, B., Potts, J., Jones, M., Cullinan, P., 2019. Laboratory animal allergy is preventable in modern research facilities. Eur. Respir. J. 53 (6).

FELASA working group on revision of guidelines for health monitoring of rodents and rabbits, Mähler Convenor, M., Berard, M., et al., 2014. FELASA recommendations for the health monitoring of mouse, rat, hamster, Guinea pig and rabbit colonies in breeding and experimental units [published correction appears in Lab Anim. 2015 Jan;49 (1), 88]. Lab. Anim. 48 (3), 178–192. https://doi.org/10.1177/0023677213516312.

Ferraz, E., Arruda, L.K., Bagatin, E., Martinez, E.Z., Cetlin, A.A., Simoneti, C.S., Freitas, A.S., Martinez, J.A., Borges, M.C., Vianna, E.O., 2013. Laboratory animals and respiratory allergies: the prevalence of allergies among laboratory animal workers and the need for prophylaxis. Clinics (Sao Paulo). 68 (6), 750–759. https://doi.org/10.6061/clinics/2013(06)05. PMID: 23778494; PMCID: PMC3674264.

Krakowiak, A., Szulc, B., Górski, P., 1999. Allergy to laboratory animals in children of parents occupationally exposed to mice, rats and hamsters. Eur. Respir. J. 14 (2), 352–356. https://doi.org/10.1034/j.1399-3003.1999.14b19.x. PMID: 10515413.

National Academies of Sciences, Engineering, and Medicine, 2017. Transportation of laboratory animals. In: Proceedings of a Workshop. The National Academies, Washington, DC. https://doi.org/10.17226/21734 (Chapter 4). Press.

Pritchett-Corning, K.R., Girod, A., Avellaneda, G., Fritz, P.E., Chou, S., Brown, M.J., 2010. Handbook of Clinical Signs in Rodents and Rabbits. Charles River Laboratories.

Nicklas, W., 1999. Microbiological standardization of laboratory animals. Berl. Munch. Tierarztl. Wochenschr. 112 (6–7), 201–210. Refer: https://pubmed.ncbi.nlm.nih.gov/10472715/.

Nicklas, W., 2002. Recommendations for the health monitoring of rodent and rabbit colonies in breeding and experimental units. Lab. Anim.

Nicklas, W., et al., 2010. FELASA guidelines for the accreditation of health monitoring programs and testing laboratories involved in health monitoring. Lab. Anim. 39.

Percie du Sert, N., et al., 2020. The ARRIVE guidelines 2.0: Updated guidelines for reporting animal research. PloS. Biol. 18.

Percy, D.H., Barthold, S.W., 2007. Pathology of Laboratory Rodents and Rabbits, Third Edition. Blackwell Publishing Professional.

Pritchett-Corning, K.R., et al., 2014. AALAS/FELASA Working Group on Health Monitoring of rodents for animal transfer. J. Am. Assoc. Lab. Anim. Sci. 53, 633–640.

Pritchett-Corning, K.R., Chou, S.T., Conour, L.A., Elder, B.J., 2011. Guidebook on Mouse and Rat Colony Management. Charles River Laboratories.

Scavizzi, F., Bassi, C., Lupini, L., Guerriero, P., Raspa, M., Sabbioni, S., 2021. A comprehensive approach for microbiota and health monitoring in mouse colonies using metagenomic shotgun sequencing. Anim. Microb. 3 (1), 53. https://doi.org/10.1186/s42523-021-00113-4. PMID: 34325744.

Silverman, J., 2002. Managing the laboratory animal facility. Third edition ed. 2002. CRC Press.

Smith, A.J., Clutton, R.E., Lilley, E., Hansen, K.E.A., Brattelid, T., 2018. Improving animal research: PREPARE before you ARRIVE. BMJ 360, k760. https://doi.org/10.1136/bmj.k760. PMID: 29472181.

Recommended reading

Arts, J.W., Kramer, K., Arndt, S.S., Ohl, F., 2012. The impact of transportation on physiological and behavioral parameters in Wistar rats: implications for acclimatization periods. ILAR J. 53 (1), E82–E98. https://doi.org/10.1093/ilar.53.1.82. PMID: 23382273.

D'Ovidio, M.C., Wirz, A., Zennaro, D., Massari, S., Melis, P., Peri, V.M., Rafaiani, C., Riviello, M.C., Mari, A., 2018. Biological occupational allergy: protein microarray for the study of laboratory animal allergy (LAA). AIMS Public Health 5 (4), 352–365. https://doi.org/10.3934/publichealth.2018.4.352. PMID: 30631779; PMCID: PMC6322995.

European Parliament Resolution of 14 February 2019 on the Implementation of Council Regulation (EC) No 1/2005 on the Protection of Animals During Transport Within and Outside the EU (2018/2110(INI)).

FELASA-AALAS Working Group on Transport of laboratory animals. http://www.felasa.eu/working-groups/working-groups-present/felasa-aalas-working-group-on-the-transport-of-laboratory-animals/.

Hedrich, H. (Ed.), 2004. The Laboratory Mouse. Academic Press.

Karicheti, V., 2018. Evaluation of shipping stress in surgically altered animals. In: Presentation at 2018 ATA Conference. Available at: https://www.animaltransportationassociation.org/annual-ata-conference/2018-orlando.

Hedrich, H. (Ed.), 2012. The Laboratory Mouse, second ed.

Lee, S., Nam, H., Kim, J., Cho, H., Jang, Y., Lee, E., Choi, E., Jin, D.I., Moon, H., 2012. Body weight changes of laboratory animals during transportation. Asian-Australas. J. Anim. Sci. 25 (2), 286–290. https://doi.org/10.5713/ajas.2011.11227. Epub 2012 Feb 1. PMID: 25049564; PMCID: PMC4093136.

Live Animal Regulations (LAR) of the International Air Transport Association (IATA). Available at: https://www.iata.org/en/publications/store/live-animals-regulations.

McCormick-Ell, J., Connell, N., 2019. Laboratory safety, biosecurity, and responsible animal use. ILAR J. 60 (1), 24–33. https://doi.org/10.1093/ilar/ilz012. PMID: 31423527.

Montonye, D.R., Ericsson, A.C., Busi, S.B., Lutz, C., Wardwell, K., Franklin, C.L., 2018. Acclimation and institutionalization of the mouse microbiota following transportation. Front. Microbiol. 28 (9), 1085. https://doi.org/10.3389/fmicb.2018.01085.

Obernier, J.A., Baldwin, R.L., 2006. Establishing an appropriate period of acclimatization following transportation of laboratory animals. ILAR J. 47 (4), 364–369. https://doi.org/10.1093/ilar.47.4.364. PMID: 16963816.

Schmitt, J.M., Wilson, D.E., Raber, J.M., 2018. Occupational safety and health. In: Weichbrod, R.H., Thompson, G.A.H., Norton, J.N. (Eds.), Management of Animal Care and Use Programs in Research, Education, and Testing, second ed. CRC Press/Taylor & Francis, Boca Raton (FL). https://doi.org/10.1201/9781315152189-14 (Chapter 14). Available from: https://www.ncbi.nlm.nih.gov/books/NBK500440/.

Shim, S., Lee, S., Kim, C., Kim, B., Jee, S., Lee, S., Sin, J., Bae, C., Woo, J.M., Cho, J., Lee, E., Choi, H., Kim, H., Lee, J., Jung, Y., Cho, B., Chae, K., Hwang, D., 2009. Effects of air transportation cause physiological and biochemical changes indicative of stress leading to regulation of chaperone expression levels and corticosterone concentration. Exp. Anim. 58 (1), 11–17. https://doi.org/10.1538/expanim.58.11. PMID: 19151507.

https://www.animaltransportationassociation.org.

Chapter 7

Recognition of pain, distress, and suffering

Cholawat Pacharinsak[1], Patrick Sharp[2], Anne Zintzsch[3], Sara Fuochi[4]
[1]*Department of Comparative Medicine, Stanford University, Stanford, CA, United States;* [2]*University of California, Merced, CA, United States;* [3]*Project on Severity Assessment of Genetically Altered Animals, University of Basel, Basel, Switzerland;* [4]*University of Bern, Bern, Switzerland*

Introduction (Cholawat Pacharinsak, Patrick Sharp, and Sara Fuochi)

Appropriate and timely recognition of signs of pain, suffering, and distress in laboratory animals is a crucial step in securing animal welfare and care through effective behavioral and environmental management and, when needed, pain treatment, thus improving ethical standards and validity of research data.

Recognition and alleviation of pain, suffering, and distress in laboratory animals is acknowledged as a crucial milestone to improve the ethical use of animals. Despite the strong will and contribution of all stakeholders involved and the huge efforts toward such a goal, the nonverbal nature of animals together with the use of, mostly, prey-species have been a challenge for professionals trying to identify reliable parameters that would allow to understand the nature and mitigate the severity of pain, distress, and suffering experienced by laboratory animals.

Despite such limits, the research and veterinary communities massive commitment to this task led to the availability of a large number of standardized, multimodal approaches to evaluate pain—but also distress and suffering—in animal models, combining mostly subjective assessments with objective methodologies, when available.

This chapter provides pain, suffering, and distress recognition methods in most relevant laboratory species, including genetically modified rodents.

Background (Cholawat Pacharinsak, Patrick Sharp, and Sara Fuochi)

The European Directive 63/2010 on the protection of animals used for scientific purposes clearly states the aim to strengthen legislation, and improve

Practical Handbook on the 3Rs in the Context of the Directive 2010/63/EU
https://doi.org/10.1016/B978-0-12-821180-9.00005-2

the welfare of those animals still needed to be used, as well as to firmly anchor the principle of the Three Rs, to replace, reduce, and refine the use of animals for scientific purposes. In this context, this chapter's focus is specifically driven by the *Directive's* approach and provide an overview of methods to recognize distress, pain, and suffering in the most commonly used laboratory animals in the EU.

A menagerie of animals is used in the course of research, teaching, and testing that constitute the broad spectrum of biomedical research. The *2019 Report on the Statistics on the Use of Animals for Scientific Purposes in the Member States of the European Union in 2015–17* along with the *2021 Summary Report on the statistics on the use of animals for scientific purposes in the Member States of the European Union and Norway in 2018* were used to determine the "most common" animal species employed. This is the most recent data provided by the 28 Member States (including the UK) and Norway. It is relevant to note that the 2019 report included categories that were previously not reported on a regular basis, suggesting an increased attention of competent authorities for:

- Fetal forms of mammals
- Animals killed solely for organs and tissues, and sentinels, unless the killing is performed under a project authorization using a method not included in Annex IV of Directive 2010/63/EU
- Animals bred and killed without being used, apart from genetically altered (GA) animals with an intended and exhibited harmful phenotype, and those genotyped with an invasive method before being killed.

The authors used the aformentioned 2019 and 2021 reports to define the "most common laboratory animals" and will limit the chapter's scope to the following species:

- Small laboratory animals: mice, rats, guinea pigs, rabbits, and zebrafish
- Large laboratory animals: dogs and pigs.

The above species comprise the large majority of the 12,093,096 animals used in 2018 in the EU and Norway. Of these 12,093,096 animals, 10,572,305 animals (87%) were used for experimental purposes while 1,520,791 animals (13%) were used for the creation and breeding of GA animals.

Furthermore, in the light of the 3Rs principle inspiring the Directive and under the practical approach, accurate recognition and mitigation of pain, suffering and distress have an enormous impact on the application of refinement as well as a potential impact on reduction, the latter specifically avoiding unwanted death.

According to EU Directive, Annex VIII, for the purposes of the final severity classification of a procedure, the application of humane endpoints should also be taken into account: several guidelines underline the relevance of identifying early or subterminal standardized conditions of pain, suffering, and distress to set humane endpoints that mitigate the severity of constraint experienced by animal models without compromising scientific data.

Relevant examples include cancer studies (Workman et al., 2010) and safety evaluations (OECD Guidance Document on the Recognition, Assessment, and Use of Clinical Signs as Humane Endpoints for Experimental Animals Used in Safety Evaluation, 2000; Stokes, 2000, 2002, 2011; Stokes et al., 2012). Finally, to the same purpose, methods including refinement of housing, husbandry, and care conditions can be used to reduce or eliminate pain, suffering, and distress, thus concurring to the application of the refinement principle.

Pain, distress, and suffering (Cholawat Pacharinsak, Patrick Sharp, and Sara Fuochi)

The EU *Directive 63/2010* leaves these three repeatedly used words quite undefined; of the three words, suffering has the least concurrence and generally tends to be open to broad definition. All three concepts though, bear a clear ethical and regulatory imperative to minimize or eliminate each condition; additionally, their deleterious impact on research and research reproducibility adds additional bias or confounding influence on research results. When dealing with animal pain, suffering, and distress, users need to be aware that such terms originally described states of the human mind, referring to human perceptions, and translating these concepts to animals is not always a straightforward task. Also, the same terms refer to an *experience* meaning something that happens to a subject that affects how the subject feels, and being each experience subjective, it is highly challenging to transfer each animal experience into an easily measurable value.

In this chapter, authors refer to the following definitions and commentaries of each condition:

Pain: "An unpleasant sensory and emotional experience associated with, or resembling that associated with, actual or potential tissue damage." (Srinivasa et al., 2020).

There are many ways to categorize pain—mild, moderate, or severe, acute versus chronic, short-lasting versus long-lasting, protective versus nonprotective—pain is always subjective and users must learn to evaluate and assess relevant interspecies differences as well as differences among individuals of the same species.

It is also relevant to emphasize the FELASA Report on pain and distress in laboratory rodents and lagomorphs (1994) drives the definition toward a more physical dimension, defining pain as "the perception or experience of nociceptive stimuli," leaving a more psychological dimension to the definition of suffering, and introducing the concept of nociception. Pain and nociception are often used interchangeably with the wrong assumption of being synonyms, while they are not. Nociception is widely defined as "the neural process of encoding noxious stimuli." Therefore, nociception is the physiological mechanism that can lead to autonomic or behavioral consequences that can further be used to assess pain—yet nociception can also occur without the sensation of pain (Turner et al., 2019).

For further details about nociception and pain in mammals and fish the authors direct readers to Chapter 3 of this manual.

Distress: A maladaptive response to a stressor yielding a new psychologic or physiologic state of prolonged stress and a failure to return to the previous, prestressor homeostasis (National Research Council, 2008).

Distress is also defined within the FELASA Report on pain and distress in laboratory rodents and lagomorphs (1994) as "a state where the animal has to devote substantial effort or resources to the adaptive response to challenges emanating from the environmental situation".

Furthermore, The NRC Committee on Pain and Distress in Laboratory Animals (1991), stated that, in most cases, laboratory animal stressors are either internal (physiological or psychological) including pain or external—environmentally induced stressors or events that stimulate homeostasis.

The combined analysis of such definitions can actually lead to further considerations having relevant practical impact on prevention and modulation of such a condition: distress does not occur without stress (prevention), and it occurs only when the biological cost of stress negatively affects biological functions critical to the animal's well-being, therefore the better the animal's opportunity and capacity to cope, the less severe the distress will be (modulation).

Suffering: many definitions of suffering were given by several authors, and once again a combination seems more appropriate to accurately define the concept.

Explanations range from the "bearing of pain or distress" (American Heritage Dictionary of the English Language) to a "physical or mental pain that a person or animal is feeling" (Cambridge English Dictionary) and to "a specific state of mind, which is not identical to, but might be a consequence of, pain and distress" (FELASA Report, 1994).

Suffering can therefore holistically be defined as "an animal's state involving pain, distress, and/or a psychologic or physical condition irrespective of chronicity" (National Research Council, 2008).

It is thus relevant to note that when referring to suffering, the focus is more driven to a psychological and mental condition of the animal. Physical conditions such as pain and distress may result in suffering, with a cause-and-effect relationship, if of sufficient intensity or duration.

Independently from the duration of the cause in itself—acute versus chronic condition—suffering is reached when pain or distress, and related detrimental effects, are no longer tolerable to the animal.

Pain assessment methodologies

Research animal pain assessment processes and techniques are defined, monitored, accomplished, and approved at multiple levels: institutional, protocol, and regulatory levels.

At the *institutional level*, the Animal Welfare Bodies should develop guidance and guidelines for researchers. Guidelines include suggesting/requiring researchers to complete training/workshops (aseptic surgery, anesthesia/analgesia, or handling training prior to performing procedures), performing a pilot study, and consulting with veterinarians. In addition, guidelines should include classification of painful procedures as mild, moderate, and severe pain together with creating pain scoring systems. Ideally, pain scoring systems should be customized for each study and included in the documentations submitted to the competent authority for the protocols' approval. When the degree of pain with novel procedures is uncertain, a pilot study should be suggested, and these results reported back to the Animal Welfare Bodies and authorities prior to the protocol's full approval or prolongation. In addition to the vital role provided by veterinarians and researchers, the critical role of nonscientists and community members must not be overlooked to augment the veterinarian and scientist perspectives. This includes their inputs both at meetings of Animal Welfare Bodies and in assessing animals in painful procedures.

At the *protocol level*, researchers should consult with laboratory animal veterinarians, veterinary anesthesiologists, or pain specialists to develop and customize pain scales or assessment methods specific for their procedures. Veterinarians can assist/suggest clear guidelines, for example, (1) when analgesic intervention can be supplemented (i.e., when pain score >3 out of 5). More importantly, when in doubt, analgesics should be administered; (2) when to follow up with pain assessments to ensure analgesics are effective; and (3) when euthanasia or an early endpoint is needed. When appropriate, one-on-one pain assessment training can be performed. Additionally, veterinarians should educate or train animal care technicians who are the frontline staff. Record keeping of pain scoring or surgery and anesthesia is crucial and Animal Welfare Bodies should review these records on a regular basis leading to improvement of the methods. Particularly, the development of procedure (or species, or strain) specific scoresheets is crucial in securing animal welfare and pain and distress control. An enormous variety of score sheets templates is available as web resources and can be easily customized based on each institution's needs. An excellent evolution of simple text-based score sheets are visual score sheets. Visual scoresheets associate the standard grid with clinical signs, scoring, and reactions providing a clear, representative picture of the clinical sign and severity, thus helping users to clearly identify the level of severity experienced by the animal and concurring to reduce interpersonal assessment biases.

At the *Regulatory level*, Competent National Authorities, while reviewing submitted research protocols, double check and define the severity degree of the research project. Competent authorities can also define, based on anticipated severity, further specific assessments including intermediate checkpoints to reassess the actual severity degree and the efficacy of measures used and described in the research proposal and ancillary documents to reduce and control pain, suffering and distress experienced by experimental animals.

Given the above, training is crucial for all the actors involved, ranging from animal caretakers to research technicians, PhD students and postdocs, principle investigators and veterinarians. In accordance with the Directive, Article 23, Member States must address personnel competence through setting minimum requirements with regards to education and training and the requirements for obtaining, maintaining, and demonstrating requisite competence for role and task specific functions. Remarkably, one specific area that must be addressed is the competent species-specific recognition of distress, pain, and suffering in the most commonly used laboratory animal species. Additionally, the Directive applies to fetal animal forms, especially in the last third of their developmental period, thus promoting the general safeguard. The Guide (NRC, 2011) restates and reinforces language from the US Government Principles for the Utilization and Care of Vertebrate Animals Used in Testing, Research, and Training (IRAC, 1985), "unless the contrary is known or established, it should be considered that procedures that cause pain in humans may also cause pain in other animals." Hence it is reasonable to consider this precaution applicable to such immature animal models as well as to adults (Rokyta, 2008).

It is also relevant to note that as interevaluator variables exist, evaluators or researchers should be trained to recognize normal behaviors and the same personnel should evaluate animals before and after procedures. More importantly, researchers should be trained and supported to use systems that work for them, so they get familiar and develop skills with the particular techniques.

In humans, the "gold standard" of assessing pain is self-reporting (Xu and Huang, 2020); however, in research animals, this technique is almost impossible. Because pain is subjective, developing criteria to objectively assess pain and measure pain severity in animals is challenging. Pain assessment in nonverbal children is similar to that in animals because it is challenging to identify and quantify; furthermore, it passes through a filter of a secondary caregiver. In children, pain assessment is comprised of physiologic (heart rate [HR], respiratory rate [RR], blood pressure [BP], cortisol, %SpO_2, behavioral (facial expression, sleep patterns, movement, etc.), and subjective [visual analog scale (increased numbers with increasing discomfort) or color analog scale (increased intensity of red with increasing discomfort, Fig. 7.1)] indicators (Stasiak et al., 2003). In animal research, although an objective pain

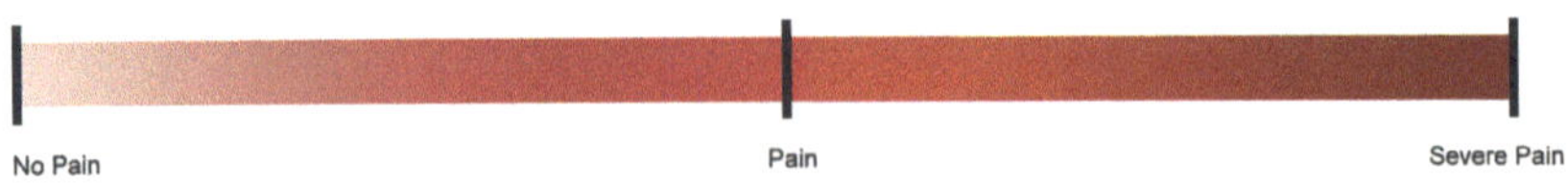

FIGURE 7.1 Color analog scales—increased intensity of red with increasing discomfort (Stasiak et al., 2003).

assessment is preferred, it is challenging to impractical, depending on the species and the pain assessment parameters (Keubler et al., 2020). Although other objective measurements (i.e., functional magnetic resonance imaging or electroencephalogram, blood cortisol levels, etc.) have been used in both humans and animals, they may not be practical for day-to-day management and may be limited to specific experimental setups (Xu and Huang, 2020).

In veterinary medicine, many pain scoring systems were initially adapted from humans; nowadays frequently employed approaches in many laboratory animal species include behavioral assessment, physiologic assessment, and clinical assessment. A special kind of evaluation is algesiometry.

Behavioral assessment: the advantage of using animal behavior as an assessment tool resides in its least intrusive nature: in the hands of a skilled, knowledgeable observer, behavior as an indicator of well-being (or lack thereof) can be reliable and powerful (NRC Committee on Regulatory Issues in Animal Care and Use, 2000). On the other hand, an accurate understanding and knowledge of the baseline, or ethogram, for the species under assessment is crucial to properly identify any deviation from normality. Ethograms can support pain or distress assessment mainly by highlighting the loss of normal behaviors or the presence of distress or pain associated behaviors (Turner et al., 2019). This kind of assessments is based on spontaneous, nonevoked behaviors. In small-sized laboratory rodents, using somatic or behavioral responses predominates. Motor or behavioral responses related to pain may be species-specific and may require accurate targeting and validation for different species. Common behavioral parameters, transversal to multiple species, include facial expression, movements (from stereotypes to abnormal ambulation), nesting behavior, burrowing, grooming, vocalization, reaction to manipulation, aggressiveness versus passivity.

A peculiar kind of behavioral assessment is represented by algesiometry (or analgesiometry) assays, used to screen analgesic activity or nociception—mostly in rodents, but assessments in dogs, cats, and rabbits are reported, too (Mogil, 2009). Examples of such screenings include well-known tests like the tail-flick test, the hot plate test, the paw withdrawal test, Randall-Selitto test, formalin injection test, and the Partial sciatic nerve ligation model. In recent years, the actual translatability of such assays has been reconsidered (Mogil, 2009), as they might represent an oversimplification of pain (Turner et al., 2019), thus suggesting that voluntary behavior in unrestrained animals should be considered as a relevant alternative in assessing pain in rodents (Tappe-Theodor and Kuner, 2014).

Physiologic assessment: several physiologic parameters can serve as indicators of distress or pain in laboratory animals. Such indicators are typically nonspecific, yet their relevance in the general assessment of animal well-being is universally acknowledged—particularly in combination with other parameters, hypothalamo-pituitary-adrenocortical axis activation is associated with acute stress (including pain). Its activation induces the release of so-called stress hormones (cortisol) and a series of physiologic effects that can be variably measurable in several laboratory species: increase in HR, RR, BP, and temperature (T). Notably, in some species, cortisol levels in blood or feces can be either higher or lower than normal in the case of prolonged pain compromising welfare (Pawluski et al., 2017; Turner et al., 2005; Jang et al., 2017).

Body weight and growth have been used as welfare indicators too. More specifically, decreased body weight, particularly without recovery of the normal body condition scoring, and slow growth are commonly used as indicators of pain and distress, as well as standard humane endpoints in many research fields (Turner et al., 2019; Workman et al., 2010; Olfert and Godson, 2000). When considering distress of laboratory animals, reproductive fitness and breeding performance are relevant indicators.

Clinical examination: Behavioral or physiologic assessments by themselves may be insufficient to properly identify signs and severity of distress and pain. A composite approach, conversely, can largely increase the accuracy of such evaluations. Clinical assessment of animals will allow integration of the aforementioned parameters within a wider, holistic assessment of the animal, including general clinical examination in accordance with best veterinary practices and/or procedure specific assessments—consult Tables 7.1–7.3 (Masson-Meyers et al., 2020). Several pain and distress scoring systems such as grimace scales (rodents, rabbits, pigs Fig. 7.2), behavioral changes/physical appearance (rodents, rabbits), Glasgow composite measuring pain scale (dogs), and Colorado State University Canine Acute pain, have been used in the laboratory animal setting.

Recognition of species-specific signs of distress, pain, and suffering of most common laboratory animals (Cholawat Pacharinsak, Patrick Sharp, and Sara Fuochi)

This part of the chapter is aimed to review the clinical and practical pain and distress assessment in the laboratory animal setting for veterinarians, researchers, technicians, and animal care staff. The tables below summarize pain signs and scoring systems for the relevant species as identified in the *Background* paragraph.

Small laboratory animals (suggested pain scoring systems are shown in Tables 7.1–7.3)

In general, researchers should be familiar with normal rodent behaviors—grooming, movement, cage exploration, nesting pattern (mice, rats)—and note that animals (i.e., mice, rats, guinea pigs, rabbits) at the bottom of the food chain may hide their pain very well because of their survival mechanisms. Therefore, when their pain signs are observable and evident, the pain level is likely moderate to severe.

TABLE 7.1 Overall pain signs. With veterinarians' consultation, pain signs should be customized to fit the need.

Parameters	Possible pain signs
General activity/movement	Overall activity, restlessness, lameness, recumbent
Appearance	Rough haircoats, piloerection, grooming, porphyrin stain (rats)
Food/water consumption	Reduce body weight, reduced urine/fecal production
Physiological alterations	Increased HR, RR, BP, body temperature, skin/mucous membrane color
Vocalization	Whining, teeth grinding, chattering (rodents)
Temperament	Quiet, aggression, avoidance, reluctant to be handled/ palpated, self-mutilation
Surgical wound	Lick, chew, redness, infection, dehiscence, swelling, discharges

Modified from Stasiak, K.L., Maul, D., French, E., Hellyer, P.W., Vandewoude, S., 2003. Species-specific assessment of pain in laboratory animals. Contemp. Top. Lab. Anim. Sci. 42, 13–20.

TABLE 7.2 Rodents (mice, rats, and guinea pigs)—pain scales.

Criteria	Score	Observation
Appearance	0	Normal, cage exploration
	1	Quiet, minimal cage exploration, hunched, move if stimulated, piloerection, quint eyes, porphyrin staining (rats)
	2	Hunched, rough hair coat, piloerection, labored breathing, flat ears, porphyrin staining (rats)
Wound appearance	0	Clean, no/very minimal swelling
	1	Moderate swelling, incision intact
	2	Severe swelling, dehiscence, infected
Body weight	0	Normal or <5% decrease
	1	<6%–15% decrease
	2	<16%–20% decrease
Total	**6**	When score >2, administer analgesics

Modified from Stasiak, K.L., Maul, D., French, E., Hellyer, P.W., Vandewoude, S., 2003. Species-specific assessment of pain in laboratory animals. Contemp. Top. Lab. Anim. Sci. 42, 13–20.

TABLE 7.3 Rabbits—pain scales.

Criteria	Score	Observation
Appearance	0	Clam, cage exploration, food/water consumption
	1	Minimal cage exploration and food/water consumption, squint eyes
	2	No cage exploration, no food/water consumption, hunched, quiet, increase respiratory rate or labored breathing, flat ears
Body temperature	0	Normal
	1	39–40°C
	2	>40°C
Wound	0	Clean, minimal swelling
	1	Moderate swelling, incision intact
	2	Severe swelling, dehiscence, infected
RR	0	0%–15% above preoperative value
	1	15%–30% above preoperative value
	2	>30% above preoperative value
Swelling	0	None
	1	Mild and noticeable
	2	Obvious
Total	**10**	When score >3, administer analgesics

Modified from Stasiak, K.L., Maul, D., French, E., Hellyer, P.W., Vandewoude, S., 2003. Species-specific assessment of pain in laboratory animals. Contemp. Top. Lab. Anim. Sci. 42, 13–20.

- *Mouse*: The laboratory mouse is the most commonly used species in biomedical research comprising the 52% of the species utilized in 2018. Part of their keystone use in biomedical research includes significant use in behavioral research; therefore, there is substantial behavioral information on the species as a whole. However, there are thousands of GA mice and multiple mouse strains making the assessment of "normal" behavior challenging. Furthermore, like many other small mammals that will be discussed, mice are preyed upon by larger mammals, avian species, and reptiles and they are, unless severely affected, good at withholding outward evidence of distress, pain, and suffering. Given the variety of mice available to today's biomedical researchers, there will be mouse lines that will

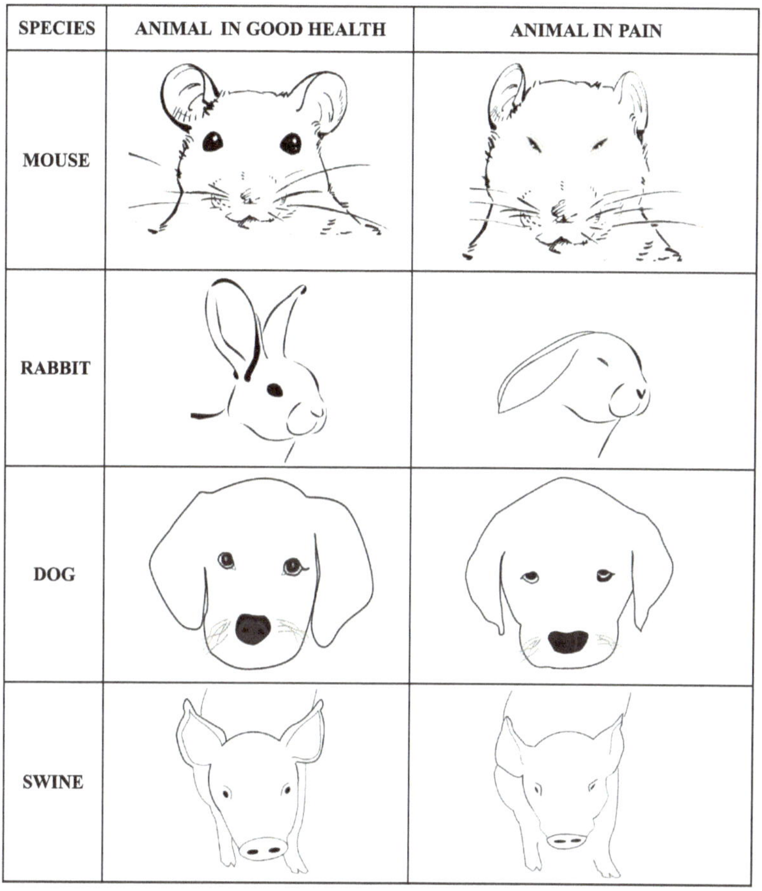

Photos from pixabay.com and outlined by Janis Atuk-Jones

FIGURE 7.2 Grimace or facial expression during pain in variety of species—note that majority of common noticeable facial signs are squinty eyes, flattened whiskers, and flattened ears.

be behavioral outliers, and it is absolutely critical for the veterinary and/or caretaker staff to be aware of the normal, yet abnormal, mouse lines and their peculiarities. As a general rule, confounding environmental, clinical, or experimental factors (light/dark cycle or noise, relevant clinical conditions like dehydration, high fat diet, treatments) may require consideration. General "sick mouse syndrome" signs include dehydration, reluctance to move, rapid breathing, hunched posture, and "spiky" fur. Aside from the clinical signs, the mouse's nest may offer additional insight to the mouse's well-being. Mice that are in good condition will readily incorporate additional material into an existing nest structure. Ultrasonic vocalization has been proposed and assessed as a method of pain assessment, with several limitations (Turner et al., 2019).

Summary of relevant clinical signs associated with pain and distress in the mouse:

- Ungroomed appearance and piloerection
- Reduced food/water intake
- Hunched posture
- Writhing
- Pressing: stretching and pressing the abdomen against cage bottom
- Recumbency
- Facial expression—facial action units (FAUs), e.g., orbital tightening, nose bulge, cheek bulge, ear position, and whisker change
- Licking activity/selfmutilation/guards limbs
- Aggressive behavior
- Reduced/altered social activities
- Reduced nest building and burrowing behavior
- Eating bedding/litters
- Vocalization

Rats: As another rodent prey species, ilk principles and characterizations apply to observing rats from a distance. Like mice, clinical observation can be made cage side. One distinct behavioral difference between rats and mice is that nest building is learned, rather than innate, in rats (Van Loo and Baumans, 2004). A rat grimace scale (RGS) has been used and it correlates with pain severity (Leung et al., 2019). RGS in some pain models (i.e., neuropathic pain) have a zenith scoring at six hours from injury and remained high for the next two days with a particularly high correlation seen with orbital tightening (Philips et al., 2017). The RGS method is reliable and consistent among evaluators. In addition, some behaviors [i.e., stretch, stretch-walk sequence, writhing, back arch, twitch, stagger/fall, abdominal licking (frequency and duration), decreased appetite, decreased postural rearing, circular ambulation, sleep-wake disturbances, and shaking] are observed with abdominal (visceral)-related models. Specifically, writhing is observed with laparotomy and intestinal mucositis models, while back arching is observed in laparotomy, intestinal mucositis, and urethral calculi models (Leung et al., 2019; Whittaker and Howarth, 2014).

Summary of relevant clinical signs associated with pain and distress in the rat:

- Ungroomed appearance and piloerection
- Chromodacryorrhea—red tears staining
- Reduced food/water intake

- Hunched posture
- Stretch, stretch-walk sequence
- Writhing
- Rebalancing weight over the limbs, staggering, and falling
- Recumbency
- Facial expression—FAUs, e.g., orbital tightening, nose bulge, cheek bulge, ear position, and whisker change
- Increased RR, shallow breathing
- Licking activity/self-mutilation/guards limbs
- Aggressive behavior
- Reduced/altered social activities
- Reduced nest building and burrowing behavior
- Eating bedding/litters
- Vocalization

Guinea pigs: As with other rodents, guinea pig pain can be difficult to detect even cage side. When painful, guinea pigs tend toward silence during handling (typically, nonpainful guinea pigs squeal, yet squealing is reported in distressed guinea pigs). Reduced activity, eye closure or squinting, reduced food/water consumption, reduced grooming, weight loss, subtle behavioral changes or body movements, i.e., weight shifting, and cage mate aggression have been reported as signs of pain. Guinea pigs are generally highly food-motivated animals eating continuously to support their high metabolism; therefore, some mild pain may not affect eating (Dunbar et al., 2016). Prior reports suggest differences between male and female guinea pigs (Oliver et al., 2017; Mogil, 2009). As with other species where pain assessment can be challenging, pain control should be based on pain procedure severity.

Summary of relevant clinical signs associated with pain and distress in the guinea pig:

- Ungroomed appearance and piloerection
- Stampede or immobility
- Anorexia (requires therapeutic measures as associated with moderate to severe conditions)
- Sialorrhea: associated to buccal/dental pain
- Vocalization: absent or squealing; spontaneous intermittent vocalization is described (Bochmann et al., 2016).

Rabbits: While not rodents, rabbits are a prey species; normally, quiet, and reserved. Recognizing abnormal subtle behaviors requires observer familiarity with normal behaviors. Autonomic system changes (i.e., HR, BP, etc.) are often challenging to assess without inducing a stress response. Therefore, behavioral changes are still considered a reliable pain sign in many species, including rabbits. In contrast to nonpainful rabbits, which usually show a short and shallow breathing pattern, painful rabbits may show deep breathing together with decreased RR and decreased nares movement (Johnston, 2005). Painful rabbits may also elicit a characteristic piercing squeal. Other signs include serous nasal discharge, epiphora, poor grooming or unkempt appearance, hiding or quietly sitting in a far/dark corner of a cage, decreased activity/movement or uncontrolled locomotion, decreased urination/defecation and neglecting cecotropes, inappetence, dehydration, and bruxism or teeth grinding (indicative of visceral or dental pain). Flinching, twitching, and shuffling may be observed (Thompson, 2014; Rocca et al., 2015). More importantly, rabbits rely on nutrients' supply from the gastrointestinal tract, and they can graze or eat continuously. Therefore, decreased food intake or anorexia may cause further gastrointestinal stasis, hepatic lipidosis, and diabetic ketoacidosis (Varga, 2014). Anorexia can be easily recognized over other behavioral changes (Johnston, 2005). The pain scale is shown in Table 7.3.

Summary of relevant clinical signs associated with pain and distress in the rabbit:

- Ungroomed appearance and piloerection
- Decreased locomotor activity and exploratory behavior
- Freezing
- Flinching, twitching, and shuffling
- Facial expression—FAUs, e.g., orbital tightening, nose bulge, cheek bulge, ear position, and whisker change
- Bruxism or teeth grinding
- Anorexia (requires prompt therapeutic measures), decreased urine, and fecal pellet output.
- Vocalization: squealing in case of acute pain, distress (fear), or even preagonic.

Zebrafish: Although fish may not possess the higher neural structures (or the alternatives) to feel pain, this does not absolve scientists from an ethical or scientific responsibility to assess and treat the noxious input considering the high potential for negatively impacting animal welfare and research results, plus evidence that fish respond characteristically when employing analgesia in research paradigms that would be painful to humans (Sneddon, 2015, 2019; Taylor et al., 2017; Deakin et al., 2019; Costa et al., 2019; Schroeder and Sneddon, 2017). Noxious stimuli in fish elicit various responses, including rubbing the affected area against or rocking back and forth using the pectoral fins on environmental substrates, guarding activity, increased gill ventilation rate, shoaling, reduced swimming (and speed), use of cover, reduced environmental exploration, spending increased time in atypical tank strata (e.g., tank top, middle, bottom), angular body curvature, temporary cessation of normal behaviors, and physical appearance. Decreased food consumption and increased RRs are also described in zebrafish (Sneddon, 2011). Evidence suggests zebrafish behavioral pain preservation between the sexes (Taylor et al., 2017; Costa et al., 2019). There is currently greater potential for automated fish behavioral (pain) assessment following various potentially painful procedures than perhaps any other species due to their ability to fully utilize all four dimensions of their environment (when not in pain) over other species (Costa et al., 2019; Deakin et al., 2019; Lopez-Luna et al., 2017a,b; Taylor et al., 2017).

Summary of relevant clinical signs associated with pain and distress in the Zebrafish:

- Abnormal swimming behavior
  - Quantitative: reduced movement and speed
  - Qualitative
- Increased RR, more rapid opercular movements
- Strong muscular movements (after needle puncture)
- Guarding behavior
- Rocking on substrate and rubbing injured area
- Decreased food consumption.

Larger laboratory animals

Because of their larger size, both autonomic nervous system (ANS) and somatic systems could be evaluated as part of a pain assessment in these animals. During pain, autonomic response parameters (HR, RR, BP, but also pupillary diameter, etc.) can be altered; of which, some altered parameters may be observable, such as tachycardia, tachypnea, hypertension, and mydriasis (note that analgesics may cause these signs, too). Therefore, these ANS signs should be assessed when possible and/or safe to do so. In the case of chronic pain, these parameter alterations may not be evident. In general, when animals experience pain, these parameters may increase approximately 15%–20% from baseline measures (Hernandez-Avalos et al., 2019). Therefore, obtaining baseline measurements before starting an experiment is highly recommended. Other indirectly altered physiologic parameters are hormonal levels (cortisol, catecholamines, etc.) or chemical mediator levels (i.e., bradykinin, prostaglandin, or cytokines). Note that these changes can result from physiological stress and may be largely variable among species or only correlate weakly to pain (De Jonckheere et al., 2015).

Dogs: Dogs are gregarious, predator animals that when well-socialized react positively and enthusiastically to human interaction. Generally, pain signs include exaggerated responses to stimulus, reluctance to move, depression, aggression, submission, anorexia, vocalization, guarding, licking injured areas, and "prayer position" (abdominal pain; raising the hindquarters with front limbs and head on the floor) (Hernandez-Avalos et al., 2019). There are several dog-specific pain scoring systems (confront Table 7.4), including the Glasgow composite measuring pain scale and Colorado State University Canine Acute pain (Reid et al., 2007; Hellyer et al., 2006). Surprisingly, despite the development of Grimace scales in many species (Mogil et al., 2020),

TABLE 7.4 Dogs—pain scales.

Criteria	Score	Observation
Comfort	0	Calm, asleep, looks comfort
	1	Mild-moderate agitation, wake, interested in surroundings
	2	Trashing
Movement	0	No movement
	1	Frequent positional change
	2	Continuous positional change
Appearance	0	Normal or mild changes: eyelids partially close, ears flattened or carried normally
	1	Moderate changes: eyes sunken or glazed, unthrifty appearance
	2	Severe changes: eye pale, enlarge pupils, guarding, legs in abnormal position or other abnormal facial expressions
Vocalization	0	None
	1	Crying, responds to calm voice and stroking
	2	Crying, no response to calm voice and stroking
HR	0	0%–15% above preoperative value
	1	15%–30% above preoperative value
	2	>30% above preoperative value
RR	0	0%–15% above preoperative value
	1	15%–30% above preoperative value
	2	>30% above preoperative value
Total	**12**	When score >4–5, administer analgesics

Modified from Pacharinsak, C., Greene, S.A., Keegan, R.D., Kalivas, P.W., 2003. Postoperative analgesia in dogs receiving epidural morphine plus medetomidine. J. Vet. Pharmacol. Therapeut. 26, 71–77. https://doi.org/10.1046/j.1365-2885.2003.00452.x and Stasiak, K.L., Maul, D., French, E., Hellyer, P.W., Vandewoude, S., 2003. Species-specific assessment of pain in laboratory animals. Contemp. Top. Lab. Anim. Sci. 42, 13–20.

no grimace scale has yet been validated for pain assessment in dogs. Facial expressions of pain can be inferred from veterinary clinical practice and include arched eyebrows, darting eyes, dilated pupils, and ears flattened against the head.

Summary of relevant clinical signs associated with pain and distress in the dog:

- Abnormal gait posture: hunched, "prayer position" for severe abdominal pain
- Guarding behavior
- Licking, biting painful area up to self-injure/mutilation if pain is severe
- Increased RR
- Decreased locomotor activity and social interactions
- Abnormal behaviors: anxiety, fear, aggressiveness, and passivity
- Attempt to escape if painful area is palpated
- Abnormal eliminatory behaviors: increased urination frequency; reduced fecal output associated with motor activity reduction
- Decreased food intake, anorexia and weight loss
- Vocalization
- Sleep disruption

Pigs: Pigs show many of the same clinical pain indicators seen in other species including vocalization, appearance, social interaction, bruxism, demeanor, gait, posture, food/water intake, and wound palpation response (Table 7.5). Perhaps the most powerful pain indicator in pigs is their unwillingness to eat, especially considering these animals are potentially as sociable as dogs when properly socialized with humans. As with any species, assessing multiple parameters provides a more robust pain assessment (Talbot et al., 2020).

TABLE 7.5 Pigs—pain signs.

Pain signs in pigs
Decreased in activity, quiet, isolated from other pigs
Kicking the abdomen (abdominal pain)
Inappetite, lose weight
Hunched posture, carpal walking (walking on knee), limping, difficult to rise or lie down
Trembling
Reduce interests in surrounding or upon humans approaching
Grunting, teeth grinding,

Modified from Thompson, L., 2014. Recognition and assessment of pain in small exotic mammals. In: Egger, C.M., Love, L., Doherty, T. (Eds.), Pain Management in Veterinary Practice. Ames, IA, John Wiley & Sons, Inc., pp. 391–397, Mozzachio, K., Tynes, V.V., 2013. Recognition and Treatment of Pain in Pet Pigs. Pain Management in Veterinary Practice, John Wiley & Sons, Ltd., Ames, IA, pp. 383–389, and Herskin Mette, S., Di Giminiani, P., 2018. Pain in Pigs: Characterisation, Mechanisms and Indicators. Advances in Pig Welfare, Woodhead Publishing Series in Food Science, Technology and Nutrition, pp. 325–355. https://doi.org/10.1016/B978-0-08-101012-9.00011-3.

Harvey-Clark et al. (2000) devised a pain scoring system to assess two postoperative analgesic protocols for pigs undergoing a left lung allograft transplant. Reyes et al. (2002) developed a pig behavioral and physiologic scoring system using an inguinal incision to assess two postoperative analgesics. Murison et al. (2009) developed a pain scoring system for a swine laryngeal transplant model, which was striking in that it combined elements of veterinary and human medicine to direct perioperative care. Castel et al. (2014) developed a pig scoring system for skin and muscle incision with retraction model employing various elements based on individual animal and social behaviors, including the impact of a human presence. Royal et al. (2013) described a behavioral and physiologic pain scoring system to assess postoperative analgesia in a swine femoral fracture model. A recent review article (Bradbury et al., 2016) focusing on applying the ARRIVE Guidelines (Kilkenny et al., 2010; Percie du Sert et al., 2020) to experimental surgery and postoperative analgesia determined that the lack of a swine-specific pain scoring system and poor pain management recommendations could hinder swine welfare. Less than 10% of the articles surveyed presented the five elements outlined in the ARRIVE Guidelines. Viscardi et al. (2017) developed a piglet grimace scale to assess castration and tail docking pain via ear position, cheek tightening/nose bulge, and orbital tightening. A thorough assessment of pain and pain-related signs in pigs was published by Herskin and Di Giminiani in 2018 (2018).

Summary of relevant clinical signs associated with pain and distress in the pig:

- Reduced food consumption
- Isolation and reduced social interaction
- Altered motor activity: increased sitting, but also reduced lying
- Facial expression—FAUs, e.g., ear position, nose bulge, cheek tightening, and orbital tightening
- Vocalization: screaming, acute pain.

Welfare assessment of genetically induced phenotypes (Anne Zintzsch)

In principle, the GA animal welfare assessment follows the same practices for any experimental animal or conspicuous animals in animal husbandry. However, with the implementation of the EU Directive 2010/63, the identification and refinement of harmful genetically induced phenotypes has become emphasized. Creating and maintaining animals with a likely harmful phenotype is considered a "procedure" and subject to authorization. Nevertheless, Bert and colleagues point out the lack of knowledge on how to objectively measure relevant abnormal states, also caused by the genetic alteration. According to the first EU report under the regulations of Directive 2010/63, about 20% of GA animals experience pain, distress, or suffering resulting from their genetic alteration (European Commission, 2020). Together with zebrafish, mice are the most commonly used GA species in biomedical research.

Guidance on key elements for a GA rodent welfare assessment has been developed by an Expert Working Group on the creation, breeding, and maintenance of GA animals commissioned by the EU (European Commission Expert Working Group, 2013) and is currently under revision. Additionally, complementing guidance has been published by Member States (Home Office, 2015; Federal Institute for Risk Assessment, 2015, 2016). Welfare assessment templates are intended to ensure consistency across research institutions. The current approach to classify a line in question as a line with a harmful phenotype primarily focusses on clinical observations at the cage-side level and noninvasive measurements easily implemented into the daily care routine. Welfare-relevant information should be collected systematically, where relevant genotypes and the entire animal's lifespan are covered. This systematic basic welfare assessment should be designed according to the line-specific prospective severity assessment and tailored to the expected phenotype. If no or little information is available, the welfare assessment should be adapted in terms of the monitoring frequency, general assessment parameters, and assessor selection (Zintzsch et al., 2020). Defining the early humane endpoints may be the most vital task to reduce pain, suffering, and distress (Schuppli et al., 2004). If a particular phenotype is expected, the assessment scheme and selected parameters should be suitable to reveal expected and unexpected phenotypes. Common assessment schemes, in general, include clinical parameters such as body weight, general appearance, behavior, and clinical signs. For example, increased water consumption is a key Huntington's disease marker suggesting vasopressin dysregulation related to hypothalamic vasopressin release, which contributed to weight loss by impacting food taste, mastication, and swallowing (Littin et al., 2008). Continuous distress monitoring is crucial in securing appropriate welfare and, if needed, pain assessment and control. Furthermore, evaluating reproduction data and mortality rates, considering age and genotype, complement the information gained in the animal facility. This information forms the basis, to implement appropriate stress-reducing measures. However, additional information from scientific studies should complement the line's overall welfare evaluation. Severity assessments are a continuous process and requiring a team approach (European Commission, 2012). When new scientific results relevant to animal welfare are revealed, lines should be reevaluated and, if needed, reclassified as a non-harmful/harmful line (Zintzsch et al., 2017). Further information may be crucial for a correct classification to cover all relevant animal welfare facets that may otherwise go unobserved in the cage environment (e.g., blindness or the loss of the sense of hearing or smell).

Documenting and exchanging GA line information via the scientific community plays a vital role in ensuring appropriate animal care and use.

While existing databases provide valuable information (The Jackson Laboratory, 2021; International Mouse Phenotyping Consortium; Infrafrontier; Eppig et al., 2015), only an amount of GA lines are included in such databases necessitating the transfer of information with the animals.

References

Bochmann, M., Knell, S., Dennler, M., Grest, P., Wenger, S., 2016. Clinical presentation, diagnosis, and treatment of Cholelithiasis in a Pet Guinea pig (*Cavia porcellus*). J. Exot. Pet Med. 25 (4), 327–331. https://doi.org/10.1053/j.jepm.2016.06.009.

Bradbury, A.G., Eddleston, M., Clutton, R.E., 2016. Pain management in pigs undergoing experimental surgery; a literature review (2012-4). Br. J. Anaesth. 116, 37–45. https://doi.org/10.1093/bja/aev301.

Castel, D., Willentz, E., Doron, O., Brenner, O., Meilin, S., 2014. Characterization of a porcine model of post-operative pain. Eur. J. Pain 18, 496–505. https://doi.org/10.1002/j.1532-2149.2013.00399.x.

Committee on Pain and Distress in Laboratory Animals Institute of Laboratory Animal Resources, National Research Council, National Academy of Sciences, 1991. Recognition and alleviation of pain and distress in laboratory animals. ILAR J. 33 (4), 71–74. https://doi.org/10.1093/ilar.33.4.71.

Costa, F.V., Rosa, L.V., Quadros, V.A., Santos, A.R.S., Kalueff, A.V., Rosemberg, D.B., 2019. Understanding nociception-related phenotypes in adult zebrafish: behavioral and pharmacological characterization using a new acetic acid model. Behav. Brain Res. 359, 570–578. https://doi.org/10.1016/j.bbr.2018.10.009.

De Jonckheere, J., Bonhomme, V., Jeanne, M., Boselli, E., Gruenewald, M., Logier, R., et al., 2015. Physiological signal processing for individualized anti-nociception management during general anesthesia: a review. Yearb. Med. Inform. 10, 95–101. https://doi.org/10.15265/IY-2015-004.

Deakin, A.G., Buckley, J., Alzu'bi, H.S., Cossins, A.R., Spencer, J.W., Al'nuaimy, W., et al., 2019. Automated monitoring of behaviour in zebrafish after invasive procedures. Sci. Rep. 9, 9042. https://doi.org/10.1038/s41598-019-45464-w.

Dunbar, M.L., David, E.M., Aline, M.R., Lofgren, J.L., 2016. Validation of a behavioral ethogram for assessing postoperative pain in Guinea pigs (*Cavia porcellus*). JAALAS 55, 29–34.

Eppig, J.T., Motenko, H., Richardson, J.E., Richards-Smith, B., Smith, C.L., 2015. The international mouse strain resource (IMSR): cataloging worldwide mouse and ES cell line resources. Mamm. Genome 26, 448–455. https://doi.org/10.1007/s00335-015-9600-0.

European Commission, 2012. National competent authorities for the implementation of directive 2010/63/eu on the protection of animals used for scientific purposes. Working document on a severity assessment framework. Brussels. Available at: http://ec.europa.eu/environment/chemicals/lab_animals/pdf/Endorsed_Severity_Assessment.pdf. (Accessed 3 December 2021).

European Commission, 2020. 2019 Report on the Statistics on the Use of Animals for Scientific Purposes in the Member States of the European Union in 2015-2017. Brussels. Available at: http://ec.europa.eu/transparency/regdoc/rep/1/2020/EN/COM-2020-16-F1-EN-MAIN-PART-1.PDF. (Accessed 26 July 2021).

European Commission Expert Working Group, 2013. Working Document on Genetically Altered Animals- Corrigendum of 24 January 2013. Available at: http://ec.europa.eu/environment/chemicals/lab_animals/pdf/corrigendum.pdf. (Accessed 9 March 2021).

Federal Institute for Risk Assessment, 2015. Severity Assessment of Genetically Altered Fish (Bony Fish, Teleost Fish). Available at: http://www.bfr.bund.de/cm/349/severity-assessment-of-genetically-altered-fish-bony-fish-teleost-fish.pdf. (Accessed 9 March 2021).

Federal Institute for Risk Assessment, 2016. Severity Assessment of Genetically Altered Mice and Rats - Version 2. BfR, Berlin. Available at: http://www.bfr.bund.de/cm/349/severity-assessment-of-genetically-altered-mice-and-rats-version-2.pdf. (Accessed 10 March 2021).

Federation of European Laboratory Animal Science Associations (FELASA), 1994. Report of the working group on pain and distress accepted by the FELASA board of management, pain and distress in laboratory rodents and lagomorphs: November 1992. Lab. Anim. 28 (2), 97–112. https://doi.org/10.1258/002367794780745308.

Harvey-Clark, C.J., Gilespie, K., Riggs, K.W., 2000. Transdermal fentanyl compared with parenteral buprenorphine in post-surgical pain in swine: a case study. Lab. Anim 34, 386–398. https://doi.org/10.1258/002367700780387750.

Hellyer, P.W., Uhrig, S.R., Robinson, N.G., 2006. Canine Acute Pain Scale and Feline Acute Pain Scale. Colorado State University Veterinary Medical Center. Available at: https://www.vasg.org/pdfs/CSU_Acute_Pain_Scale_Canine.pdf. (Accessed 29 July 2021).

Hernandez-Avalos, I., Mota-Rojas, D., Mora-Medina, P., Martínez-Burnes, J., Casas Alvarado, A., Verduzco-Mendoza, A., et al., 2019. Review of different methods used for clinical recognition and assessment of pain in dogs and cats. Int. J. Vet. Sci. Med. 7, 43–54. https://doi.org/10.1080/23144599.2019.1680044.

Herskin Mette, S., Di Giminiani, P., 2018. Pain in Pigs: Characterisation, Mechanisms and Indicators. Advances in Pig Welfare, Woodhead Publishing Series in Food Science, Technology and Nutrition, pp. 325–355. https://doi.org/10.1016/B978-0-08-101012-9.00011-3.

Home Office, 2015. Severity Classification of Genetically Altered Animals under the Animals (Scientific Procedures) Act 1986. Available at: http://assets.publishing.service.gov.uk/government/uploads/system/uploads/attachment_data/file/785514/asp-severity-classification-of-genetically-altered-animals.pdf. (Accessed 9 March 2021).

Infrafrontier. European mouse mutant archive. INFRAFRONTIER mouse disease models. Available at: http://www.infrafrontier.eu/infrafrontier-research-infrastructure/organisation/european-mouse-mutant-archive. (Accessed 9 March 2021).

International Mouse Phenotyping Consortium. Available at: http://www.mousephenotype.org/. (Accessed 9 March 2021).

IRAC [Interagency Research Animal Committee], 1985. US Government Principles for the Utilization and Care of Vertebrate Animals Used in Testing, Research, and Training. Federal Register, May 20. Office of Science and Technology Policy, Washington. Accessed 11 October 2021.

The Jackson Laboratory, 2021. Mouse Genome Database (MGD) at the Mouse Genome Informatics Website. Bar Harbor Maine. Available at: http://www.informatics.jax.org/mgihome/other/citation.shtml. (Accessed 9 March 2021).

Jang, S.J., Kang, S.S., Son, S.J., Lee, J.Y., Kim, G., Choi, S.H., 2017. Cortisol levels and gastrointestinal disorders after stressful surgery in rabbits. In Vivo 31 (4), 637–640. https://doi.org/10.21873/invivo.11105.

Johnston, M.S., 2005. Clinical approaches to analgesia in ferrets and rabbits. Seminars Avian Exot. Pet Med. 14 (4), 229–235.

Keubler, L.M., Hoppe, N., Potschka, H., Talbot, S.R., Vollmar, B., Zechner, D., et al., 2020. Where are we heading? Challenges in evidence-based severity assessment. Lab. Anim 54, 50–62. https://doi.org/10.1177/0023677219877216.

Kilkenny, C., Browne, W.J., Cuthill, I.C., Emerson, M., Altman, D.G., 2010. Improving bioscience research reporting: the arrive guidelines for reporting animal research. PLoS Biol. 8, e1000412. https://doi.org/10.1371/journal.pbio.1000412.

Leung, V.S.Y., Benoit-Biancamano, M.-O., Pang, D.S.J., 2019. Performance of behavioral assays: the rat grimace scale, burrowing activity and a composite behavior score to identify visceral pain in an acute and chronic colitis model. PAIN Rep. 4, e718. https://doi.org/10.1097/pr9.0000000000000712.

Littin, K., Acevedo, A., Browne, W., Edgar, J., Mendl, M., Owen, D., et al., 2008. Towards humane end points: behavioural changes precede clinical signs of disease in a Huntington's disease model. Proc. Biol. Sci. 275, 1865–1874. https://doi.org/10.1098/rspb.2008.0388.

Lopez-Luna, J., Al-Jubouri, Q., Al-Nuaimy, W., Sneddon, L.U., 2017a. Impact of analgesic drugs on the behavioural responses of larval zebrafish to potentially noxious temperatures. Appl. Anim. Behav. Sci. 188, 97–105.

Lopez-Luna, J., Al-Jubouri, Q., Al-Nuaimy, W., Sneddon, L.U., 2017b. Reduction in activity by noxious chemical stimulation is ameliorated by immersion in analgesic drugs in zebrafish. J. Exp. Biol. 220, 1451–1458. https://doi.org/10.1242/jeb.146969.

Masson-Meyers, D.S., Andrade, T., Caetano, G.F., Guimaraes, F.R., Leite, M.N., Leite, S.N., Frade, M., 2020. Experimental models and methods for cutaneous wound healing assessment. Int. J. Exp. Pathol. 101 (1–2), 21–37. https://doi.org/10.1111/iep.12346.

Mogil, J.S., 2009. Animal models of pain: progress and challenges. Nat. Rev. Neurosci. 10, 283–294. https://doi.org/10.1038/nrn2606.

Mogil, J.S., Pang, D., Silva Dutra, G.G., Chambers, C.T., 2020. The development and use of facial grimace scales for pain measurement in animals. Neurosci. Biobehav. Rev. 116, 480–493. https://doi.org/10.1016/j.neubiorev.2020.07.013.

Mozzachio, K., Tynes, V.V., 2013. Recognition and Treatment of Pain in Pet Pigs. Pain Management in Veterinary Practice. John Wiley & Sons, Ltd., Ames, IA, pp. 383–389.

Murison, P.J., Jones, A., Mitchard, L., Burt, R., Birchall, M.A., 2009. Development of perioperative care for pigs undergoing laryngeal transplantation: a case series. Lab. Anim. 43, 338–343. https://doi.org/10.1258/la.2009.008101.

National Research Council, Committee on Regulatory Issues in Animal Care and Use, 2000. Definition of pain and distress and reporting requirements for laboratory animals. In: Proceedings of the Workshop Held June 22, 2000. National Academies Press (US). Commission on Life Sciences, Washington (DC). Available at: https://www.ncbi.nlm.nih.gov/books/NBK99526/. (Accessed 27 July 2021).

National Research Council, 2008. Recognition and Alleviation of Distress in Laboratory Animals. The National Academies Press, Washington, DC. https://doi.org/10.17226/11931. (Accessed 11 August 2021).

Olfert Ernest, D., Godson Dale, L., 2000. Humane endpoints for infectious disease animal models. ILAR J. 41 (2), 99–104. https://doi.org/10.1093/ilar.41.2.99.

Oliver, V.L., Athavale, S., Simon, K.E., Kendall, L.V., Nemzek, J.A., Lofgren, J.L., 2017. Evaluation of pain assessment techniques and analgesia efficacy in a female Guinea pig (*Cavia porcellus*) model of surgical pain. JAALAS 56, 425–435.

Organization for Economic Co-operation and Development (OECD/OCDE) Environment Directorate Joint Meeting of the Chemicals Committee and the Working Party on Chemicals, Pesticides and Biotechnology: Guidance Document on The Recognition, Assessment, and Use of Clinical Signs as Humane Endpoints for Experimental Animals Used In Safety Evaluation. Paris. Available at: https://www.oecd.org/officialdocuments/publicdisplaydocumentpdf/?cote=env/jm/mono%282000%297&doclanguage=en. (Accessed 26 July 2021).

Pacharinsak, C., Greene, S.A., Keegan, R.D., Kalivas, P.W., 2003. Postoperative analgesia in dogs receiving epidural morphine plus medetomidine. J. Vet. Pharmacol. Therapeut. 26, 71–77. https://doi.org/10.1046/j.1365-2885.2003.00452.x.

Pawluski, J., Jego, P., Henry, S., Bruchet, A., Palme, R., Coste, C., Hausberger, M., 2017. Low plasma cortisol and fecal cortisol metabolite measures as indicators of compromised welfare in domestic horses (*Equus caballus*). PLoS One 12 (9), e0182257. https://doi.org/10.1371/journal.pone.0182257.

Percie Du Sert, N., Ahluwalia, A., Alam, S., Avey, M.T., Baker, M., Browne, W.J., et al., 2020. Reporting animal research: explanation and elaboration for the ARRIVE guidelines 2.0. PLoS Biol. 18, e3000411. https://doi.org/10.1371/journal.pbio.3000411.

Philips, B.H., Weisshaar, C.L., Winkelstein, B.A., 2017. Use of the rat grimace scale to evaluate neuropathic pain in a model of cervical radiculopathy. Comp. Med. 67, 34–42.

Reid, J., Nolan, A.M., Hughes, J.M.L., Lascelles, D., Pawson, P., Scott, E.M., 2007. Development of the short-form Glasgow composite measure pain scale (CMPS-SF) and derivation of an analgesic intervention score. Anim. Welf. 16 (Suppl. 1), 97-104(8).

Reyes, L., Tinworth, K.D., Li, K.M., Yau, D.F., Waters, K.A., 2002. Observer-blinded comparison of two nonopioid analgesics for postoperative pain in piglets. Pharmacol. Biochem. Behav. 73, 521–528. https://doi.org/10.1016/S0091-3057(02)00820-1.

Rocca, G.D., Catanzaro, A., Salvo, A.D., Goldberg, M.E., 2015. Diagnosis of pain in small companion animals. Am. J. Anim. Vet. Sci. 10, 57–66.

Rokyta, R., 2008. Fetal pain. Neuro. Endocrinol. Lett. 29 (6), 807–814. PMID: 19112406.

Royal, J.M., Settle, T.L., Bodo, M., Lombardini, E., Kent, M.L., Upp, J., et al., 2013. Assessment of postoperative analgesia after application of ultrasound-guided regional anesthesia for surgery in a swine femoral fracture model. JAALAS 52, 265–276.

Schroeder, P.G., Sneddon, L.U., 2017. Exploring the efficacy of immersion analgesics in zebrafish using an integrative approach. Appl. Anim. Behav. Sci. 187, 93–102. https://doi.org/10.1016/j.applanim.2016.12.003.

Schuppli, C.A., Fraser, D., Mcdonald, M., 2004. Expanding the three Rs to meet new challenges in humane animal experimentation. Altern. Lab. Anim. 32, 525–532. https://doi.org/10.1177/026119290403200507.

Sneddon, L.U., 2011. Pain perception in fish. J. Conscious. Stud. 18 (9–10), 209–229.

Sneddon, L.U., 2015. Pain in aquatic animals. J. Exp. Biol. 218, 967–976. https://doi.org/10.1242/jeb.088823.

Sneddon, L.U., 2019. Evolution of nociception and pain: evidence from fish models. Philos. Trans. R. Soc. Lond. Ser. B Biol. Sci. 374, 20190290. https://doi.org/10.1098/rstb.2019.0290.

Srinivasa, N.R., Carr, D.B., Cohen, M., Finnerup, N.B., Flor, H., Gibson, S., Keefe, F.J., Mogil, J.S., Ringkamp, M., Sluka, K.A., Song, X.-J., Stevens, B., Sullivan, M.D., Tutelman, P.R., Takahiro, U., Vader, K., 2020. The revised international association for the study of pain definition of pain: concepts, challenges, and compromises. Pain 161 (9), 1976–1982. https://doi.org/10.1097/j.pain.0000000000001939.

Stasiak, K.L., Maul, D., French, E., Hellyer, P.W., Vandewoude, S., 2003. Species-specific assessment of pain in laboratory animals. Contemp. Top. Lab. Anim. Sci. 42, 13–20.

Stokes, W.S., 2000. Reducing unrelieved pain and distress in laboratory animals using humane endpoints. ILAR J. 41 (2), 59–61. https://doi.org/10.1093/ilar.41.2.59.

Stokes, W.S., 2002. Humane endpoints for laboratory animals used in regulatory testing. ILAR Journal 43 (Suppl. 1), S31–S38.

Stokes, W.S., 2011. Best practices for the use of animals in toxicological research and testing. Ann. N. Y. Acad. Sci. 1245, 17–20.

Stokes, W.S., Strickland, J., 2012. Validation and regulatory acceptance of dermatotoxicology methods: recent progress and the role of NICEATM and ICCVAM. In: Wilhelm, K.P., Zhai, H., Maibach, H. (Eds.), Dermatotoxicology, eighth ed. Informa Healthcare, London, pp. 475–487.

Talbot, S.R., Biernot, S., Bleich, A., Van Dijk, R.M., Ernst, L., Häger, C., et al., 2020. Defining body-weight reduction as a humane endpoint: a critical appraisal. Lab. Anim 54, 99–110. https://doi.org/10.1177/0023677219883319.

Tappe-Theodor, A., Kuner, R., 2014. Studying ongoing and spontaneous pain in rodents – challenges and opportunities. Eur. J. Neurosci. 39, 1881–1890. https://doi.org/10.1111/ejn.12643.

Taylor, J.C., Dewberry, L.S., Totsch, S.K., Yessick, L.R., Deberry, J.J., Watts, S.A., et al., 2017. A novel zebrafish-based model of nociception. Physiol. Behav. 174, 83–88. https://doi.org/10.1016/S0031-9384(17)30067-7.

Thompson, L., 2014. Recognition and assessment of pain in small exotic mammals. In: Egger, C.M., Love, L., Doherty, T. (Eds.), Pain Management in Veterinary Practice. John Wiley & Sons, Inc., Ames, IA, pp. 391–397.

Turner, A.I., Hemsworth, P.H., Tilbrook, A.J., 2005. Susceptibility of reproduction in female pigs to impairment by stress or elevation of cortisol. Domest. Anim. Endocrinol. 29 (2), 398–410.

Turner, P., Pang, D., Lofgren, J., 2019. A review of pain assessment methods in laboratory rodents. Comp. Med. 69 (6). https://doi.org/10.30802/AALAS-CM-19-000042, 451-467(17).

Van Loo, P.L., Baumans, V., 2004. The importance of learning young: the use of nesting material in laboratory rats. Lab. Anim 38, 17–24. https://doi.org/10.1258/00236770460734353.

Varga, M., 2014. Textbook of Rabbit Medicine, second ed. Elsevier Ltd, New York.

Viscardi, A.V., Hunniford, M., Lawlis, P., Leach, M., Turner, P.V., 2017. Development of a piglet grimace scale to evaluate piglet pain using facial expressions following castration and tail docking: a pilot study. Front. Vet. Sci. 4, 51. https://doi.org/10.3389/fvets.2017.00051.

Whittaker, A.L., Howarth, G.S., 2014. Use of spontaneous behaviour measures to assess pain in laboratory rats and mice: how are we progressing? Appl. Anim. Behav. Sci. 151, 1–12. https://doi.org/10.1016/j.applanim.2013.11.001.

Workman, P., Aboagye, E.O., Balkwill, F., Balmain, A., Bruder, G., Chaplin, D.J., Double, J.A., Everitt, J., Farningham, D.A., Glennie, M.J., Kelland, L.R., Robinson, V., Stratford, I.J., Tozer, G.M., Watson, S., Wedge, S.R., Eccles, S.A., Committee of the National Cancer Research Institute, 2010. Guidelines for the welfare and use of animals in cancer research. Br. J. Cancer 102 (11), 1555–1577. https://doi.org/10.1038/sj.bjc.6605642.

Xu, X., Huang, Y., 2020. Objective pain assessment: a key for the management of chronic pain. F1000Research 9 (F1000 Faculty Rev), 1–7. https://doi.org/10.12688/f1000research.20441.1.

Zintzsch, A., Noe, E., Reißmann, M., Ullmann, K., Krämer, S., Jerchow, B., et al., 2017. Guidelines on severity assessment and classification of genetically altered mouse and rat lines. Lab. Anim 51, 573–582. https://doi.org/10.1177/0023677217718863.

Zintzsch, A., Noe, E., Grimm, H., 2020. Navigating uncertainties: how to assess welfare and harm in genetically altered animals responsibly-a practical guideline. Animals (Basel) 10. https://doi.org/10.3390/ani10050857.

Recommended reading

Commission Staff Working, July 14, 2021. Document Summary Report on the Statistics on the Use of Animals for Scientific Purposes in the Member States of the European Union and Norway in 2018 Brussels. Available at: https://ec.europa.eu/environment/chemicals/lab_animals/pdf/SWD_%20part_A_and_B.pdf. (Accessed 26 July 2021).

Leonard, R.B., 1985. Primary afferent receptive field properties and neurotransmitter candidates in a vertebrate lacking unmyelinated fibers. Prog. Clin. Biol. Res. 176, 135–145.

Liles, J.H., Flecknell, P.A., 1993. The effects of surgical stimulus on the rat and the influence of analgesic treatment. Br. Vet. J. 149, 515–525. https://doi.org/10.1016/S0007-1935(05)80036-3.

National Research Council, 2011. Guide for the Care and Use of Laboratory Animals, eighth ed. The National Academies Press, Washington, DC. https://doi.org/10.17226/12910.

Raja, S.N., Carr, D.B., Cohen, M., Finnerup, N.B., Flor, H., Gibson, S., Keefe, F.J., Mogil, J.S., Ringkamp, M., Sluka, K.A., Song, X.J., Stevens, B., Sullivan, M.D., Tutelman, P.R., Ushida, T., Vader, K., 2020. The revised International Association for the Study of Pain definition of pain: concepts, challenges, and compromises. Pain 161 (9), 1976–1982. https://doi.org/10.1097/j.pain.0000000000001939.

Reilly, S.C., Quinn, J.P., Cossins, A.R., Sneddon, L.U., 2008. Behavioural analysis of a nociceptive event in fish: comparisons between three species demonstrate specific responses. Appl. Anim. Behav. Sci. 114, 248–259.

Rose, J.D., 2002. The neurobehavioral nature of fishes and the question of awareness and pain. Rev. Fish. Sci. 10, 1–38.

Sneddon, L.U., 2002. Anatomical and electrophysiological analysis of the trigeminal nerve in a teleost fish, *Oncorhynchus mykiss*. Neurosci. Lett. 319, 167–171. https://doi.org/10.1016/S0304394001025848.

Sneddon, L.U., 2009. Pain perception in fish: indicators and endpoints. ILAR J. 50, 338–342. https://doi.org/10.1093/ilar.50.4.338.

Snow, P.J., Plenderleith, M.B., Wright, L.L., 1993. Quantitative study of primary sensory neurone populations of three species of elasmobranch fish. J. Comp. Neurol. 334, 97–103. https://doi.org/10.1002/cne.903340108.

Chapter 8

Harm to research animals, severity categories, and humane endpoints

Aurora Brønstad
Department of Clinical Medicine, University of Bergen, Bergen, Norway

Harm and suffering in research animals

Despite continuous initiatives to refine experimental procedures to minimize suffering to experimental animals, some suffering or harm is still inevitable (Hendriksen et al., 2010). For ethical reasons, pain, suffering, or distress to research animals can only be ethically justified based on the potential benefits expected to be obtained by the experiment in the lack of alternative methods to achieve the same objectives. There must be a reasonable proportionality between harm to the animals and the benefits expected to be achieved (Luy, 2015).

Bateson was the first to describe a model of the relationship between harm to animals and benefits of animal studies (Bateson, 1986). The EU Directive 2010/63 ("the Directive") says that the project evaluation shall include a harm−benefit analysis.

> *(d) a harm-benefit analysis of the project, to assess whether the harm to the animals in terms of suffering, pain and distress is justified by the expected outcome taking into account ethical considerations, and may ultimately benefit human beings, animals or the environment.*
>
> EuropeanCommission (2010) Article 38.

What is "harm"?

In the EU Directive 2010/63, "pain, suffering, and distress" are the commonly terms used to refer to "harm" as a condition that must be avoided, unless it is scientifically justified by needs necessary to achieve certain scientific, knowledge, or safety objectives (EuropeanCommission, 2010).

Practical Handbook on the 3Rs in the Context of the Directive 2010/63/EU
https://doi.org/10.1016/B978-0-12-821180-9.00012-X

Sometimes there are conflicts between a researcher's and the common public's perception of what is ethically relevant with regard to welfare in animals (Buning Tde, 2004). Mellor and Reid in 1994 defined a categorization of harm based on impacts on the domains of the five freedoms—originally defined as a set of criteria to assess welfare in farm animals (FAWC, 1992). Mellor and Reid argued that the five freedoms encompass the impacts on animals in a broader perspective more aligned with modern regulations, guidelines, as well as laypersons perspective, than only concern about suffering (Mellor and Reid, 1994).

A working group established by The American Association for Laboratory Animal Science (AALAS) and the Federation of European Laboratory Animal Science Associations (FELASA) published in 2016 a review paper on harm–benefit analysis (HBA) (Bronstad et al., 2016) where harm was defined as any negative impact on welfare experienced by the animal (Bronstad et al., 2016). According to this definition, harm is reflecting any condition that puts animals in a situation where they cannot cope to maintain basic natural or life sustainable activities based on the domains of five freedoms (FAWC, 1992; Mellor and Reid, 1994; Laber et al., 2016) throughout the whole period of the animal's life (Hendriksen et al., 2010).

Ability to experience harm depends on sentience and to the degree that the animal is aware of its situation. Followingly, harm is only relevant for autonomously living, sentient animals (Hendriksen et al., 2010). Based on the assumption that early life stages and younger organisms experience less distress (Krzykwa et al., 2018), harm is not relevant for a cell or an early stage of an embryo.

An example of a distinction where harm is relevant can be illustrated by the production and use of genetically modified animals. The harm they inflict is primarily what is involved in the generation (superovulation, embryo transfer surgery, and vasectomy) and then any burden due to the genetic manipulation. The harm by the following maintenance breeding depends on whether the genetic alteration causes a harmful or nonharmful phenotype. A harmful phenotype (Zintzsch et al., 2017) might need extra protection or recourses to obtain good welfare.

Another example is the use of animals for terminal procedure or for use of organs and tissues after euthanasia. Use of animal in research only for harvesting organs or tissues is not even regulated in many countries. Killing animals for their organs and tissues is not considered a procedure by the Directive (Article 3.1), and thus is not subject to HBA according to the Directive (EuropeanCommission, 2010). However killing of animals is still an ethical concern, especially if not all parts of the animal is utilized for a scientific benefit (Bronstad and Sandoe, 2019). Programs for exchange and better utilization of animal tissues are encouraged, and the Directive in *Article 18* says:

> *Member States shall facilitate, where appropriate, the establishment of programmes for the sharing of organs and tissues of animals killed.*

Some initiatives like the Animatch (Animatch) and Animal and Tissue Share (Research Animal Resources) have been set up for exchange of organs and tissues from lab animals. Use of surplus animals from breeding for necessary practical training is another example to reduce the number of animals in science, education, and testing.

Harm to animals is usually discussed in the context of a harm-benefit analysis of evaluation of a particular project. The benefit puts the expected harm into a particular context and is the key for justifying an animal study (Pfister et al., 2012). The framework of acceptance has been described in the regulations of the Directive and typical requirements include application of the 3Rs and some judgment of the likelihood that the outcome of each project will contribute to the core scientific information that ultimately produces benefit (Bronstad et al., 2016).

The harm-benefit analysis provides an ethical framework and is an essential part of the ethical review (Bronstad et al., 2016) and is based on utilitarian consequence ethics i.e., "the proper course of an action is the one that maximizes utility" (Hepple et al., 2005; Rollin, 2013). A systematic approach to analysis of harm as suggested by the AALAS-FELASA Working group on harm—benefit analysis (Laber et al., 2016) is helpful to identify all "harms." Identification of harm factors should stimulate to seek alternative, less harmful approaches by applying the 3R principles (Russell and Burch, 1959), more animal-centric practices (Sørensen et al., 2020a), and more humane endpoints (Hendriksen et al., 2010). Together with an analysis of benefits, HBA questions and clarifies if all harms are necessary for achieving certain benefits.

The domain "animal" included every living species that is not a plant or a fungus; however, when the regulation refers to "animal," this usually means a vertebrate. The 2010 EU Directive also include cephalopods in the definition of an "animal" (EuropeanCommission, 2010) based on updated knowledge for this group of animals (Fiorito et al., 2015; Andrews et al., 2013; Cooke et al., 2019).

Avoidable and unavoidable suffering

Following up the discussion above, there is a distinction between avoidable and unavoidable harm. Despite initiatives to continuously refine experimental procedures to minimize suffering to experimental animals, some suffering or harm is still inevitable (Hendriksen et al., 2010). The Directive is clear that unnecessary pain, suffering, and distress must be avoided (European Commission, 2010). Pain, suffering, and distress that are not strictly necessary to achieve legitimate research objectives cannot be ethically justified when live, sentient animals are involved. However, as animals are often used as disease models or subjects for human safety, pain, suffering, and distress are implicit when justified because of scientific needs.

Going beyond what is necessary to achieve a scientific objective, however, is regarded as unethical, inhumane, and such suffering must be avoided. Unnecessary harm can be avoided or minimized by, for example, implementing earlier and more refined practices like animal-centric care and management (Sørensen et al., 2020a) or by defining more humane endpoints (Morton, 1998; Hendriksen et al., 2010). Severe clinical conditions usually follow a cascade of events. If an animal is too weak and not able to take water, it becomes dehydrated, causing haemoconcentration, increased blood viscosity with negative strain on cardiac function, and finally heart failure (Morton, 1998). Such cascades of events should be avoided by defining earlier end- and action-points (Hendriksen et al., 2010).

Animals in tumor studies usually experience tumor growth, as study of the tumor is the objective of the study. Large tumors might impair basic functions like food intake and grooming and cause discomfort or pain (Wallace, 2000). The disease might also cause complications because of spread to vital organs like lungs, liver, or brain (Helgers et al., 2020; Wallace, 2000). Diabetic animal models might experience increased thirst, increased urination, and need for increased cage changes because of increased urination. In vaccine development, especially the control animals will develop the most severe disease compare to the experimental group—assumed that the vaccine provides the desired protective effect (Vlach et al., 2000; Warn et al., 2003). Further harmful conditions might include things like hunger and thirst because powerless animals become too weak to take food and water. Suffering pain after invasive procedures or stress because of handling by untrained personnel are other sources of harm that should be considered for refinement. Sometimes animals are harmed because of unexpected events—for example, fighting wounds because of male mice fighting or other unexpected adverse effects of procedures. If these harms are not necessary, means should be used to avoid them of ethical, scientific, and legal reasons.

A project proposal for a study with animals should include a plan for how to act in all such events—and an upper limit for how much of these accumulated experiences that can be justified or claimed to be necessary. Old practices are unreliable—updated knowledge and technology must be taken into consideration when new series of studies are planned.

Direct, contingent, and cumulative suffering

Direct suffering is defined as suffering caused by procedures. A collection of examples of severity of different procedures has been defined by an expert working group by the European Commission (European Commission Expert Working Group on Severity Assessment, 2012).

For animal studies—all necessary procedures must be described. An assessment of the impact on the animals, as well as how negative impact on the animal can be minimized by (refinement) is a fundamental constituent of every study proposal. Severity assessment for an oncology study, experimental autoimmune encephalitis, arthritis, and stroke are included as examples in a document on severity assessment provided by the European Commission (European Commission Expert Working Group on Severity Assessment, 2012).

Contingent suffering refers to other suffering or distress implicit by being in capture and housed under experimental conditions not always meeting the animal's natural needs. It also covers transportation or by being handled by humans as a part of husbandry and care practices. A simple thing as being handled may cause stress especially in prey animals like rodents (Arras et al., 2007) and handling (Gouveia and Hurst, 2013; Ghosal et al., 2015; Gouveia and Hurst, 2017) and housing practices are targets for refinement in the same way as for experimental procedures. The Directive *(Article 4.3)* says:

> *Member States shall ensure refinement of breeding, accommodation and care, and of methods used in procedures, eliminating or reducing to the minimum any possible pain, suffering, distress or lasting harm to the animals.*

For wild animals that must be captured for studies or for fish that need to be taken out of water—their natural habitat to be handled—with the risk of serious damage—handling will significantly add to the overall severity for that animal (Hawkins et al., 2011).

Cumulative harm—relates to the whole series of factors through the whole life experience of the animal that cause negative impact on animal's well-being (Honess and Wolfensohn, 2010).

An overview of cumulative suffering in animal studies is summarizing all contingent interventions and procedures that the animal experience as a consequence of being a research animal in an "activity map" (Brønstad, 2011), or do a systematic harm assessment as suggested by Laber et al. (2016) including all interventions through the life of the animal.

Severity classification

Severity of animal studies—prospective estimate, reporting retrospective assessment

Severity limits can be described by the degree of deviation from a normal, healthy, naive animal (Morton, 1998). The EU Directive Annex VIII has defined three different severity categories based on the level of impact on the animal, i.e., mild, moderate, and severe (European Commission, 2010). In addition a fourth category, "non-recovery" has been defined for procedures which are performed entirely under general anaesthesia from which the animal shall not recover consciousness, but will be killed while it is under anaesthesia (European Commission, 2010).

There are several good reasons of having a severity classification system. First, to continuously focus on the application of the 3Rs when planning studies in animals, especially for severe studies. The severity system improves transparency on actual animal welfare costs and facilitates communication between those using, caring, and monitoring research animals building a common language. All sources of pain, suffering, and distress should be identified in the planning and design phase of the study and means to minimize negative effect on the animals (refinement) must be considered. Input on refinement from the Animal Welfare Body and Designated Veterinarian may be helpful to reduce severity by applying less aggravating alternatives. This planning phase should also uncover if there are training needs. There might be a need for specific training of those that will do the daily observation of the animals with focus on what to observe and how to report and respond. Specific assessment protocols or score sheets (Hendriksen et al., 2010) might be helpful in the monitoring of the animals. The more severe impact on the animals—the more frequent and rigorous monitoring of animals is required (Morton, 1998; Wallace, 2000). Factors to take into consideration when defining severity level include the procedure, the level of invasiveness, level of necessary restraint, duration of the procedure, whether treatment to reduce pain, suffering, or distress can be applied and to what level the animal's normal behavior is affected. This evaluation must also consider the species in questions, how well are they suited to the experimental conditions we provide them, and how do they cope with contact with humans. In addition, it must be evaluated if there is a need to repeat procedures during the study, will the animal learn and feel fear, and will the animal be allowed to rest and recover between procedures (European Commission, 2010). All these factors sum up to the cumulative severity for that animal.

Upper and lower limits

The severity classification in the Directive defines lower and upper limits for what is defined as a procedure, as well as an upper limit for what should never been authorized.

The lower level is defined as pain, suffering, distress, or lasting harm equivalent to, or higher than, that caused by the introduction of a needle in accordance with good veterinary practice—Article 3.1 (European Commission, 2010). Similarly, an upper limit—that should be prohibited—is defined as procedures that result in severe pain, suffering, or distress, which is likely to be long-lasting and that cannot be ameliorated (European Commission, 2010).

According to Hendriksen et al. (2010), the following observations reflects levels of upper limit of suffering that should not be exceeded.

- Inability to eat and/or drink
- Rapid or continuous weight loss
  - 20% during few days
  - 15% as a chronic loss over time

- Dehydration
- General decrease in grooming behavior
  - Recognized as dirty hair coat, red tears, and nasal orifices, soiled personal and peri oral regions especially in young animals
- Indications of pain, distress, or suffering
- Severe or continuing respiratory distress.

These conditions will lead to an unbearable (Olsson et al., 2020), life-threatening condition for the animal if no corrective actions are taken at an early stage (Toth, 2000).

Examples of the mild, moderate, and severe categories

A complete overview of procedures belonging to the different severity categories has been defined in Annex VIII of the EU Directive (European Commission, 2010). An expert working group provided examples on how to allocate animal procedures to the right category (European Commission Expert Working Group on Severity Assessment, 2012). All example provided are based on the assumption that best practices and refinement measures are strictly followed (European Commission Expert Working Group on Severity Assessment, 2012).

According to the Directive Annex VIII, "*mild severity*" is defined as procedures on animals as a result of which the animals are likely to experience short-term mild pain, suffering, or distress. That means procedures that cause no significant impairment on the well-being or general condition of the animals. Mild procedures also include noninvasive imaging of animals with appropriate sedation or anaesthesia; breeding of genetically altered animals which is expected to result in a phenotype with mild effects and short term (<24 h) restraint in metabolic cages have been defined as mild procedures.

Moderate procedures have been defined as procedures where the animals are likely to experience short-term moderate pain, suffering, or distress, or long-lasting mild pain, suffering, or distress. Examples include surgery that involves penetration of a body cavity under general anaesthesia and appropriate analgesia, associated with postsurgical risk of pain, suffering, or impairment of general condition. Moderate procedures also include breeding of genetically altered animals which are expected to result in a phenotype with moderate effect on animal welfare. The moderate severity class also includes creation of genetically altered animals through surgical procedure and use of metabolic cages involving moderate restriction of movement over a prolonged period (up to 5 days).

Severe procedures on animals are procedures likely cause severe pain, suffering, or distress, or long-lasting moderate pain, suffering, or distress. That includes experiments where death is the endpoint, fatalities are to be expected, and alarming pathophysiological states are induced. The severe procedures also include use of metabolic cages involving distressing restriction of movement over a prolonged period, inescapable electric shock (e.g., to produce learned helplessness), complete isolation for prolonged periods of social species, immobilization stress to induce conditions like gastric ulcers or cardiac failure in rats, or forced swim or exercise tests with exhaustion as the end point.

Also, if several mild procedures are performed in the same animal, this might cause the project to be categorized as moderate (a more severe category). The same principle applies for animals experiencing a series of moderate procedures—in such cases, the grand score for that project can be severe.

The examples listed here are not complete but can be used as guidance and comparison when studies are planned. Guidance documents that include examples that are relevant for aquatics (Hawkins et al., 2011) and cephalopods (Cooke et al., 2019) are available. There is also a publication and severity assessment for genetically altered mice (Zintzsch et al., 2017).

The fourth category that is classified by the Directive is the "*non-recover*" class. *Non-recovery* is used for studies where animals are anesthetized without any prior intervention or procedures and the whole study is completed while the animal is in anaesthesia and animal killed without recovering from anaesthesia. In this way, the animal will not have other experiences of procedures beside the injection of the needle or other method to induce anaesthesia.

Prospective severity assessment

Studies in animals need to be assigned a severity class prospectively *before* projects are authorized by the competent authority. The prospective severity described in a project proposal is a best guess based on the guiding documents, experience, and skills of the research team (European Commission Expert Working Group on Severity Assessment, 2009). This estimate of severity applies for the whole groups of animals in a study.

The actual severity experienced by each animal during the course of a procedure shall be reported annually after the study and reported in the statistical information made publicly available.

Reporting actual severity

The actual severity of procedures will be reported by Member States in the annual statistical returns. This reflects the highest severity experienced by the animal because of the procedure(s).

The actual severity does not necessarily equal the prospective severity, as experiments can show to be both less and more severe than anticipated. Experimental groups in testing a cure for a disease can advantage from that cure with less grave consequences, while control animals may have no such advantage, and the outcome might be more severe for them.

The actual severity also gives direction on criteria among others if an animal can be reused in another procedure as reuse depends on severity of the preceding procedure(s). Animals may only be reused provided that the severity of the previous procedure was "mild" or "moderate."

A common severity classification system promotes harmonization between European countries. Suffciently trained and competent staff are absolute requirements to assess animal welfare during the study, and to report actual severity correctly. An observational strategy and a common recording system should capture all the necessary data in a consistent format to be able to harmonize the severity category system across the European countries. A free, online training for the severity assessment has been available on the Education and Training Platform for Laboratory Animal Science (ETPLAS).

Retrospective severity assessment

The prospective severity classification has impact on requirement for retrospective assessment (RA). The retrospective severity assessment is helpful to identify improvements on the 3Rs as well as evaluating if the model is suitable to achieve expected objectives. All projects using nonhuman primates and projects involving procedures classified as "severe" shall undergo retrospective assessment (Article 39). The competent authority may decide that also other studies shall undergo retrospective assessment. This can be relevant for new animal models, pilots, newly created genetically altered animals, new test substances or classes of compounds, or animal use for education and training.

The retrospective assessment provides an opportunity to review if the outcome of the study met the defined objective(s) as well as evaluated if the estimated severity prospectively matches the actual severity. These experiences should be used for consideration of 3R alternatives and best practices for planning similar studies in the future. So, the retrospective assessment is a tool for an iterative process of learning and continuous improvement.

The retrospective assessment should be carried out as soon as the experiment is completed, and both the scientific and animal care and welfare team should be involved and share experiences from the study.

More information on retrospective assessment can be found in the publication on Project Evaluation and Retrospective Assessment from the European Commission (Evaluation and Retrospective Assessment, 2013).

Humane endpoints

Planning animal studies with early, less serious endpoints is an important step against less severe animal studies. The concept of humane endpoints was introduced in Europe by OECD (2000) and the European Directive Article 13.3 (European Commission, 2010) states that death shall be avoided as an endpoint and be replaced by earlier, humane endpoints.

"Alternative to death" reflects a very narrow definition of humane endpoints (Ashall and Millar, 2014), and it has been questioned if all earlier endpoints can be really considered "humane" (Franco et al., 2012). Other authors therefore propose a more broad definition of humane endpoint as a *concept for continuous refinement* of animal studies (Hendriksen et al., 2010). This latter broad definition will be basis for the rest of this chapter as it is better aligned with the 3R principles (Balls et al., 1995; Evaluation and Retrospective Assessment, 2013; Russell and Burch, 1959). Continuous improvements are embedded in the Directive (European Commission, 2010)—Article 4.3:

> *Member States shall ensure refinement of breeding, accommodation and care, and of methods used in procedures, eliminating or reducing to the minimum any possible pain, suffering, distress or lasting harm to the animals.*

What is a "humane endpoint"?

Humane endpoints are defined as clear, predictable, and irreversible criteria, which can be used as a substitute for a more severe experimental outcome (Mei et al., 2019; Hendriksen et al., 2010).

Studies to better understand diseases and hopefully develop better treatment, testing efficacy or safety of substances are the objectives for many studies in animals. The Directive Article 13 states that unnecessary pain, suffering, distress, or lasting harm must be avoided (European Commission, 2010). Potential or actual pain, suffering or distress that is regarded necessary to achieve a study objective must be justified based on a harm–benefit assessment as a part of project evaluation (Laber et al., 2016, Bronstad et al., 2016; European Commission, 2010). That means that the negative experience of the animal must be justified by a greater benefit by doing the study. Pain, suffering, and distress above what is required to achieve legitimate scientific objectives are regarded unnecessary and inhumane (Morton, 1998). Animal studies are authorized on the condition that pain, suffering, and distress should not exceed the necessary, justified level. If the actual negative impact on the animal passes the level predicted in the prospective severity assessment, defining the conditions for authorization of the study (Morton, 1998; Ashall and Millar, 2014) or when the condition for the animal reaches the level of unbearable (Olsson et al., 2020), authorities need to be notified and may set further conditions for continuing the study. Just as important—there must be plans in place to eliminate or minimize pain, suffering, or distress for the animals.

There are also scientific reasons for applying earlier endpoints. Chronic diseases gradually progress in the level of pain, suffering, and distress for the animal, it gradually become more severe (Fig. 8.1). Animals in advanced stages of disease are so physiologically deranged or disrupted that they do not

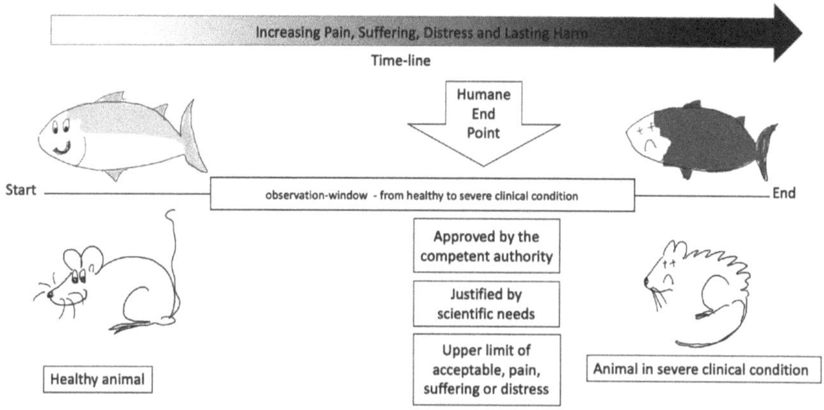

FIGURE 8.1 Chronic diseases gradually progress in the level of pain, suffering, and distress for the animal, it gradually become more severe. The window of possibilities for humane endpoints can theoretically be set at any time between the first detectable deviation from normality to the first indication of severe suffering. The level of necessary pain, suffering, or distress must be ethically justified and legally authorized be the competent authority.

provide reliable scientific information (Morton, 1998). Pain and distress might confound animal studies and interfere with scientific outcomes such as immunological or hormonal responses. Severe clinical symptoms reflect alterations in the normal physiology and reduced ability to adapt to maintain homeostasis. The sense of being self-threatened over time causes severe distress in the animal. The animal experiences pain, loss of control, fear or anxiety, or lack of social support (Olsson et al., 2020). The serious condition of moribund or dying reflects an irreversible situation with death as the only outcome (Toth, 2000). Autolytic changes in tissues take place as soon as death is imminent and if discovered by the next scheduled inspection, the carcass and tissues are already damaged or even cannibalized by cagemates. Timely termination by euthanasia has the advantage that investigative staff were better prepared for euthanasia and tissue collection (Collymore et al., 2018).

Besides the European Directive, other internationally recognized documents like the Guide for the Care and Use of Laboratory Animals address the importance of humane endpoints which better explains the scientific importance of applying early endpoints.

> *Studies that may result in severe or chronic pain or significant alterations in the animals' ability to maintain normal physiology, or adequately respond to stressors, should include descriptions of appropriate humane endpoints.*
>
> Garber et al. (2010).

Clinical symptoms or signs of poor welfare develop when the animal cannot any longer compensate for the progressing disease. This is especially critical for pray animals—like rodents and several species of fish. Prey animals will mobilize resources to compensate for and try to hide early signs of disease (Arras et al., 2007). Hiding weakness for a prey animal is important to avoid attention from potential predators as this might make their lives at risk. The implication is that the animal has already suffered from coping distress for a while, before staff are able to recognize the problem. The observation window from no clinical signs to obvious, severe clinical signs can therefore be narrow for many species. Several signs humans are able to observe during ordinary inspections—reflect serious conditions in many species, for example:

> *Sunken flanks, neglected grooming, piloerection, hunched back, immobility are clear evidence of severely impaired often moribund health status in mice.*
>
> Jirkof et al. (2010).

Therefore, we should search to define earlier predictors of pain, suffering, and distress and use them to define more humane endpoints.

The concept of humane endpoints is about setting earlier endpoints and follows up with mitigating actions to minimize, reduce, or eliminate unnecessary suffering is in accordance with the principle of refinement (Russell and Burch, 1959).

Death and near-death as an endpoint

Replacement of death as an endpoint is explicitly stated in the EU Directive Article 13 and has been translated into national regulations in European countries.

> *Death as the end-point of a procedure shall be avoided as far as possible and replaced by early and humane end-points.*
>
> European Commission (2010).

The Directive also states that if death is unavoidable as an endpoint, the study should be designed to result in the deaths of as few animals as possible.

The Canadian guideline (Olfert et al., 1998) has also included the severe state of dying—the moribund state as—a category that should be avoided as endpoint.

> *... moribund animal is one that is close to death and may be comatose or unresponsive to stimuli, exhibit dyspnea or other severe breathing problems, hypothermia, prostration, etc. However, before the animal gets to the point of being moribund, detailed observations of the animal can help to set an earlier endpoint and thereby reduce the actual cost to the animal, in terms of pain and/or distress.*

Of animal welfare reasons also, the moribund state should be avoided as an endpoint equally as avoiding of death (Franco et al., 2012; Toth, 2000). It is questioned if the moribund stage can be regarded as a "humane" endpoint—as it is very likely that the animals have been subjected to pain, suffering, and distress before it reached this stage and so it questioned if these near-death endpoints are really "humane" (Franco et al., 2012).

Define and apply appropriate humane endpoints; establish suitable criteria to identify when the humane endpoint has been reached

"Harm" to research animals was defined as compromised welfare based on the impact of the five freedom domains (Laber et al., 2016; Bronstad et al., 2016; Mellor and Reid, 1994; FAWC, 1992). Absence of good welfare can followingly be used to assess harm and used to define endpoints that should trigger follow up action points to restore better welfare. Such welfare indicators are often based on observations like deviating morphology, behavioral or physiological parameters (Hendriksen et al., 2010).

Endpoints can systematically be defined based on how they impair welfare indicators and be classified as either:

- Morphological welfare indicators
- Behavioral welfare indicators
- Physiological welfare indicators

Morphological welfare indicators must be based on knowledge about the specific species and life stage. Malformations that cause disabilities and problems to maintain the animal's normal functions and responses are relevant endpoint indicators. They are especially relevant in developmental studies and characterization of new genetically lines, where malformations may be an (unexpected) outcome. Examples include skeletal deformities (Wang et al., 1996; Martini et al., 2020) and hypoplasia of organs (Shawlot and Behringer, 1995; Min et al., 1998; Pichel et al., 1996). Malformation can cause critical vital malfunction such as food intake caused by jaw or teeth deformities. Changes in connective tissue may limit growth and development (Zeltz et al., 2014). Lip or cleft palate deformities (Oh et al., 2010) cause problems already in the suckling stage for a neonate mammal. Protocols suggesting systematic approaches to categorization of new phenotypes (Thon et al., 2010) and assessment of severity of genetic altered mice (Zintzsch et al., 2017) are available. Malformation of vertebra column in a fish causes extra strain with swimming both for search food and escape from threats (Martini et al., 2020; Noble et al., 2012) putting the animal under chronic coping stress. Other examples of morphological welfare indicators may include damage to skin or eyes, ulcers, emaciation, change in body condition, body shape, or fitness factor (Martini et al., 2020; Noble et al., 2012). The Fishwell project was made to define morphological welfare indicators in Atlantic salmon (Stien et al., 2020).

Several morphological welfare indicators are simple to register, and the optimum condition is usually well described. Others will demand more laboratory testing like CT/MR for examination of internal organs or necropsy of sentinel animals. Collaboration with experienced comparative pathologists is strongly recommended in defining morphological welfare indicators (Cardiff, 2009).

A main weakness of using morphological welfare indicators is that they are *retrospective*. That means that the condition that caused morphological deviation has happened back in time and the animal might have struggled with disabilities before evidence are detected by regular observation. Also, absence of morphological deviations is not a guaranty for good welfare and presence of a morphological deviation is not necessarily representing a welfare issue either—as long as the animal copes well under the conditions provided.

Behavioral welfare indicators are detected by observation of the animal. A significant advantage is that they often reflect responses in *real time*. Changes in behavior may be the first visible response to aversive conditions and an early warning. One example is response to daily husbandry practices like feeding. Animals fed at regular intervals will be hungry and come to be fed when the food is served. Observing the animals during daily husbandry practices gives valuable information of the physical condition—and such observations should be considered to define endpoints. Animals that do not respond as usual when the operator occurs by the cage, pen, tank, or net should be followed up for underlying cause so appropriate actions can be applied.

Nest building is a natural behavior in mice. Healthy mice are highly motivated to build nests when they have access to nesting material (Arras et al., 2007), and incomplete or lack of nest building in mice that have access to nesting material may be an indicator of pain, suffering, or distress and should be checked for underlying causes. Also reduced burrowing behavior in mice (Jirkof et al., 2010; Deacon, 2006; Zhang et al., 2020b) or the normal vertical activity (rearing) in rats (Matson et al., 2007) have been used to assess pain and distress in laboratory rodents. Other behaviors like hanging in the cage lid is a strongly motivated, though not essential, behavior in mice, and deviations from such behaviors can be used to assess pain and discomfort (Zhang et al., 2020a).

Not all behavioral welfare indicators reflect real-time problems. Stereotypic behavior can be a coping response to environmental distress that might have been undetected for some time (Latham, 2010; Mason and Rushen, 2006). In case of stereotypies in individual animals, attention should also be made to other animals in the same environment; they might also be suffering distress even if stereotypies are not manifested yet. Stereotypic behavior might also have underlying physiological causes (Garner, 2006)—and this might be of especially interest when defining endpoints for newly generated genetically altered animals.

The general activity is also a useful endpoint indicator. Healthy rodents show exploratory behavior rearing onto hind legs and sniffing. Activity can be reduced because the animals are weakened, exhausted, or in pain. Increased activity might be a response to irritation, stress, or a treat that upset the animals. Lethargy (apathy, drowsiness) is characterized by reduced sensitivity to stimuli, often occurring in the final stage of illness—a situation called moribund. Willingness to move only when stimulated will be an earlier endpoint than no willingness to move when stimulated—but still this will reflect an alarming situation for the animal.

Self-grooming is important for many species and one of the most frequently performed behavioral activities in rodents using up to 25%–40% of the awake time budget (Spruijt et al., 1988). Rodent fur should be smooth and dense and a ruffled fur is therefore an important early indicator of poor welfare (Baumans, 2010; Koolhaas, 2010). Other early indicators of pain or distress in rodents include reduced food intake and isolation from group mates (Baumans, 2010; Koolhaas, 2010).

When resting rodents are curled up—and when housed in groups, they group together to better maintain preferred temperature and avoid thermal stress (Gaskill and Garner, 2017). Finding a rodent alone—laying stretched out—is an indicator of impaired welfare.

Observing swimming pattern in fish gives valuable information of the physical condition of the fish, however there are big differences between species and life stages. Zebrafish typically swim in the middle of the tank and swimming at the tank bottom is interpreted as an indicator of pain (Deakin et al., 2019). Salmon Parr are territorial, live in the demersal zone, and show a cross-current swimming pattern in the rivers (Huntingford et al., 2012). During the Parr–Smolt transformation, they reduce cross-current swimming and show tendency to move downstream against sea water (Huntingford et al., 2012). In the lab in the tank, salmon smolt typically swim in the same direction in a coordinated manner (they are schooling). Loss of equilibrium, position in tank or water, and gill movements (hyperventilation, hypoventilation) are useful welfare indicators in fish. Flared gills occur when fish have trouble breathing caused by low O_2 level in water or infection on the gills. Such conditions must immediately be followed up by identifying the cause and take corrective actions.

For rats (Roughan and Flecknell, 2006), mice (Wright-Williams et al., 2007), rabbits (Leach et al., 2009), and zebrafish (Sneddon, 2019) specific pain behaviors have been described. Striking painful body part against surface has been used as an indicator of pain in rats (Roughan and Flecknell, 2006) and rabbits (Leach et al., 2009) and behavior changes caused by noxious stimuli have been demonstrated in several aquatic species including cephalopods (Crook et al., 2014; Crook and Walters, 2011), hermit crabs (Magee and Elwood, 2016), and several species of fish including rainbow trout (Sneddon et al., 2003) and zebra fish (Thomson et al., 2020; Deakin et al., 2019). Pain assessment by facial grimace scoring has been validated and showed to be a reliable method for pain assessment in several species (Langford et al., 2010; Keating et al., 2012; Sotocinal et al., 2011; Hageri et al., 2017; Viscardi et al., 2017), and some traits seems to be conserved across species.

Observations by humans may—depending on species and the adaption to human presence or interactions—cause observation bias. Some animals modulate behavior to mask symptoms. This is especially a problem in prey animals (Arras et al., 2007) as predation is a strong selective force in evolution. Most species maintain a set of natural defense mechanisms (instincts) even when they are raised in captivity protected from natural predators. In some cases, natural defense mechanisms may be modulated by use of training, like with positive reinforcement or by more animal-centric approaches to care and management (Sørensen et al., 2020a). Some animals avoid predation by occupying areas or habitats that are not suitable for their predators or they are active in times where predators are not. For mice—the nest function as a hiding place and offering the animal a better opportunity to self-environmental control reducing the stress in the animal. For animals in other habitats like fish in water, observation and evaluation of individual animals can be difficult as animals are not easily to access. Panic response in fish reflects that the fish are stressed. It can be poor water quality, fear of a predator presence, or just the animal caretaker dressing up in other colors than they are used to.

Reversed activity cycles must also be taken into consideration in evaluation of behavioral welfare indicators, as nocturnal animal will not show their full behavioral repertoire during day time (Hendriksen et al., 2010). There is a risk of missing important behavioral cues of impaired welfare when animals are observed in their normal resting time. An enriched environment, providing remedies necessary to meet animals' natural needs will be necessary to be able to detect abnormal behavior.

By use of technology and remote recording like telemetry (Arras et al., 2007; Vlach et al., 2000; Kramer and Kinter, 2003), GPS registrations and similar automated technology for (Laure Bégout et al., 2012; Zhang et al., 2020a; Deakin et al., 2019; Roughan et al., 2009; Sotocinal et al., 2011; Wright-Williams et al., 2013) behavior observation can be possible without direct interfering with the animal.

A main weakness with behavioral welfare indicators is that interpretation of deviant behavior is not always clear. The use of deviant behavior assumes that all personnel (researchers, technicians, veterinarians, and care staff) are equally able to recognize behavioral deviations as well that all animals respond in a similar manner (Hendriksen et al., 2010). Basic knowledge of natural behavior for the species and life stage is a fundamental condition for using behavioral welfare criteria as endpoints.

In evaluation of animal behavior, the minimum the researcher should be able to answer are.

- How do the species naturally move?
- How do they naturally feed, what do they eat and how much?
- How do the species natural respond to disturbances, potential dangers, or interactions with conspecific as well as humans?

Modified from Jobling, M., 2012. Fish in aquaculture environments. In: Huntingford, F., Jobling, M., Kadri, S.L. (Eds.), Aquaculture and Behavior, Chichester, West Sussex Ames, Iowa, Wiley (Jobling, 2012).

Assessment points that form the basis for evaluating adverse effects should include:

- Appearance
- Posture
- Spontaneous behavior
- Provoked behavior

Modified from Hendriksen, C., Morton, D., Cussler, K., 2010. Use of humane endpoints to minimise suffering. In: Howard, B., Nevalainen, T., Perreta, G. (Eds.), The COST Manual of Laboratory Animal Care and Use, CRC Press Taylor & Francis.

Physiological welfare indicators and surrogate endpoints: Physiological welfare indicators and surrogate endpoints include clinical observation, like breathing pattern, hearth frequency or results from laboratory test of blood, hormone levels, saliva, urine, or other samples. They have potential as early indicators or disease markers (surrogate endpoints) predicting a cascade of events leading to a severe condition. Many of them are sensitive, have validated methods and range so comparison with the normal condition is reliable. "Surrogate endpoints" (Morton, 1998) are used to predict an outcome to early terminate a study before the animal develops severe distress (Workman et al., 2010). Examples of physiological parameters as surrogate endpoints can be a specific blood pressure measurement that predicts death, a defined tumor volume that consistently precedes painful ulceration (Workman et al., 2010) or blood glucose used as prelethal surrogates (Morton, 1998).

Subjective evaluation might be a weakness with many clinical observations. The fact that many of the physiological endpoints are quantifiable has the advantage that they are *less vulnerable for subjective evaluation* (Hendriksen et al., 2010, Hendriksen, 2011).

A main challenge is the collection of endpoint data. It often involves manual handling, restrain, and sampling. Such procedures which can be a burden for many species, for example, for fish that must be taken out of water or for animals that are not adapted to human interaction or handling. Training and habituation can be used for many species like dogs, NHP, and pigs to reduce the stress related to handling and sampling (Sørensen et al., 2020b).

New technological solutions may provide alternative less intrusive solutions for data collection. Surface hypothermia using infrared noncontact thermometers has been used as a predictor for death in infectious studies in mice (Warn et al., 2003). Reduced body temperature measured by telemetry (Vlach et al., 2000) or by infrared measurement (Warn et al., 2003) used as an early endpoint for sepsis in studies of infectious disease among others.

Method, frequency, and volume of samples must be taken into consideration in study design—as this might affect the level of stress and physiology of the animal (Workman et al., 2010).

Morphological, behavioral, or physiological endpoints

Morphological, behavioral, and physiological endpoints as described in this section have their strength and weaknesses regarding evaluating impact on welfare or emerging pain, suffering, or distress. However, by combining more of them using a triangulation approach, like combining body weight and behavioral change like burrowing, (Zhang et al., 2020b), or body weight/ breathing patterns/activity (Collymore et al., 2018), they can provide useful indicators for early humane endpoints and be a game changer in defining earlier and more humane endpoints.

Generic, specific, or unexpected events

Another systematic way to asses endpoints is to classify as generic endpoints, project specific endpoints, or endpoints for unexpected events (Ashall and Millar, 2014).

Generic endpoint parameters reflect unspecific welfare issues (morphologic, behavioral, or physiological) like loss of appetite, lack of self-grooming (Garner, 2006), activity level, change in body weight or body condition (Easterly et al., 2001), indicators of pain, distress, or poor welfare. They give an indication that the animal is not doing well, but do not specifically relate to a study specific condition. Lack of highly motivated species specific behavior like burrowing (Jirkof et al., 2010; Deacon, 2006; Zhang et al., 2020b), nest building, or cage lid hanging (Zhang et al., 2020a) in mice can be used to determine earlier endpoint indicators in studies as generic endpoints. Generic endpoints are relevant for all studies of gradually progressing diseases.

Using weight loss as an endpoint parameter

Weight loss, typically 10%–20% is a commonly used generic endpoint parameter. There are many good reasons for using weight loss as an endpoint. It is not a risk of subjective bias, and it is also easy to measure in land living animals also without too much restrain. Some species can even be trained to walk voluntary on the weight (Sørensen et al., 2020a). Other species like aquatics, however, weighing is not easily performed without handling and bringing the animals out of its natural habitat that will be stressful.

One criticism is that weight loss as an isolated parameter is not very sensitive for the animal's actual condition. Maintaining body weight or weight gain can be caused by conditions like tumor growth or accumulation of fluid in the peritoneal cavity (ascites) (Wallace, 2000). Stable weight in an animal in a growth phase is a poor sign as they are expected to increase weight according to the weight-gain curve. Weight gain can also be a problem as obesity increases the risk of other disease, complicates self-grooming, and increases the load on the limbs and cardiac function. Also, obese animals take up more space and there might be a need to split groups in more cages or pens unless this is accounted for on from the start of the study already. Overweight is especially a problem in long-term studies. Also animals with a significant loss of body weight might be lively and have a good quality of life (Morton, 1998).

Body weight should be evaluated together with body score evaluation (Ullman-Cullere and Foltz, 1999; Easterly et al., 2001). The body score in mice is usually defined in five categories from emaciated (score 1) to obese (score 5) and each category is well defined. However, there is a risk of subjective evaluation so body score should be combined with body weight (objective) or other objective indicators. Body condition scoring can be used more objectively by micro-CT imaging (Easterly et al., 2001). Studies relying exclusive on weight loss as the only criterion for euthanasia as an endpoint might result in an unnecessary loss of animals (Talbot et al., 2020).

Weight loss is reflecting a "problem in the past," i.e., the animal might have reduced appetite or been under a catabolic process already for a while before we are able to detect it. Another complementary endpoint parameter could therefore be to measure food intake as an indication of appetite in real time. However reduced food intake of preferred food did not seem to be a useful measure in guinea pigs (Collymore et al., 2018).

Project specific endpoints are endpoints that reflect disease progression of the phenomenon of interest in the study. For a cancer study, tumor size, number of tumors, metastases, tumor's location and interaction with other functions (mobility, food intake) can be relevant project-specific endpoints. For surgery—wound complications like redness, swelling, infection, and broken sutures are relevant and should be included as potential endpoints with defined action points for follow up. See example in Table 8.2. Preclinical screening of ALS mice showed that loss of motoric function was used as an early sensitive and rapid indicator for the initial phase of denervation of muscle fibers (Mead et al., 2011) and reduced activity in the home cage running wheel as an early diagnostic sign (Bennett et al., 2014). For diabetes studies, examples can include blood glucose, urination, cage changes need (because of increased urination), or water consumption.

Project-specific endpoints should be defined for each particular study. Any indicator used as surrogates for death must be defined for each study and one should not be tempted to reuse endpoints from other random studies or blindly rely on global endpoints (Nemzek et al., 2004).

An overview of indicators for study-specific endpoints and alternatives to death and other severe endpoints is summarized in Table 8.1.

Endpoints for unexpected events relate to pain, suffering, or distress caused by other unrelated illness, accidents, or unexpected adverse effects of the study that necessitate human intervention to avoid continued unnecessary and unjustified pain, suffering, or distress (Ashall and Millar, 2014; Hendriksen et al., 2010; Morton, 1998). Unwanted events might be fighting and fighting ulcers that an animal is hurt under procedures or husbandry practices. It might also be that severity exceeds the level anticipated at the start of the experiment

TABLE 8.1 Refinement of study-specific endpoints and alternatives to death. This table shows examples of study-specific endpoints and alternatives to death and other severe endpoints as an important pathway to refinement of animal studies.

Study objectives	Endpoint	Reference
Infections, immune system		
Infectious diseases	Acute phase responses Body temperature Weight loss	Olfert and Godson (2000)
Rheumatoid arthritis in rodents	Swollen digits, knuckle, midfoot, and ankle/wrist scoring, ulceration, gait and posture analysis	Hawkins et al. (2015)
Shock by cecal ligations and puncture (CLP)	Loss of ability to ambulate Body weight and temperature	Nemzek et al. (2004)
Tuberculosis vaccine studies	Bodyweight, breathing pattern, and activity during handling combined predict need for euthanasia one week ahead	Collymore et al. (2018)
Bacterial virulence	Hypotermiadose50 (HD50) replaces LD50	Soothill et al. (1992)
Sepsis	Surface temperature and sickness score at 24 h after the first injection by machine learning	Mei et al. (2019)
Inoculation with infectious organisms	Temperature and body weight	Trammell and Toth (2011)
Cancer, tumor studies		
Intracranial glioma in rat	Body weight algorithm	Helgers et al. (2020)
Experimental liver metastases	Laparoscopy	Kobaek-Larsen et al. (2004b)
Colon cancer in rats	Black feces or presence of fresh blood	Kobaek-Larsen et al. (2004a)
Early cancer detection	Ultrasound	Wang (2003)
Severe irradiations studies in NHP	Rapid decrease of body temperature for three consecutive days combined with general behavior score	Bertho et al. (2020)

Continued

TABLE 8.1 Refinement of study-specific endpoints and alternatives to death. This table shows examples of study-specific endpoints and alternatives to death and other severe endpoints as an important pathway to refinement of animal studies.—cont'd

Infections, immune system		
Study objectives	**Endpoint**	**Reference**
Total body irradiation in mouse	Body weight, temperature by telemetry	Koch et al. (2016)
Multiple body systems		
Study objectives	**Endpoint**	**Reference**
Murine model for cholestasis	Burrowing activity	Zhang et al. (2020b)
Antiurolithiasis activity of test compounds by zinc disc implantation	Imaging (X-ray radiographs) of the bladder deposits	Singh et al. (2010)
Ocular herpes virus infection	Combined body weight reduction (>0.05 g/day) and body temperature less than 34.5°C measured by SC implanted temperature probe. Predict death in 98% of mice.	Hankenson et al. (2013)
Central Nervous System (CNS)		
Amyotrophic lateral sclerosis (ALS)	Decline in motoric function	Mead et al. (2011)
	Home cage running wheel	Bennett et al. (2014)
Stroke	Weight change on the first and third day after treatment. Core temperature change on the third day after treatment (male, stroke model), or with neuroscore, weight change, and core temperature change on the second day after treatment (female, stroke model) by machine learning.	Mei et al. (2019)
Toxicity		
Toxicity testing	Heart rate, snout–vent length, eye size, and pericardial area	Krzykwa et al. (2018)
Surgery		
Postlaparotomy	Telemetric recording of heart rate	Arras et al. (2007)

TABLE 8.2 Defining scores, grades, and actions points. Example of evaluation of surgical wound.

	Clinical evaluation	Score	Action
Wound	Closed wounds, no erythema	0	New observation next day
	Slight erythema around wound, no edema/swelling	3	Frequent observation two times per day
	Moderate erythema, edema	6	Consider painkiller or antibiotics based on veterinary recommendation
	Severe erythema, swelling/ edema, open wound	8	Provide painkiller, conservative wound care, or reoperation of wound
	No effect of treatment	12	Humane killing of animal

(Hendriksen et al., 2010) like more serious adverse effects of a test substance than expected.

Unexpected events might include disasters like fire or flood, earthquake, technical breakdown, power supply in the building, contagious disease outbreak, a pandemic situation, or others.

The guide says (Garber et al., 2010):

> *Well-planned experiments with clearly delineated scientific and humane endpoints will help to ensure that a contingency plan is in place for problems that may arise during the study.*

Depending on the level of impact, this event might or may not influence the study, and it must be made a case-by-case decision if the study can continue or if it should be terminated. Continuation of the study may be on the cost of animal harm and reliability of the study results. Termination and restarting the experiment will be on the cost of animal lives.

Actions to be taken when an endpoint is reached and consider possible options for refining methods to finish at an earlier endpoint

When an animal reaches a certain defined endpoint, it must be followed up by actions to reduce or eliminate pain, suffering distress or lasting harm. "Action points" describe the actions we are going to take to mitigate the condition and avoid further pain, suffering, distress, or lasting harm when or if an animal reaches certain stages—or endpoints—in a study.

In many cases, for many species, this will be to kill the animal to avoid further suffering. However, with earlier endpoint, other alternative action points should also be considered, as said in the Directive (European Commission, 2010) on care an accommodation:

Article 33 (d) arrangements are made to ensure that any defect or avoidable pain, suffering, distress or lasting harm discovered is eliminated as quickly as possible.

That could include medical treatment to relieve pain, nausea, correct blood glucose, and fluid therapy etc. Other action points can include more frequent checks, providing soft or alternative food, provide enrichment (for example, for behavioral problems or fighting) or consultation by the veterinarian. The animal care staff and the designated veterinarian should be involved in planning endpoint and action points as they can provide useful information (Nemzek et al., 2004). In all cases, the study director must consider whether these actions might bias the study in an unfortunate manner.

Only when the condition cannot be ameliorated, the animal must be killed as the end of a procedure as a final endpoint.

An animal shall be killed when it is likely to remain in moderate or severe pain, suffering, distress or lasting harm (Article 17.2)

Score sheets—clinical assessment and follow-up action points

Score sheets—endpoints and action points

A score sheet ("observation sheets" or "welfare assessment sheets") is a protocol for systematically recording of key clinical observations and other test results of the animals (Hendriksen et al., 2010; Morton, 1998). These observation and test results show to which degree the animals physiologically or mental state deviate from the normal and are used to determine when the animal reaches predefined endpoints that trigger certain actions. The score sheet helps to identify when the condition for an animal exceeds, what has been defined and approved (by the competent authority) as the humane endpoint in the light of a harm–benefit analysis (the harm to the animals must be balanced against potential benefits):

... a harm-benefit analysis of the project, to assess whether the harm to the animals in terms of suffering, pain and distress is justified by the expected outcome (article 28).

European Commission (2010).

... the IACUC is obliged to weigh the objectives of the study against potential animal welfare concerns.

Garber et al. (2010).

A score sheet can also be used to document when the condition for an animal exceeds, what has been defined and approved by the competent authority as the humane endpoint.

When defining scores for clinical signs, they are typically categorized in distinct categories from 0—or no impact to higher numbers depending on how severe the situation is for the animal and the highest score for the most severe condition—often that cannot be relieved or scientifically justified.

The facial grimace score (FGS) for pain (Sotocinal et al., 2011; Langford et al., 2010), for example, a score 0 is no signs of pain using the FGS, a score 1 is reflecting moderately present, and score 2 an obviously present sign of pain.

I theory any of morphological, behavioral, or physiological welfare indicators can be used to define endpoints and action points in a score sheet. Some might be sensitive to subjective evaluation by the observer unless very clear, strict categories are defined. It is advisable to consider the relevance for the particular study, how feasible it is to collect data during regular observation, and to what degree can we define them in an objective manner to avoid observation bias. Ambiguous score sheets including many observation points may look impressive—but there is also a risk they are completely impractical in use. Score sheets should therefore be regularly revised to include a limited number of relevant parameters (Nemzek et al., 2004).

Binary scoring

An alternative to score observations in numerous categories of increasing seriousness, as for the FGS (Langford et al., 2010; Sotocinal et al., 2011) using 0, 1, and 2—altogether three categories—is to use binary scoring. Observation points are not scored and categorized of increasing seriousness—but they are scored as either present or not present. For example, you observe:

- Indication of pain or no indiaction of pain as +/−, 1/0 or y/n
- Normal behaviour or abnormal behavior as +/−, 1/0 or y/n
- Normal food intake or changed food intake as +/−, 1/0 or y/n

The convention is that minus (or 0 or "n") indicate the normal situation while + (or 1 or "y") indicate that the animal is outside the normal range (Morton, 1998).

Binary scoring represents a simplification that may reduce risk of misinterpretation or subjective evaluation—which can be the case when too many poorly defined score categories are used. Binary scoring is therefore less sensitive to observer bias. The principal investigator must, define how many 1/+ that defines actions to follow up and if any particular of them elicits immediate actions.

Many or few endpoint indicators in the score sheet?

Some authors stress that endpoints can rarely be generalized (Morton, 1998). Using too many and too general endpoints for scoring is labor intensive.

Researchers might be required to evaluate variables that might appear arbitrary or unrelated to the animals condition (Toth, 2000). Compliance may be increased if the measurement of the variables are related to research objectives (Toth, 2000). Other authors underline that humane endpoint cannot rely on a single variable, suggesting to use a combination of variables (Bertho et al., 2020; Zhang et al., 2020b) and recommend a combination of assessment parameters for a robust welfare assessment in laboratory animals (Honess and Wolfensohn, 2010) and early predictor of death (Zhang et al., 2020b).

Refinement of clinical assessment criteria and score sheets

Score sheets should not be a static "copy-and-paste" document from study to study—but should be constantly developed and updated with further experience (Morton, 1998). Hendriksen et al. (2010) propose that endpoints should be regularly reassessed as a refinement strategy to improve welfare for research animals and tailor-made them to the specific study.

Conditions that have been considered crucial for validity of humane endpoints are a low frequency of both false positive and false negative predictions (Hendriksen, 2011). False positive means aborting an animal from a study (by killing) while the animal would have survived the observation period. False negative endpoints mean that the animal already died before it reached the predefined endpoint (Hendriksen, 2011). False positive endpoints invalidate the potency of the study, while false negative endpoints compromise animal welfare (Hendriksen, 2011).

Refinement of clinical assessment criteria should also include analysis of the performance of each single as well as combinations of multiple parameters. For example, body weight reduction and reduced burrowing (Zhang et al., 2020b) or body weight and body temperature (Hankenson et al., 2013; Trammell and Toth, 2011) combined (e.g., the product of two parameters) of more than one physiological parameter to derive a surrogate indicator for a higher risk of death showed to be a better predictor of death than each one of them alone.

Machine learning, a technique used to identify underlying patterns from given datasets, to produce reliable and repeatable predictions, has been used to refine endpoints in mice (Mei et al., 2019). By analysis of large datasets, it is possible to separate animals that are at a higher risk of death based on surrogate indicators and even identify sex differences for that risk (Mei et al., 2019). Machine learning can also be used to set up alarms and alerts to the person responsible for monitoring the animals, and thereby reduce the risk of failure to respond in time.

Score sheets should be regularly reassessed so that criteria that are never observed at all should be excluded (Zhang et al., 2020b). Also, criteria that called for immediate euthanasia should be removed from the list of humane endpoints (Zhang et al., 2020b; Mei et al., 2019)—as it might be questioned if they are really "humane" (Franco et al., 2012). Rather criteria that call for immediate euthanasia should be listed as "emergency procedures for severe conditions and unwanted incidents." Animals dying from unexpected incidents should not contribute to refined endpoints criteria in future studies as their endpoint is irrelevant as a predictor for study-specific endpoints (Mei et al., 2019).

Control animals

When defining endpoints for studies, special consideration must be made to control animals. Especially negative controls (not receiving the drug candidate of interest) will not have the advantage of a potential treatment of protection effect of the test substance. These animals will therefore likely be the first to reach the endpoint state—assumed that the test substance is effective. This applies for both the generic and project specific endpoints.

Unregulated species and life stages

Regulations and guidelines define borders of which animal species and life stages that are included. Legally, humane endpoints are relevant for studies in these species and life stages. However, the regulations are based on historical knowledge and new knowledge about sentience of species and life stages is evolving. The inclusion and protection of cephalopods in the 2010 EU Directive (European Commission, 2010) is an example of a new group of nonvertebrate animals that become protected based on scientific knowledge. So, attention also to potential pain, suffering, and distress for early life stages, like fish larva, before they start-feed, or drosophila may be worth considering—even if they are not currently covered by regulations.

How humane is your endpoint?

Humane endpoint is a professional concept that is defined both in regulations as well as in literature and guidelines for animal research (European Commission, 2010; Garber et al., 2010; Hendriksen et al., 2010; Ashall and Millar, 2014). The most widely used, but also most narrow definition, as by the Directive (European Commission, 2010) and OECD guidelines (OECD, 2000) promotes humane endpoints as an *alternative to death* (Ashall and Millar, 2014). However humane endpoints as defined by Hendriksen et al. (2010)—are a *strategy to minimize pain, suffering, and distress* in animal studies. Humane endpoints together with proper pain relief and animal-centric practices for handling, are all keys *to design less inhumane studies* in accordance with the 3Rs in Russell and Burch's—"The principles of humane experimental technique" (Russell and Burch, 1959).

The term "humane" reflects an altruistic position identified with kindness, care, compassion, or sympathy for others—especially humans but also animals. The term "humane" is also used in connection with humane killing of animals—which in veterinary practice means mercy killing performed by specially trained personnel, generally to prevent suffering from incurable or painful conditions (Hatch, 1988).

"Humane" endpoints can similarly be perceived as studies identified with end- and action-points assuring kindness, care, compassion, or sympathy for the animals. The much-used statement "unnecessary pain, suffering, and distress" cause cognitive dissonance in the context of humane endpoints—especially among nonprofessional, lay persons. Pain, suffering, or distress might be approved if necessary, to reach certain scientific objectives, justified by potential legitimate benefits and when there are no alternatives. However, accepting pain, suffering, or distress—even when justified—is contradictory to beliefs, ideas, or values of common people's understanding of "humane." How "humane" endpoints are—is not a fixed dimension, but relative to the context, study objective and what alternatives that are available or that have been considered. If brining in the "humane" dimension causes confusion and contributes to blurring communication on how animals are used in research, it might be suggested to only refer to "endpoints"—leaving out the humanity. However, to use "humane" to describe the process of continuously refining to aim against "more humane" practices in animal studies still makes sense.

The Directive's definition applies for avoiding death as an endpoint (European Commission, 2010), but does not explicitly apply for the moribund state. However, the CCAC guideline says:

> *... moribund animal is one that is close to death and may be comatose or unresponsive to stimuli, exhibit dyspnea or other severe breathing problems, hypothermia, prostration, etc. However, before the animal gets to the point of being moribund, detailed observations of the animal can help to set an earlier endpoint and thereby reduce the actual cost to the animal, in terms of pain and/ or distress. CCAC.*
>
> Olfert et al. (1998).

It is therefore questioned if endpoints—like moribund that involve a serious and threatening status for the animal can be regarded as "humane".

The Guide for the care and use of laboratory animals better link the experimental and humane endpoints and better connect it to refinement (Garber et al., 2010).

> *The experimental endpoint of a study occurs when the scientific aims and objectives have been reached. The humane endpoint is the point at which pain or distress in an experimental animal is prevented, terminated, or relieved. The use of humane endpoints contributes to refinement by providing an alternative to experimental endpoints that result in unrelieved or severe animal pain and distress, including death. For many invasive experiments, the experimental and humane endpoints are closely linked.*

How humane is my endpoints?—Key check question

- Will my endpoint be perceived as humane by the common public?
- Will the aim of the study be perceived as a legitimate justification for the endpoints defined?
- Do I have room for further refinement of my study so I can define more humane endpoints?

References

Andrews, P.L.R., Darmaillacq, A.S., Dennison, N., Gleadall, I.G., Hawkins, P., Messenger, J.B., Osorio, D., Smith, V.J., Smith, J.A., 2013. The identification and management of pain, suffering and distress in cephalopods, including anaesthesia, analgesia and humane killing. J. Exp. Mar. Biol. Ecol. 447, 46–64.

ANIMATCH. https://www.animatch.eu. (Accessed 10 February 2021).

Arras, M., Rettich, A., Cinelli, P., Kasermann, H.P., Burki, K., 2007. Assessment of post-laparotomy pain in laboratory mice by telemetric recording of heart rate and heart rate variability. BMC Vet. Res. 3, 16.

Ashall, V., Millar, K., 2014. Endpoint matrix: a conceptual tool to promote consideration of the multiple dimensions of humane endpoints. ALTEX 31, 209–213.

Balls, M., Goldberg, A.M., Fentem, J.H., Broadhead, C.L., Burch, R.L., Festing, M.F., Frazier, J.M., Hendriksen, C.F., Jennings, M., Van Der Kamp, M.D., Morton, D.B., Rowan, A.N., Russell, C., Russell, W.M., Spielmann, H., Stephens, M.L., Stokes, W.S., Straughan, D.W., Yager, J.D., Zurlo, J., VAN Zutphen, B.F., 1995. The Three Rs: The Way Forward: the Report and Recommendations of ECVAM Workshop 11.

Bateson, P., 1986. When to experiment on animals. New Sci. 109, 30–32.

Baumans, V., 2010. The laboratory mouse. In: Hubrecht, R., Kirkwood, J.K. (Eds.), The UFAW Handbook on the Care and Management of Laboratory and Other Research Animals, eighth ed. Chichester: Wiley-Blackwell. edited for UFAW.

Bennett, E.J., Mead, R.J., Azzouz, M., Shaw, P.J., Grierson, A.J., 2014. Early detection of motor dysfunction in the SOD1G93A mouse model of Amyotrophic Lateral Sclerosis (ALS) using home cage running wheels. PLoS One 9, e107918.

Bertho, J.M., Ancelet, S., Denais-Lalieve, D., Dudoignon, N., 2020. Humane endpoints in severe irradiation experiments using non-human primates: a retrospective analysis. Scand. J. Lab. Anim. Sci. 46, 1–15.

Brønstad, A., 2011. Activity maps – a tool for evaluation of accumulated harm to animals. In: Bergen, U.O. (Ed.), Poster and Abstarct 8th World Congress on Alternatives and Animal Use in the Life Sciences.

Bronstad, A., Sandoe, P., 2019. Examining compliance with ethical standards for animal research: is there a need for refinement? A qualitative study from northern Europe. Lab. Anim. 23677219841080.

Bronstad, A., Newcomer, C.E., Decelle, T., Everitt, J.I., Guillen, J., Laber, K., 2016. Current concepts of harm-benefit analysis of animal experiments - report from the AALAS-FELASA working group on harm-benefit analysis - Part 1. Lab. Anim. 50, 1–20.

Buning TDE, C., 2004. Practical difficulties in balancing harms and benefits in the modern use of laboratory animals: biotechnology. Altern. Lab. Anim. 32 (Suppl. 1B), 459–463.

Cardiff, R.D., 2009. How to phenotype a mouse. Dis. Model. Mech. 2, 317–321.

Cooke, G.M., Anderson, D.B., Begout, M.L., Dennison, N., Osorio, D., Tonkins, B., Kristiansen, T., Fiorito, G., Galligioni, V., Ponte, G., Andrews, P.L., 2019. Prospective severity classification of scientific procedures in cephalopods: report of a COST FA1301 Working Group survey. Lab. Anim. 53, 541–563.

Collymore, C., Kent, L., Ahn, S.K., Xu, W.X., Li, M., Liu, J., Turner, P.V., Banks, E.K., 2018. Humane endpoints for Guinea pigs used for *Mycobacterium tuberculosis* vaccine research. Comp. Med. 68, 41–47.

Crook, R.J., Walters, E.T., 2011. Nociceptive behavior and physiology of Molluscs: animal welfare implications. ILAR J. 52, 185–195.

Crook, R.J., Dickson, K., Hanlon, R.T., Walters, E.T., 2014. Nociceptive sensitization reduces predation risk. Curr. Biol. 24, 1121–1125.

Deacon, R.M.J., 2006. Burrowing in rodents: a sensitive method for detecting behavioral dysfunction. Nat. Protoc. 1, 118–121.

Deakin, A.G., Buckley, J., ALZU'bi, H.S., Cossins, A.R., Spencer, J.W., AL'nuaimy, W., Young, I.S., Thomson, J.S., Sneddon, L.U., 2019. Automated monitoring of behaviour in zebrafish after invasive procedures. Sci. Rep. 9.

Easterly, M.E., Foltz, C.J., Paulus, M.J., 2001. Body condition scoring: comparing newly trained scorers and micro-computed tomography imaging. Lab. Anim. (NY) 30, 46–49.

ETPLAS, Education and Training Platform for Laboratory Animal Science. Education [Online]. https://etplas.eu/learn/eu-12-the-severity-assessment-framwork/. (Accessed 10 February 2021).

European Commission, 2010. In: Directive 2010/63/EU of the European Parliament And of the Council of 22 September 2010 on the Protection of Animals Used for Scientific Purposes. European Commission.

European Commission Expert Working Group On Severity Assessment, 2009. Expert Working Group on Severity Classification of Scientific Procedures Performed on Animals.

European Commission Expert Working Group On Severity Assessment, 2012. Working Document on a Severity Assessment Framework. http://ec.europa.eu/environment/chemicals/lab_animals/pdf/Endorsed_Severity_Assessment.pdf.

Evaluation and Retrospective Assessment, E., 2013. Project Evaluation and Retrospective Assessment - Expert Working Group Report, Brussels.

FAWC, 1992. The Five Freedoms. Farm Animal Welfare Council, London.

Fiorito, G., Affuso, A., Basil, J., Cole, A., DE Girolamo, P., D'angelo, L., Dickel, L., Gestal, C., Grasso, F., Kuba, M., Mark, F., Melillo, D., Osorio, D., Perkins, K., Ponte, G., Shashar, N., Smith, D., Smith, J., Andrews, P.L., 2015. Guidelines for the care and welfare of cephalopods in research -A consensus based on an initiative by CephRes, FELASA and the Boyd group. Lab. Anim. 49, 1–90.

Franco, N.H., Correia-Neves, M., Olsson, I.A.S., 2012. How "humane" is your endpoint?-refining the science-driven approach for termination of animal studies of chronic infection. PLoS Pathog. 8.

Garber, J.C., Barbee, R.W., Bielitzki, J.T., Clayton, L.A., Donovan, J.C., Hendriksen, C.F.M., Kohn, D.F., Lipman, N.S., Locke, P.A., Quimby, F.W., Turner, P.V., Wood, G.A., WURbel, H., 2010. Guide for the care and use of laboratory animals: eighth edition. In: Academies, T.N. (Ed.), National Academies Press Committee for the Update of the Guide for the Care and Use of Laboratory Animals. National Research Council.

Garner, J.P., 2006. Perseveration and stereotypy –Systems-level insights from clinical psychology. In: Mason, G., Rushen, J. (Eds.), Stereotypic Animal Behaviour: Fundamentals and Applications to Welfare, 2 ed. CABI, Wallingford, UK.

Gaskill, B.N., Garner, J.P., 2017. Stressed out: providing laboratory animals with behavioral control to reduce the physiological effects of stress. Lab. Anim. (NY) 46, 142–145.

Ghosal, S., Nunley, A., Mahbod, P., Lewis, A.G., Smith, E.P., Tong, J., D'alessio, D.A., Herman, J.P., 2015. Mouse handling limits the impact of stress on metabolic endpoints. Physiol. Behav. 150, 31–37.

Gouveia, K., Hurst, J.L., 2013. Reducing mouse anxiety during handling: effect of experience with handling tunnels. PLoS One 8, e66401.

Gouveia, K., Hurst, J.L., 2017. Optimising reliability of mouse performance in behavioural testing: the major role of non-aversive handling. Sci. Rep. 7, 44999.

Hageri, C., Biernot, S., Buettner, M., Glage, S., Keubler, L.M., Held, N., Bleich, E.M., Otto, K., Muller, C.W., Decker, S., Talbot, S.R., Bleich, A., 2017. The Sheep Grimace Scale as an indicator of post-operative distress and pain in laboratory sheep. PLoS One 12.

Hankenson, F.C., Ruskoski, N., Van Saun, M., Ying, G.S., Oh, J., Fraser, N.W., 2013. Weight loss and reduced body temperature determine humane endpoints in a mouse model of ocular herpesvirus infection. JAALAS 52, 277–285.

Hatch, R.C., 1988. Euthanatizing agents. In: Booth, N.H., Mcdonald, L.E. (Eds.), Veterinary Pharmacology and Therapeutics, sixth ed. AS: Iowa State University Press.

Hawkins, P., Dennison, N., Goodman, G., Hetherington, S., Llywelyn-Jones, S., Ryder, K., Smith, A.J., 2011. Guidance on the severity classification of scientific procedures involving fish: report of a Working Group appointed by the Norwegian Consensus-Platform for the Replacement, Reduction and Refinement of animal experiments (Norecopa). Lab. Anim. 45, 219–224.

Hawkins, P., Armstrong, R., Boden, T., Garside, P., Knight, K., Lilley, E., Seed, M., Wilkinson, M., Williams, R.O., 2015. Applying refinement to the use of mice and rats in rheumatoid arthritis research. Inflammopharmacology 23, 131–150.

Helgers, S.O.A., Talbot, S.R., Riedesel, A.K., Wassermann, L., Wu, Z., Krauss, J.K., Hager, C., Bleich, A., Schwabe, K., 2020. Body weight algorithm predicts humane endpoint in an intracranial rat glioma model. Sci. Rep. 10, 9020.

Hendriksen, C., 2011. Humane endpoints in vaccine potency testing. Proc. Vaccinol. 5 (5), 221–226.

Hendriksen, C., Morton, D., Cussler, K., 2010. Use of humane endpoints to minimise suffering. In: Howard, B., Nevalainen, T., Perreta, G. (Eds.), The COST Manual of Laboratory Animal Care and Use. CRC Press Taylor & Francis.

Hepple, S.B., Baldwin, T., Brazier, M., Brownsword, R., Frse, C.K., Professor, K., Harries, R.R., Lipton, P., Southwark, P., Plant, R., Raff, M., Ross, N., Sewell, H., Smith, P., Strathern, D.M., Williamson, A., 2005. Ethical Issues Raised by Animal Reserach. The Ethics of Research Involving Animals. Latimer Trend & Company Ltd Estover Road, Plymouth PL6 7PY (Nuffield Council of Bioethics).

Honess, P., Wolfensohn, S., 2010. The extended welfare assessment grid: a matrix for the assessment of welfare and cumulative suffering in experimental animals. Altern. Lab. Anim. 38, 205–212.

Huntingford, F., Kadri, S.L., Jobling, M., 2012. Introduction: aquaculture and behavior. In: Huntingford, F., Jobling, M., Kadri, S.L. (Eds.), Aquaculture and Behavior. Wiley, Chichester, West Sussex Ames, Iowa.

Jirkof, P., Cesarovic, N., Rettich, A., Nicholls, F., Seifert, B., Arras, M., 2010. Burrowing behavior as an indicator of post-laparotomy pain in mice. Front. Behav. Neurosci. 4.

Jobling, M., 2012. Fish in aquaculture environments. In: Huntingford, F., Jobling, M., Kadri, S.L. (Eds.), Aquaculture and Behavior. Wiley, Chichester, West Sussex Ames, Iowa.

Keating, S.C.J., Thomas, A.A., Flecknell, P.A., Leach, M.C., 2012. Evaluation of EMLA cream for preventing pain during tattooing of rabbits: changes in physiological, behavioural and facial expression responses. PLoS One 7.

Kobaek-Larsen, M., Fenger, C., Ritskes-Hoitinga, J., 2004a. Secondary effects induced by the colon carcinogen azoxymethane in BDIX rats. Apmis 112, 319–329.

Kobaek-Larsen, M., Rud, L., Oestergaard Soerensen, F., Ritskes-Hoitinga, J., 2004b. Laparoscopy of rats with experimental liver metastases: a method to assess new humane endpoints. Lab. Anim. 38, 162–168.

Koch, A., Gulani, J., King, G., Hieber, K., Chappell, M., Ossetrova, N., 2016. Establishment of early endpoints in mouse total-body irradiation model. PLoS One 11, e0161079.

Koolhaas, J.M., 2010. The laboratory rat. In: Hubrecht, R., Kirkwood, J.K. (Eds.), The UFAW Handbook on the Care and Management of Laboratory and Other Research Animals, eighth ed. Chichester: Wiley-Blackwell. edited for UFAW.

Kramer, K., Kinter, L.B., 2003. Evaluation and applications of radiotelemetry in small laboratory animals. Physiol. Genom. 13, 197–205.

Krzykwa, J.C., Olivas, A., Jeffries, M.K.S., 2018. Development of cardiovascular and neurodevelopmental metrics as sublethal endpoints for the fish embryo toxicity test. Environ. Toxicol. Chem. 37, 2530–2541.

Laber, K., Newcomer, C.E., Decelle, T., Everitt, J.I., Guillen, J., Bronstad, A., 2016. Recommendations for addressing harm-benefit analysis and implementation in ethical evaluation - report from the AALAS-FELASA working group on harm-benefit analysis - Part 2. Lab. Anim. 50, 21–42.

Langford, D.J., Bailey, A.L., Chanda, M.L., Clarke, S.E., Drummond, T.E., Echols, S., Glick, S., Ingrao, J., Klassen-Ross, T., Lacroix-Fralish, M.L., Matsumiya, L., Sorge, R.E., Sotocinal, S.G., Tabaka, J.M., Wong, D., Van Den Maagdenberg, A.M.J.M., Ferrari, M.D., Craig, K.D., Mogil, J.S., 2010. Coding of facial expressions of pain in the laboratory mouse. Nat. Methods 7. 447-U52.

Latham, N., 2010. Brief introduction to welfare assessment: a "toolbox" of techniques. In: Hubrecht, R., Kirkwood, J.K. (Eds.), The UFAW Handbook on the Care and Management of Laboratory and Other Research Animals, eighth ed. Wiley-Blackwell, Chichester. edited for UFAW.

Laure Bégout, M., Kadri, S., Huntingford, F., Damsgård, B., 2012. Tools for studying the behaviour of farmed fish. In: Huntingford, F., Jobling, M., Kadri, S.L. (Eds.), Aquaculture and Behavior. Wiley, Chichester, West Sussex Ames, Iowa.

Leach, M.C., Allweiler, S., Richardson, C., Roughan, J.V., Narbe, R., Flecknell, P.A., 2009. Behavioural effects of ovariohysterectomy and oral administration of meloxicam in laboratory housed rabbits. Res. Vet. Sci. 87, 336–347.

Luy, J., 2015. The principle of proportionality:the concept and its application to animal protection issues such as the "three Rs" and the "harm-benefit analysis". ALTEX Proc. 4, 16–23.

Magee, B., Elwood, R.W., 2016. Trade-offs between predator avoidance and electric shock avoidance in hermit crabs demonstrate a non-reflexive response to noxious stimuli consistent with prediction of pain. Behav. Process. 130, 31–35.

Martini, A., Huysseune, A., Witten, P.E., Boglione, C., 2020. Plasticity of the skeleton and skeletal deformities in zebrafish (*Danio rerio*) linked to rearing density. J. Fish. Biol.

Mason, G., Rushen, J., 2006. Stereotypic Animal Behaviour: Fundamentals and Applications to Welfare. CABI, Wallingford, UK.

Matson, D.J., Broom, D.C., Carson, S.R., Baldassari, J., Kehne, J., Cortright, D.N., 2007. Inflammation-induced reduction of spontaneous activity by adjuvant: a novel model to study the effect of analgesics in rats. J. Pharmacol. Exp. Therapeut. 320, 194–201.

Mead, R.J., Bennett, E.J., Kennerley, A.J., Sharp, P., Sunyach, C., Kasher, P., Berwick, J., Pettmann, B., Battaglia, G., Azzouz, M., Grierson, A., Shaw, P.J., 2011. Optimised and rapid pre-clinical screening in the SOD1(G93A) transgenic mouse model of amyotrophic lateral sclerosis (ALS). PLoS One 6, e23244.

Mei, J., Banneke, S., Lips, J., Kuffner, M.T.C., Hoffmann, C.J., Dirnagl, U., Endres, M., Harms, C., Emmrich, J.V., 2019. Refining humane endpoints in mouse models of disease by systematic review and machine learning-based endpoint definition. ALTEX 36, 555–571.

Mellor, D., Reid, C., 1994. Concepts of animal well-being and predicting the impact of procedures on experimental animals. In: Baker, R.M., Jenkin, G., Mellor, D.J. (Eds.), Improving the Well-Being of Animals in the Research Environment. Australia: Australian and New Zealand Council for the Care of Animals in Research and Teaching.

Min, H., Danilenko, D.M., Scully, S.A., Bolon, B., Ring, B.D., Tarpley, J.E., Derose, M., Simonet, W.S., 1998. Fgf-10 is required for both limb and lung development and exhibits striking functional similarity to *Drosophila* branchless. Genes Dev. 12, 3156–3161.

Morton, D., 1998. Humane Endpoints in Animal Experimentation for Biomedical Research: Ethical, Legal and Practical Aspects.

Nemzek, J.A., Xiao, H.Y., Minard, A.E., Bolgos, G.L., Remick, D.G., 2004. Humane endpoints in shock research. Shock 21, 17–25.

Noble, C., Jones, H.A.C., Damsgard, B., Flood, M.J., Midling, K.O., Roque, A., Saether, B.S., Cottee, S.Y., 2012. Injuries and deformities in fish: their potential impacts upon aquacultural production and welfare. Fish Physiol. Biochem. 38, 61–83.

(OECD), O. F. E. C.-O. A. D, 2000. OECD Guidance Document on the Recognition, Assessment and Use of Clinical Signs as Humane Endpoints for Experimental Animals Used in Safety Evaluation. Paris, France.

Olfert, E.D., Godson, D.L., 2000. Humane endpoints for infectious disease animal models. ILAR J. 41, 99–104.

Olfert, E., Latt, R., Macallum, E., Mccutcheon, K., Rainnie, D., Schunk, M., 1998. Guidelines on Choosing Appropriate Endpoints in Experiments Using Animals for Research, Teaching and Testing. Canadian Council of Animal care. http://www.ccac.ca/Documents/Standards/Guidelines/Appropriate_endpoint.pdf.

Oh, W.J., Westmoreland, J.J., Summers, R., Condie, B.G., 2010. Cleft palate is caused by CNS dysfunction in Gad1 and Viaat knockout mice. PLoS One 5.

Olsson, A.S., Nicol, C.J., Niemi, S.M., Sandoe, P., 2020. From unpleasant to unbearable—why and how to implement an upper limit to pain and other forms of suffering in research with animals. ILAR J.

Pfister, R., Gyger, M., Bugnon, P., Desbaillets, I., Fellay, J.-M., ILLGEN-Wilcke, B., Kohler, I., Zeller, W., 2012. Editorial ethical guidelines: what about the evaluation of the benefits to humans? Lab. Anim. 46, 176.

Pichel, J.G., Shen, L., Sheng, H.Z., Granholm, A.C., Drago, J., Grinberg, A., Lee, E.J., Huang, S.P., Saarma, M., Hoffer, B.J., Sariola, H., Westphal, H., 1996. Defects in enteric innervation and kidney development in mice lacking GDNF. Nature 382, 73–76.

Research Animal Resources. Animal & Tissue Share (Online), University of Minnesota. https://www.researchservices.umn.edu/services-name/research-animal-resources/research-support/animal-tissue-share. (Accessed 10 February 2021).

Rollin, B.E., 2013. The moral status of invasive animal reserach. In: Gilbert, S., Kaebnick, G.E., Murray, T.H. (Eds.), Animal Research Ethics: Evolving Views and Practices. US.

Roughan, J.V., Flecknell, P.A., 2006. Training in behaviour-based post-operative pain scoring in rats - an evaluation based on improved recognition of analgesic requirements. Appl. Anim. Behav. Sci. 96, 327–342.

Roughan, J.V., Wright-Williams, S.L., Flecknell, P.A., 2009. Automated analysis of postoperative behaviour: assessment of HomeCageScan as a novel method to rapidly identify pain and analgesic effects in mice. Lab. Anim. 43, 17–26.

Russell, W.M.S., Burch, R.L., 1959. The Principles of Humane Experimental Technique. Methuen, London.

Shawlot, W., Behringer, R.R., 1995. Requirement for Lim1 in head-organizer function. Nature 374, 425–430.

Singh, P.K., Patil, C.R., Harlalka, G.V., Gaud, N.P., 2010. Zinc disc implantation model of urinary bladder calculi and humane endpoints. Lab. Anim. 44, 226–230.

Sneddon, L.U., 2019. Evolution of nociception and pain: evidence from fish models. Phil. Trans. Biol. Sci. 374.

Sneddon, L.U., Braithwaite, V.A., Gentle, M.J., 2003. Novel object test: examining nociception and fear in the rainbow trout. J. Pain 4, 431–440.

Soothill, J.S., Morton, D.B., Ahmad, A., 1992. The HID50 (hypothermia-inducing dose 50): an alternative to the LD50 for measurement of bacterial virulence. Int. J. Exp. Pathol. 73, 95–98.

Sørensen, D.B., Cloutier, S., Gaskill, B.N., 2020a. Animal-centric care and management. In: Sørensen, D.B., Pedersen, A., Bailey, R.E. (Eds.), Animal-centric Care and Management: Enhancing Refinement in Biomedical Research, 1 ed. CRC Press.

Sørensen, D.B., Pedersen, A., Bailey, R.E., 2020b. Animal-centric Care and Management : Enhancing Refinement in Biomedical Research. CRC Press.

Sotocinal, S.G., Sorge, R.E., Zaloum, A., Tuttle, A.H., Martin, L.J., Wieskopf, J.S., Mapplebeck, J.C.S., Wei, P., Zhan, S., Zhang, S.R., Mcdougall, J.J., King, O.D., Mogil, J.S., 2011. The Rat Grimace scale: a partially automated method for quantifying pain in the laboratory rat via facial expressions. Mol. Pain 7.

Spruijt, B.M., Welbergen, P., Brakkee, J., Gispen, W.H., 1988. An ethological analysis of excessive grooming in young and aged rats. Ann. N. Y. Acad. Sci. 525, 89–100.

Stien, L.H., Bracke, M., Noble, C., Kristiansen, T., 2020. Assessing fish welfare in aquaculture. In: T., K., Fernö, A., Pavlidis, M., Van de Vis, H. (Eds.), The Welfare of Fish. Springer, Switzerland.

Talbot, S.R., Biernot, S., Bleich, A., Van Dijk, R.M., Ernst, L., Hager, C., Helgers, S.O.A., Koegel, B., Koska, I., Kuhla, A., Miljanovic, N., Muller-Graff, F.T., Schwabe, K., Tolba, R., Vollmar, B., Weegh, N., Wolk, T., Wolf, F., Wree, A., Zieglowski, L., Potschka, H., Zechner, D., 2020. Defining body-weight reduction as a humane endpoint: a critical appraisal. Lab. Anim. 54, 99–110.

Thomson, J.S., Deakin, A.G., Cossins, A.R., Spencer, J.W., Young, I.S., Sneddon, L.U., 2020. Acute and chronic stress prevents responses to pain in zebrafish: evidence for stress-induced analgesia. J. Exp. Biol. 223.

Thon, R.W., Rritskes-Hoitinga, M., Gates, H., Prins, J.B., 2010. Phenotyping of genetically modified mice. In: Hubrecht, R., Kirkwood, J.K. (Eds.), The UFAW Handbook on the Care and Management of Laboratory and Other Research Animals, eighth ed. Wiley-Blackwell, Chichester. edited for UFAW.

Toth, L.A., 2000. Defining the moribund condition as an experimental endpoint for animal research. ILAR J. 41, 72–79.

Trammell, R.A., Toth, L.A., 2011. Markers for predicting death as an outcome for mice used in infectious disease research. Comp. Med. 61, 492–498.

Ullman-Cullere, M.H., Foltz, C.J., 1999. Body condition scoring: a rapid and accurate method for assessing health status in mice. Lab. Anim. Sci. 49, 319–323.
Viscardi, A.V., Hunniford, M., Lawlis, P., Leach, M., Turner, P.V., 2017. Development of a piglet grimace scale to evaluate piglet pain using facial expressions following castration and tail docking: a pilot study. Front. Vet. Sci. 4.
Vlach, K.D., Boles, J.W., Stiles, B.G., 2000. Telemetric evaluation of body temperature and physical activity as predictors of mortality in a murine model of staphylococcal enterotoxic shock. Comp. Med. 50, 160–166.
Wallace, J., 2000. Humane endpoints and cancer research. ILAR J. 41, 87–93.
Wang, L.V., 2003. Ultrasound-mediated biophotonic imaging: a review of acousto-optical tomography and photo-acoustic tomography. Dis. Markers 19, 123–138.
Wang, Y., Schnegelsberg, P.N., Dausman, J., Jaenisch, R., 1996. Functional redundancy of the muscle-specific transcription factors Myf5 and myogenin. Nature 379, 823–825.
Warn, P.A., Brampton, M.W., Sharp, A., Morrissey, G., Steel, N., Denning, D.W., Priest, T., 2003. Infrared body temperature measurement of mice as an early predictor of death in experimental fungal infections. Lab. Anim. 37, 126–131.
Workman, P., Aboagye, E.O., Balkwill, F., Balmain, A., Bruder, G., Chaplin, D.J., Double, J.A., Everitt, J., Farningham, D.A., Glennie, M.J., Kelland, L.R., Robinson, V., Stratford, I.J., Tozer, G.M., Watson, S., Wedge, S.R., Eccles, S.A., Committee of the National Cancer Research, I, 2010. Guidelines for the welfare and use of animals in cancer research. Br. J. Cancer 102, 1555–1577.
Wright-Williams, S.L., Courade, J.P., Richardson, C.A., Roughan, J.V., Flecknell, P.A., 2007. Effects of vasectomy surgery and meloxicam treatment on faecal corticosterone levels and behaviour in two strains of laboratory mouse. Pain 130, 108–118.
Wright-Williams, S., Flecknell, P.A., Roughan, J.V., 2013. Comparative effects of vasectomy surgery and Buprenorphine treatment on faecal corticosterone concentrations and behaviour assessed by manual and automated analysis methods in C57 and C3H mice. PLoS One 8.
Zeltz, C., Lu, N., Gullberg, D., 2014. Integrin alpha11beta1: a major collagen receptor on fibroblastic cells. In: Gullberg, D. (Ed.), Adv Exp Med Biol. 2014/07/16 (Dordrecht).
Zhang, H., Lecker, I., Collymore, C., Dokova, A., Pham, M.C., Rosen, S.F., Crawhall-Duk, H., Zain, M., Valencia, M., Filippini, H.F., Li, J., D'souza, A.J., Cho, C., Michailidis, V., Whissell, P.D., Patel, I., Steenland, H.W., Virginia Lee, W.J., Moayedi, M., Sterley, T.L., Bains, J.S., Stratton, J.A., Matyas, J.R., Biernaskie, J., Dubins, D., Vukobradovic, I., Bezginov, A., Flenniken, A.M., Martin, L.J., Mogil, J.S., Bonin, R.P., 2020a. Cage-lid hanging behavior as a translationally relevant measure of pain in mice. Pain.
Zhang, X., Kumstel, S., Tang, G., Talbot, S.R., Seume, N., Abshagen, K., Vollmar, B., Zechner, D., 2020b. A rational approach of early humane endpoint determination in a murine model for cholestasis. ALTEX 37, 197–207.
Zintzsch, A., Noe, E., Reissmann, M., Ullmann, K., Kramer, S., Jerchow, B., Kluge, R., Gosele, C., Nickles, H., Puppe, A., Rulicke, T., 2017. Guidelines on severity assessment and classification of genetically altered mouse and rat lines. Lab. Anim. 51, 573–582.

Recommended reading

Mantis shrimp (Online). Available from: http://en.wikipedia.org/wiki/Mantis_shrimp. (Accessed).
"Humane" Merriam-Webster.com Dictionary (Online), Merriam-Webster. Available from: https://www.merriam-webster.com/dictionary/humane. (Accessed 14 December 2020).

Chapter 9

Anaesthesia, analgesia, and killing

Marino Campagnol
Laboratory Animal Consultant, Italy

Introduction

The aim of this chapter is to provide all personnel involved in laboratory animal sciences key elements and points of reflection for the optimal application of *refinement* in research projects.

This chapter has been written as a key checklist to be used as baseline aimed to provide a common ground for discussions between the principal investigator and the designated veterinarian.

Recommended sources of information and relevant texts are provided in this chapter.

Anaesthesia and analgesia

Definitions in anaesthesia and analgesia

For deepening of this topic, the following readings are recommended:

- Grimm, K.A., Lamont, L.A., Tranquilli, W.J., Greene, S.A., Robertson S.A., 2015. Veterinary Anesthesia and Analgesia: The Fifth Edition of Lumb and Jones, Wiley-Blackwell (Chapter 1). ISBN-13: 978-1-118-52623-1.
- IASP Taxonomy Working Group. https://s3.amazonaws.com/rdcms-iasp/files/production/public/Content/ContentFolders/Publications2/ClassificationofChronicPain/Part_III-PainTerms.pdf.
- Raja, S.N., Carr, D.B., Cohen, M., Finnerup, N.B., Flor, H., Gibson, S., Keefe, F.J., Mogil, J.S., Ringkamp, M., Sluka, K.A., Song, X.J., Stevens, B., Sullivan, M.D., Tutelman, P.R., Ushida, T., Vader, K., 2020. The revised International Association for the Study of Pain definition of pain: concepts, challenges, and compromises. Pain 161 (9), 1976–1982. https://doi.org/10.1097/j.pain.0000000000001939. PMID: 32694387; PMCID: PMC7680716.

Practical Handbook on the 3Rs in the Context of the Directive 2010/63/EU
https://doi.org/10.1016/B978-0-12-821180-9.00011-8

Pain is "an unpleasant sensory and emotional experience associated with, or resembling that associated with, actual or potential tissue damage." Therefore, pain is "a psychological concept and not a physical measure" and that the experience of pain had to be distinguished from noxious stimulation.

Nociception is the neural process of encoding noxious stimuli (a stimulus that is damaging or threatens damage to normal tissues). Consequences of encoding may be autonomic (e.g., elevated blood pressure) or behavioral (motor withdrawal reflex or more complex nocifensive behavior). Pain sensation is not necessarily implied.

Nociceptive pain is the pain that arises from actual or threatened damage to nonneural tissue and is due to the activation of nociceptors. The term is used to describe pain occurring with a normally functioning somatosensory nervous system to contrast with the abnormal function seen in neuropathic pain.

Nociceptive stimulus is an actually or potentially tissue-damaging event transduced and encoded by nociceptors.

Analgesia is the absence of pain in response to stimulation which would normally be painful. The term is generally reserved for describing a state in a conscious patient.

Sedation is a state characterized by central depression accompanied by drowsiness and some degree of centrally induced relaxation. The patient is generally unaware of its surroundings but can become aroused and is responsive to noxious stimulation. Sedatives are not recommended by themselves to immobilize a patient during times which painful stimuli are likely to occur (L&J).

Hypnosis is a condition of artificially induced sleep, or a trance resembling sleep, resulting from moderate depression of the CNS from which the patient is readily aroused.

Narcosis is a drug-induced state of deep sleep from which the patient cannot be easily aroused. Narcosis may or may not be accompanied by antinociception, depending on the techniques and drugs used.

Local analgesia (anaesthesia) is a loss of pain sensation in a circumscribed body area.

Regional analgesia (anaesthesia) is insensibility to pain in a larger, though limited, body area usually defined by the pattern of innervation of the effected nerve(s) (e.g., paralumbar nerve blockade and anaesthesia).

General anaesthesia is drug-induced unconsciousness that is characterized by controlled but reversible depression of the CNS and perception. In this state, the patient is not arousable by noxious stimulation. Sensory, motor, and autonomic reflex functions are attenuated to varying degrees, depending upon the specific drug(s) and technique(s) used. Surgical general anaesthesia is the state/plane of anaesthesia that provides unconsciousness, amnesia, muscular relaxation, and hypoalgesia sufficient for painless surgery.

In 1950, Rees and Gray suggested a definition of general anaesthesia that included narcosis, analgesia, and relaxation. The components of general anaesthesia may be controlled by an accurate titration of drugs with the least possible disturbance of physiological functions (cardiovascular, respiratory, and different metabolic functions), and with the smallest burden upon each of the different mechanism of metabolism and excretion of the used drugs.

Criteria for selection of anaesthesia protocols and drugs

For deepening of this topic, the following reading is recommended: Paul Flecknell, 2015. "Laboratory Animal Anesthesia, fourth edition, Academic Press" (Chapter 1). ISBN-13: 978-0128000366.

Criteria for selection of anaesthesia protocols and drugs are the following:

- Influence on research
- Species, age, sex
- Strain/breed, phenotype (Genetic drift? Hybrid strains?)
- Physical status (Pregnancy? Previous procedures?)
- Procedure and its duration
- Nature, location, and severity of pain
- Available equipment
- Knowledge and experience of personal and familiarity with techniques.

Balanced anaesthesia is achieved by the simultaneous use of multiple drugs and techniques. Drugs are targeted to attenuate specifically individual components of the anaesthetic state, that is, amnesia, antinociception, muscle relaxation, and alteration of autonomic reflexes (Fig. 9.1).

The aim of the balanced anaesthesia is "to close the circle" in compliance with the requirements of the procedure.

- For a safe blood draw only, sedation will be necessary.
- For the use of imaging techniques, a complete immobilization and then a profound sedation will be necessary, that is hypnosis.

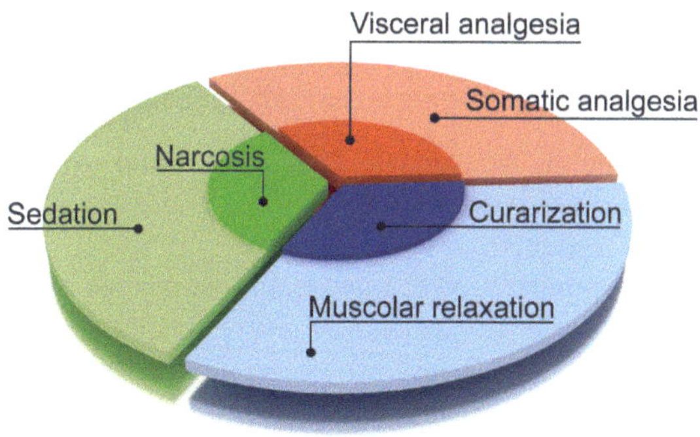

FIGURE 9.1 Anaesthesia circle.

- For a surgical procedure, narcosis, surgical muscle relaxation, and analgesia of the anatomical areas undergoing the surgical procedure should be necessary.
- For a thoracic surgery or in case the complete muscle relaxation is necessary, it will be necessary to close the whole circle and have narcosis, somatic and visceral analgesia, and curarization.

It should be considered that drugs acting on gamma-aminobutyric acid (GABA) receptors have no analgesic effect and therefore they must be associated with analgesics. Drugs acting on the GABA system are, for example, tribromoethanol, benzodiazepines (diazepam, midazolam, zolazepam, etc), barbiturates (pentobarbital, tiopenthal, etc), propofol, etomidate, alfaxalone, inhalational anaesthetics (isoflurane, sevoflurane, etc). When narcosis is induced, it is as if we temporarily "cut pharmacologically the head" of the animal.

For this reason, the Directive 2010/63/EU, Article 14 (Anaesthesia) paragraph 4. "An animal, which may suffer pain once anaesthesia has worn off, shall be treated with "*preemptive and postoperative analgesics* or other appropriate pain-relieving methods provided that it is compatible with the purpose of the procedure."

- For a preemptive analgesia, the time the drug takes to show its effect before the anaesthesia needs to be considered.
- For a postoperative analgesia, it should be considered the duration of pain, depending on the procedure.

In order to decrease the adverse effects of anaesthetics, the synergy between different active substances should be exploited to compensate the deficiencies of a single drug, and to improve the overall analgesic effect (e.g., NSAID + opioid, NSAID + local anaesthetic infiltration); an optimal multimodal analgesia is based on the synergistic combination of analgesic drugs from different classes and analgesic techniques that target different mechanisms of pain.

The analgesic therapy must consider:

- the period when the animal is more active (i.e., rodents at night)
- preferring the orally administration (training in the days before the surgery).

It is possible to administer analgesics inside a snack which animals were accustomed to in the days before the surgery (i.e.,: marshmallow for swine) or dissolved in a jelly (rodents).

With regard of administration to rodents and guinea pigs, analgesics can be dissolved in drinking water (provided that drug palatability does not represent an issue); carprofen has been proved to be stable when dissolved in drinking water and there are meloxicam liquid ad hoc medical products orally administered both for swine and for pets.

Which technique should be utilized by laboratory animal type?

- Species (the effect of anaesthetics and analgesics and their dosage varies according to the species).
- Strain (there are differences, among various strains of a species, that can be as big as those among different species).
- Inbred/outbred (in an anaesthetic protocol, inbred animals have similar response, whereas outbred animals show a much greater interindividual variability in responses).
- Sex (the response to the anaesthetic and analgesic protocol is different in the two sexes).
- Age (e.g., although a young animal has a higher metabolism than an adult animal, it can still miss the specific enzymes necessary for the drug metabolism. By contrast, an old animal has a slower metabolism and can incur in drug accumulation). Administration in young (e.g., suckling) animals can be difficult.
- Fetal and larval forms (in case of a pregnant female, once two-thirds of pregnancy have passed, we will consider anaesthesia for fetuses too. The passage of the drug to the fetus is mediated by placenta and it varies according to the active substance).
- Genetically modified (the genetic modifications, such as the genetic drift, can influence the metabolism and the pharmacologic effects of anaesthetics).
- Genetically modified with harmful phenotype: the possibility to ease the animal suffering with drugs and the possible impact on the research need to be considered.

What do we need to ease?

- Pain: its nature, intensity, and duration need to be considered.
- Suffering: it includes also the mental status.
- Distress: the possibility to ease distress pharmacologically needs to be also considered.

When a technique needs to be utilized?

- Procedure: the anaesthetic protocol must be established on the basis of the procedure animals will be subjected to, i.e., for magnetic resonance imaging, hypnosis, or narcosis only will be sufficient with no need for analgesic component, while for abdominal surgery, visceral analgesia in addition to the somatic one of the abdominal wall needs to be considered. In addition, the anaesthetic protocol shall consider also the duration of the

procedure, bearing in mind that intramuscular or intraperitoneal injectable drugs are less manageable in case of long duration.
- Postprocedure: analgesia must cover the entire period the pain is present; for this reason, it is extremely important to quantify this period the best possible.
- Even if data collection completion coincides with the project end, appropriate analgesic therapy must be applied as long as it is necessary.
- Harmful phenotype: the possibility to pharmacologically ease the suffering at the end of data collection or for the duration of the entire experimentation should be evaluated.

Please, bear in mind that "When, Why, and How" a technique needs to be applied must be described within the Project.

Pain must be:

1. Foreseen
   By procedure
   By species > breed/strain > sex
2. Identified
   With the recognition considering species-specific signs of pain
   With the continuous surveillance
3. Treated
   With drugs
   With nonpharmacological methods

Which drug of choice (anaesthetic and analgesic)?

By type of animal:

- Species
- Breed/strain (Genetic drift? Hybrid strains?)
- Sex
- Age
- Physiological/pathological condition

By pharmacodynamic:

- Interaction with the procedure
- Side effects
- Synergism with other analgesics

By pharmacokinetics:

- Duration and frequency of administration
- Onset of the pharmacological effect
- When does the drug stop to work?
- Route of administration

By prices and availability:

- Operating costs
- Availability of the equipment
- Availability of the skills
- Availability of staff

Nonpharmacological control of pain and distress

"An animal, which may suffer pain once anaesthesia has worn off, shall be treated with preemptive and postoperative analgesics *or other appropriate pain-relieving methods* provided that it is compatible with the purpose of the procedure." (Directive 2010/63/EU, Article 14).

In surgical protocols, pain and distress can be largely prevented by nonpharmacological preventive measures; appropriate environment/housing, and husbandry practices are also included.

- Acclimatization of at least seven days (all species with the exception of nonhuman primates which need a longer acclimatization period).
- Housing and husbandry taking into account physiological and ethological needs of the animal species.
- Correct manipulation techniques.

Animals, especially rodents, map the environment they live in. The cage change on the day before the surgery as well as a different cage after the surgery will result in a stressful condition.

The groups of animals should not be mixed after the surgery; in fact, mixing animal groups is a cause of distress, whereas keeping animals in a well-known environment within the usual group members is a refinement technique with an impact on analgesia efficacy.

The skills of the surgeon are an important element in this equation: the attention to the basic principles of surgery, i.e., asepsis, homeostasis, respect of the animal tissues, reduction of foreign material to the strictly minimum necessary, will be helpful in accelerating the postoperatory recovery period.

Postoperatory check of animals

- The check frequency depends on the procedure, being once a day the minimum. Checking time should consider the ethological characteristics of the animal species (e.g., rodents sleep during the day and are active at night).
- Animal observation from a distance, possibly from outside the room, minimize animal behavior modification due to the presence of the observer.
- Animals should be observed quietly for a long enough period to allow a thorough check.

- Any change in animal behavior should be noticed and reported to the designated veterinarian and the responsible of the project.
- It should be kept in mind that animals can suffer on Saturday and Sunday too!

Notes on the anaesthetic inhalers

For deepening of this topic, the following readings are recommended:

Grimm, K.A., Lamont, L.A., Tranquilli, W.J., Greene, S.A., Robertson, S.A., 2015. Veterinary Anesthesia and Analgesia: The Fifth Edition of Lumb and Jones, Wiley-Blackwell (Chapters 3 and 16). ISBN-13: 978-1-118-52623-1.

Inhalation anaesthetics are widely used in laboratory animals because their pharmacokinetic characteristics enable a rapid adjustment of anaesthesia depth. The special inhalation apparatus consists of a source of oxygen and a patient breathing circuit. The breathing circuit usually consists of an endotracheal tube or face mask, a carbon dioxide (CO_2) exhaling system, and a gas reservoir.

The vaporizer should be periodically calibrated. In fact, a noncalibrated vaporizer can affect the anaesthetic effect.

A good control of inhalation anaesthetic concentration allows a more effective and safer use than commonly used injectable anaesthetic agents.

The minimum alveolar concentration (MAC) of volatile agents is a term used to describe the potency of anaesthetic drugs; it corresponds to the end-tidal concentration between the value enabling and that preventing voluntary movements in response to noxious stimuli.

MAC is affected by several factors:

- species
- strain
- age
- pregnancy
- intensity of the noxious stimuli
- premedicants, analgesics, and other anaesthetics.

Because circadian rhythms can cause small changes in MAC, anaesthesia should be consistently administered during the same time periods to reduce variability.

To maintain the anaesthesia level, an end-tidal concentration 1.5 times the MAC needs to be applied (Table 9.1).

All the anaesthetic agents should be used in combination with a gas scavenging system; the gas anaesthesia apparatus must be subject to periodical maintenance for any possible gas leakage, since chronic exposure to low levels of these agents can cause health problems to operators.

TABLE 9.1 Minimum alveolar concentration indicative values.

Species	Isoflurane MAC	Isoflurane MAC 1,5	Sevoflurane MAC	Sevoflurane MAC 1,5
Rat	1,4	2,1	2,5	3,75
Mouse	1,35	2	2,5	3,75
Rabbit	2	3	3,7	5,55
Dog	1,3	1,95	2,2	3,3
Swine	1,75	2,6	2,6	4
Sheep	1,58	2,4	2,74	4,1
Macaca fascicularis	1,28	1,9	2	3

Beyond operators' health and safety issues, inhalation anaesthetics can also contribute to air pollution; therefore, gas evacuation procedures and environmental pollution monitoring should be put in place.

Nitrous oxide (N_2O) is a gas with analgesic properties used in combination with oxygen; N_2O is the main responsible of air pollution in surgery as it is not absorbed/eliminated by charcoal canister absorbers.

Besides the tidal volume (i.e., the volume of gas that is moved in and out of the lungs at each breath) and the animal's respiratory rate, the flow of the gas (usually oxygen) that conveys inhalation anaesthetics should be proportional to the method used for their administration. An excessively flow may result in an increased risk of environmental pollution and in money waste.

Calculation of inhalation anaesthetic consumption

For deepening of this topic, the following reading is recommended:

Grimm, K.A., Lamont, L.A., Tranquilli, W.J., Greene, S.A., Robertson, S.A., 2015. Veterinary Anesthesia and Analgesia: The Fifth Edition of Lumb and Jones (Chapter 16), Wiley-Blackwell. ISBN-13: 978-1-118-52623-1.

For the calculation of anaesthetic consumption, the following formulas are adopted:

Isoflurane > 1 mol = 123 mL of liquid isoflurane
123 mL liquid isoflurane = 24.000 mL isoflurane vapor (1 mol = 24.000 mL for all gases)
Wit Gas Flow at 1 L/min and 2% isoflurane:
Isoflurane makes up % of 1000 mL/min = 20 mL isoflurane vapor (remainder gas flow = 980 mL O_2)

In 1 h, isoflurane vapor used = 20 mL × 60 min = 1.200 mL
In 1 h, isoflurane used = 1.200 mL/24.000 (1 mol isoflurane) = 0.05 mol
Mol > ml: 0.05 × 123 mL = 6,15 mL
1 L/min - 2% isoflurane - 1 h = 6,15 mL
3 L/min - 3% isoflurane - 1 h = 27,68 mL
5 L/min - 5% isoflurane - 5 min = 6,4 mL
Sevoflurane > 1 mol = 132 mL of liquid isoflurane
132 mL liquid sevoflurane = 24.000 mL sevoflurane vapor (1mol = 24.000 mL for all gases)
Wit Gas Flow at 1 L/min and 3,5% sevoflurane:
Sevoflurane makes up % of 1000 mL/min = 35 mL isoflurane vapor (remainder gas flow = 965 mL O^2)
In 1 h, sevoflurane vapor used = 35 mL × 60 min = 2.100 mL
In 1 h, sevoflurane used = 2.100 mL/24.000 (1 mol sevoflurane) = 0,0875mol
Mol > ml: 0,0875 × 132 mL = 11,55 mL
1 L/min - 3,5% sevoflurane - 1 h = 11,55 mL
3 L/min - 4% sevoflurane - 1 h = 39,6 mL
5 L/min - 8% sevoflurane - 5 min = 11 mL

Fresh gas flows required to prevent CO_2 rebreathing during spontaneous ventilation

For deepening of this topic, the following reading is recommended:

Dugdale, A.A., Beaumont, G., Bradbrook, C., Gurney, M., 2020. Veterinary Anaesthesia: Principles to Practice: With Website: Principles to Practice, second ed., Wiley-Blackwell (Chapter 9). ISBN-13: 9781119246770 (Table 9.2).

With breathing system T-piece (Mapleson C, E, F) and Bain (Mapleson D), the fresh gas flows required to prevent CO_2 rebreathing during spontaneus ventilation is:

2-4 (tracheotube) or 5 (mask) × Minute Ventilation.

Fresh gas flows required to prevent CO_2 rebreathing in a 30 g mouse with a T-piece and mask is:

5 (mask) × 180 (breaths per minute) × 8 (tidal volume) × 0,030 (Kg weight) = 216 mL/min.

Fresh gas flows required to prevent CO_2 rebreathing in a 250 g rat with a T-piece and mask is:

5 (mask) × 80 (breaths per minute) × 8 (tidal volume) × 0,250 (Kg weight) = 800 mL/min.

TABLE 9.2 Average respiratory rates and tidal volumes in some laboratory animal species (awake status).

Species	Respiratory rate (Breaths/minute)	Tidal volume (mL/kg)
Mouse	180	8
Rat	80	8
Rabbit	50	6
Guinea pig	90	4
Dog	20–25	15
Swine >50 Kg	15–30	10
Sheep	20–30	8
Macaca fascicularis	20–30	7–8

Note: the respiratory rate is decreased by anaesthesia. Minute ventilation = Respiratory rate × Tidal volume.

Methods of killing animals

For deepening of this topic, the following readings are recommended:

- AVMA Guidelines for the Euthanasia of Animals: 2020 Edition. (Accessed on July 25, 2021), pp. 1–17. Available online: https://www.avma.org/sites/default/files/2020-02/Guidelines-on-Euthanasia-2020.pdf.
- Hawkins, P., Prescott, M.J., Carbone, L., Dennison, N., Johnson, C., Makowska, I.J., Marquardt, N., Readman, G., Weary, D.M., Golledge, H.D., 2016. A Good Death? Report of the Second Newcastle Meeting on Laboratory Animal Euthanasia. Animals (Basel) 6 (9), 50. https://doi.org/10.3390/ani6090050. PMID: 27563926; PMCID: PMC5035945.
- Herrmann, K., Flecknell, P., 2018. The application of humane endpoints and humane killing methods in animal research proposals: a retrospective review. Altern. Lab. Anim. 46 (6), 317–333. https://doi.org/10.1177/026119291804600606. PMID: 30657328.

Euthanasia (from Greek: εὐθανασία; "good death": εὖ, eu; "well" or "good" + θάνατος, thanatos; "death") is the practice of intentionally ending life to relieve pain and suffering.

Euthanasia as a matter of human disposition occurs when death is a welcome event and continued existence is not an attractive option for the animal as perceived by the owner and veterinarian (AVMA Guidelines for the Euthanasia of Animals: 2020 Edition).

Reasons for euthanasia include incurable (and especially painful) conditions or diseases, lack of resources to continue supporting the animal, or laboratory test procedures. Euthanasia methods are designed to cause the minimum pain and distress possible.

As the notion of euthanasia is linked to the one of "end of not solvable suffering" and of "death preferable to life" with regards to laboratory animals, this word is to be used only as final act of the humane end point.

Laboratory animals are killed also for reasons other than those included the concept of euthanasia. In these cases, killing is a more appropriate term and, among the methods available, only those inducing "a good death" should be used (humane killing).

It should be kept in mind that both euthanasia and killing are processes that are considered completed when the death is confirmed.

The choice of the employed method is the result of the balance between scientific and ethic needs according to the relevant legislation.

Killing per se is not considered a procedure. Similarly, killing animals only for obtaining tissues and organs is not considered a procedure.

Methods in Annex IV of the Directive 2010/63 do not apply mandatorily when an animal needs to be killed in emergency circumstances for animal health reasons, public health, public security, or environmental reasons.

Animals should be killed only by competent and trained personnel by using a method which is appropriate for the species.

Animal handling and the killing method should minimize distress experienced by the animal prior to loss of consciousness.

Whenever possible, killing should not be performed in areas where other animals are present, unless for well-justified reasons, e.g., in the case of a badly injured animal where moving the animal may cause additional suffering.

A good death must be "good" for the animal, "good" for the scientific aim, and "good" for operators and observers.

In evaluating methods of humane killing, the following criteria should be considered:

- Safety of personnel
- Compatibility with animal species, age, and health status
- Compatibility with intended animal use and purpose
- Legal requirements
- Availability of specific equipment
- Capability to induce loss of consciousness and death with the minimum pain and distress possible
- Irreversibility
- Method reliability
- Possible emotional impact on operators and observers
- Time required to loss of consciousness onset
- Drug availability and human abuse potential

The killing should be completed by one of the following methods:

(a) Confirmation of permanent cessation of the circulation
(b) Brain destruction
(c) Cervical dislocation
(d) Exsanguination
(e) Confirmation of the rigor mortis

Methods other than those listed in the table on Annex IV may be used:

(a) On unconscious animals, providing the animal does not regain consciousness before death;
(b) On animals used in agricultural research, when the aim of the project requires that the animals are kept under similar conditions to those under which commercial farm animals are kept; these animals may be killed in accordance with the requirements laid down in Annex I to Council Regulation (EC) No 1099/2009 of September 24, 2009 on the protection of animals at the time of killing.

Killing methods

For deepening of this topic, the following reading is recommended:
Directive 2010/63/EU, Annex IV.

Rodents

- Anaesthetic overdose: where appropriate, sedation should be previously induced.
- CO_2: the chamber should be filled gradually. This method is not suitable for fetal and neonate rodents.
- Cervical dislocation: this method is suitable only for rodents under 1 kg body weight. Rodents over 150 g body weight should be previously sedated.
- Decapitation: this method should be used if other methods are not possible.
- Inert gases (e.g., Ar, N_2).

Rabbits

- Anaesthetic overdose: where appropriate, sedation should be previously induced.
- Captive bolt.
- Cervical dislocation: this method is suitable only for rabbits under 1 kg body weight. Rabbits over 150 g body weight should be previously sedated.

- Head concussion/percussive blow: this should be used only with rabbits under 5 kg body weight.
- Electrical stunning: this method requires special equipment.

Dogs, cats, ferrets, and foxes

- Anaesthetic overdose: where appropriate, sedation should be previously induced.
- Head concussion/percussive blow: this method is suitable only on neonates.
- Electrical stunning: this method requires special equipment.
- Shooting with a free bullet with appropriate rifles, guns, and ammunition. This method can be used in open field conditions by experienced shooters when other methods are not applicable.

Large mammals

- Anaesthetic overdose: where appropriate, sedation should be previously induced.
- Captive bolt.
- Concussion/percussive blow to the head.
- Inert gases (Ar, N_2): this method is suitable only for pigs.
- Shooting with a free bullet with appropriate rifles, guns, and ammunition. This method can be used in open field conditions by experienced shooters when other methods are not applicable.

All systems for the humane killing of farm animals should be designed to ensure that animals are not caused unnecessary distress. Careful handling by experienced staff, with minimum disruption to normal practices, will minimize distress to the animals, before they are humanely killed.

Killing should not be performed in areas where other animals are present, unless in the case of euthanasia of a badly injured animal where additional suffering may be caused by moving the animal (Commission Recommendation of June 18, 2007 on guidelines for the accommodation and care of animals used for experimental and other scientific purposes).

Fish

- Anaesthetic overdose: where appropriate, sedation should be previously induced.
- Head concussion/percussive blow.
- Electrical stunning: this method requires special equipment.

Most fish species can be killed by:

- Anaesthetic overdose using appropriate administration route and anaesthetic agent for the size and species. When killed by immersion, fish should be left in the anaesthetic solution for at least five minutes following the cessation of opercular movement and/or vestibulo-ocular reflex.
- Head concussion.

Death should be confirmed, for example, by physical destruction of the brain or exsanguination (Commission Recommendation of June 18, 2007 on guidelines for the accommodation and care of animals used for experimental and other scientific purposes).

Amphibians

- Anaesthetic overdose: where appropriate, sedation should be previously induced.
- Head concussion/percussive blow.
- Electrical stunning: this method requires special equipment.

Birds

Anaesthetic overdose: where appropriate, sedation should be previously induced.

Carbon dioxide

- Cervical dislocation: this method is suitable only for birds under 1 kg body weight. Birds over 150 g body weight should be previously sedated.
- Head concussion/percussive blow: this method is suitable only for birds under 5 kg body weight.
- Decapitation: this method is suitable only for birds under 250 g.
- Electrical stunning: this method requires special equipment.
- Inert gases (Ar, N_2).

The preferred method of killing for juvenile and adult birds is an overdose of anaesthetic using an appropriate agent and administration route. This is preferable to CO_2 inhalation, as CO_2 may be aversive. As diving birds and some others, for example, mallard ducks, can slow their heart rates and hold their breath for long periods, care should be taken when killing such species by inhalation to ensure that they do not recover. Ducks, diving birds, and very young chicks should not be killed using CO_2 (Commission Recommendation of June 18, 2007 on guidelines for the accommodation and care of animals used for experimental and other scientific purposes).

Reptiles

- Anaesthetic overdose: where appropriate, sedation should be previously induced.
- Captive bolt: this method is suitable only for large reptiles.
- Head concussion/percussive blow.
- Shooting with a free bullet with appropriate rifles, guns, and ammunition. This method can be used in open field conditions by experienced shooters when other methods are not applicable.

An appropriate method of killing is by an overdose of a suitable anaesthetic (Commission Recommendation of June 18, 2007 on guidelines for the accommodation and care of animals used for experimental and other scientific purposes).

Notes about the use of carbon dioxide

For deepening of this topic, the following readings are recommended:

- Hawkins, P., Playle, L., Golledge, H., Leach, M., Banzett, R., Coenen, A., Cooper, J., Danneman, P., Flecknell, P., Kirkden, R., et al., February 27–28, 2006. Newcastle Consensus Meeting on Carbon Dioxide Euthanasia of Laboratory Animals, Newcastle upon Tyne, UK, pp. 1–17. Available online: https://www.nc3rs.org.uk/sites/default/files/documents/Events/First%20Newcastle%20consensus%20meeting%20report.pdf. (Accessed on 25 July 2021).
- Hawkins, P., Prescott, M.J., Carbone, L., Dennison, N., Johnson, C., Makowska, I.J., Marquardt, N., Readman, G., Weary, D.M., Golledge, H.D., August 23, 2016. A Good Death? Report of the Second Newcastle Meeting on Laboratory Animal Euthanasia. Animals (Basel) 6 (9), 50. PMID: 27563926; PMCID: PMC5035945. https://doi.org/10.3390/ani6090050.

The use of injectable anxiolytics before CO_2 exposure cannot be considered a refinement but rather an additional source of stress.

If animals have been previously exposed to inhalant anaesthetics, the use of isoflurane or sevoflurane to induce anaesthesia prior to CO_2 is not a refinement.

The use of isoflurane and sevoflurane can be considered a refinement when killing naïve rodents by inducing anaesthesia prior to introducing CO_2.

Where appropriate, CO_2 administration in the home cage can be considered to eliminate the stress of being moved.

CO_2 is aversive for mice and rats when levels rise to approximately 12%–18%.

Time to unconsciousness with CO_2 is dependent on the displacement rate, container volume, and concentration used.

CO_2 exposure by using a gradual filling method can reduce pain due to nociceptor activation by carbonic acid prior to onset of unconsciousness; a displacement rate from 30% to 70% of the chamber volume/min is recommended for rodents.

A rapidly increasing concentration (flow rate>50% of the chamber volume/minute) induces unconsciousness in only 26 to 48 seconds.

Induction of consciousness loss at CO_2 concentrations <80% may produce postmortem pulmonary and upper respiratory tract lesions that vary among different mouse strains. For this reason, the method of euthanasia needs to be adjusted on a case-by-case basis.

How to calculate the appropriate CO_2 flow rate to fill the euthanasia chamber in 1 minute?

Chamber Volume: Length × Height × Width (Centimeters)/1000 = Volume (Liters).

Calculate y% CO_2 Displacement Rate (minimum CO_2 flow rate).

Chamber Volume × 0.y CO_2 Displacement Rate = Liters CO_2 Flow/Minute (Flow Rate).

Example:

- chamber volume: 8 L
- 20% displacement rate

8 L × 0.2 CO_2 = 1.6 L CO_2 flow/minute.

To achieve the same result in 30 seconds, double the liters.

Recommended reading

AVMA Guidelines for the Euthanasia of Animals: 2020 Edition, pp. 1–17. Available online: https://www.avma.org/sites/default/files/2020-02/Guidelines-on-Euthanasia-2020.pdf. (Accessed on 25 July 2021).

Dugdale, A.A., Georgina, B., Carl, B., Matthew, G., 2020. Veterinary Anaesthesia: Principles to Practice: With Website: Principles to Practice, second ed. Wiley-Blackwell, ISBN 9781119246770.

Fish, R., Danneman, P.J., Brown, M., Karas, A., 2008. Anesthesia and Analgesia in Laboratory Animals, second ed. American College of Laboratory Animal Medicine) Academic Press, ISBN 9780123738981.

Grimm, K.A., Lamont, L.A., Tranquilli, W.J., Greene, S.A., Robertson, S.A., 2015. Veterinary Anesthesia and Analgesia: The Fifth Edition of Lumb and Jones. Wiley-Blackwell, ISBN 978-1-118-52623-1.

Hawkins, P., Prescott, M.J., Carbone, L., Dennison, N., Johnson, C., Makowska, I.J., Marquardt, N., Readman, G., Weary, D.M., Golledge, H.D., 2016. A good death? Report of the second Newcastle meeting on laboratory animal Euthanasia. Animals (Basel) 6 (9), 50. https://doi.org/10.3390/ani6090050. PMID: 27563926; PMCID: PMC5035945.

Hawkins, P., Playle, L., Golledge, H., Leach, M., Banzett, R., Coenen, A., Cooper, J., Danneman, P., Flecknell, P., Kirkden, R., et al., February 27–28, 2006. In: Newcastle Consensus Meeting on Carbon Dioxide Euthanasia of Laboratory Animals; Newcastle upon Tyne, UK, pp. 1–17. Available online: https://www.nc3rs.org.uk/sites/default/files/documents/Events/First%20Newcastle%20consensus%20meeting%20report.pdf (Accessed 25 July 2021).

Herrmann, K., Flecknell, P., 2018. The application of humane endpoints and humane killing methods in animal research proposals: a retrospective review. Altern. Lab. Anim. 46 (6), 317–333. https://doi.org/10.1177/026119291804600606. PMID: 30657328.

IASP Taxonomy Working Group. https://s3.amazonaws.com/rdcms-iasp/files/production/public/Content/ContentFolders/Publications2/ClassificationofChronicPain/Part_III-PainTerms.pdf.

Merskey, H., 2008. History and definition of chronic pain. In: Rashiq, S., Schopflocher, D., Taenzer, P., Jonsson, E. (Eds.), Chronic Pain: A Health Policy Perspective. Wiley-VCH Verlag GmbH & Co. KGaA, pp. 19–28.

Paul, F., 2015. Laboratory Animal Anaesthesia, fourth ed. Academic Press, ISBN 978-0128000366.

Raja, S.N., Carr, D.B., Cohen, M., Finnerup, N.B., Flor, H., Gibson, S., Keefe, F.J., Mogil, J.S., Ringkamp, M., Sluka, K.A., Song, X.J., Stevens, B., Sullivan, M.D., Tutelman, P.R., Ushida, T., Vader, K., 2020. The revised International Association for the Study of Pain definition of pain: concepts, challenges, and compromises. Pain 161 (9), 1976–1982. https://doi.org/10.1097/j.pain.0000000000001939. PMID: 32694387; PMCID: PMC7680716.

Chapter 10

Replacement

Chantra Eskes
Services & Consultation on Alternative Methods (SeCAM), Magliaso, Switzerland

Preface

The EU Directive on the protection of animals used for experimental and other scientific purposes (Directive 86/609/EEC replaced by Directive 2010/63/EU) is a forerunner in implementing the Three Rs (Replace, Reduce, and Refine) principle, and has, as final goal, the full replacement of procedures on live animals for scientific and educational purposes as soon as it is scientifically possible to do so.

Replacement alternatives are defined by Directive 2010/63/EU as *"a scientifically satisfactory method or testing strategy, not entailing the use of live animals"*. If in the past replacement methods were usually seen as one-to-one replacements of an animal procedure, it is currently well accepted that a combination of methods using, e.g., testing strategies is needed to achieve the full replacement of an animal procedure. Methods and approaches that can partially or fully replace the use of animal procedures encompass *in silico* approaches, *in chemico* and *in vitro* assays, integrated approaches combining different information sources, new testing tools such as high-throughput screening and methods that aim to improve the mechanistic understanding.

When used for regulatory purposes, the scientific validity of a method can be demonstrated through a validation process whereby the relevance and reliability of the method for a particular purpose is assessed. For newly developed and used methods, the scientific quality of a method can be achieved by making use of, e.g., Good In Vitro Method Practices (GIVIMP).

The present chapter addresses existing definitions, approaches, scientific quality prospects, and the international acceptance of replacement alternatives. Although further efforts are needed to overcome existing knowledge and technological limitations, replacement alternatives provide an opportunity to improve the scientific quality of obtained results by providing relevant and reproducible outcomes, while at the same time furthering the implementation of the Three Rs principle.

Practical Handbook on the 3Rs in the Context of the Directive 2010/63/EU
https://doi.org/10.1016/B978-0-12-821180-9.00008-8

Introduction

The use of animals for the benefit of humans, other animals, environment, and gain of scientific knowledge is a contentious issue that has raised and still raises ethical and societal concerns around the world since decades. Already back in 1959, the principle of the Three Rs, i.e., to replace, reduce, and refine the use of animals for experimentation purposes, has been developed by Russell and Burch, to ensure humane experimental procedures are conducted on animals (Russel and Burch, 1959).

The European Union has been one of the pioneers in the world to implement the principles of the Three Rs into its legislative framework by adopting the directive on the protection of animals used for experimental and other scientific purposes already back in 1986 (Directive 86/609/EEC). This directive, later replaced in 2010 by Directive 2010/63/EU (EU, 2010), makes a firm legal requirement to apply the Three Rs, and the Three Rs have also become embedded in the national legislation of all Member States of the EU. Furthermore, this EU directive emphasizes the phasing out of animal testing whereby the full replacement of all animal use for scientific purposes has been set as the ultimate goal by the Directive 2010/63/EU (Preamble text considerations number (10), (11), (38); article 4.1). For example, recital 10 of the EU Directive 2010/63/EU states that:

> *"While it is desirable to replace the use of live animals in procedures by other methods not entailing the use of live animals, the use of live animals continues to be necessary to protect human and animal health and the environment. However,* ***this Directive represents an important step towards achieving the final goal of full replacement of procedures on live animals for scientific and educational purposes as soon as it is scientifically possible to do so.*** *To that end, it seeks to facilitate and promote the advancement of alternative approaches. It also seeks to ensure a high level of protection for animals that still need to be used in procedures."*

The EU Directive 2010/63/EU further defines the replacement of an animal experimental procedure in Article 4 as follows:

> *"Member States shall ensure that, wherever possible, a scientifically satisfactory method or testing strategy, not entailing the use of live animals, shall be used instead of a procedure".*

When it is not yet possible to replace the use of an animal in a procedure, all efforts must be made to reduce and refine the animal experiments. In case valid replacement alternatives are available, the animal method shall therefore not be used.

This chapter addresses the current understanding of what a replacement method represents and provides examples of different types of replacement methods that take into account the latest developments of scientific knowledge and technological progress. It also addresses the meaning of "a scientifically satisfactory method" and the importance to take into consideration quality concepts. Finally, it provides information on the regulatory acceptance of replacement methods at the international level.

What does a replacement method mean?

William Russell and Rex Burch developed the concept of the Three Rs already back in the 1950s as the principles of humane experimental technique (Russel and Burch, 1959). The authors indicated a clear Three R hierarchy, with replacement being the highest priority, then reduction, then refinement. According to Russel and Burch, the replacement of animal experimentation related to "*any scientific method employing non-sentient material which may replace methods which use conscious living vertebrates*". Reduction referred to "*lowering the number of animals used to obtain information of a given amount and precision*", and refinement related to the "*decrease in the incidence or severity of procedures applied to those animals which have to be used*". Furthermore, Russell and Burch made a distinction between absolute and relative replacement. In the case of absolute replacement, animals were not required at any stage. For relative replacement, animals were still required, though in the actual experiments they were not exposed to distress. For example, an experiment performed on an anaesthetized animal, which was killed under the anaesthetic at the end of the experiment, was considered to be a replacement method.

The definitions of the Three Rs have evolved since Russell and Burch first described them, some of which differ significantly from the original definition (Tannenbaum and Bennett, 2015). They are generally understood as follows (Declaration of Bologna, 2009):

- **Replacement alternatives**: Methods which permit a given purpose to be achieved without conducting experiments or other scientific procedures on animals.
- **Reduction alternatives**: Methods for obtaining comparable levels of information from the use of fewer animals in scientific procedures, or for obtaining more information from the same number of animals.
- **Refinement alternatives**: Methods which alleviate or minimizese potential pain, suffering, and distress, and which enhance animal well-being.

However, different interpretations of the definition of replacement still exist. For example, the UK National Center on the Three Rs (NC3Rs) proposes a contemporary definition of replacement methods as "*accelerating the development and use of models and tools, based on the latest science and technologies, to address important scientific questions without the use of animals*" (https://www.nc3rs.org.uk/the-3rs).

Furthermore, if replacement has often beeen seen as a one-to-one replacement of an animal experimentation, it is currently well accepted that more than one method is necessary to fully replace an animal experimentation.

On that regard, partial replacement methods are considered to be alternative methods that can partially replace an animal experiment but requires the use of additional alternative methods as part of a strategy approach for achieving full replacement of the animal test (Manou et al., 2005). In this case, partial replacement indicates that an individual non-animal test may not fully replace the animal test, but only certain aspects of that test, and a combination of non-animal tests is required to fully replace the animal test. If one non-animal test can replace the animal test, it is called full or one-to-one replacement.

However, the terms of full and partial replacement have also been used in the sense that full replacement relates to avoiding the use of any research animals, whereas partial replacement includes the use of some animals that, based on current scientific knowledge are not considered capable of experiencing suffering, such as *Drosophila* and nematode worms and immature forms of vertebrates. Finally, partial replacement may be considered as the use of primary cells (and tissues) taken from animals killed solely for this purpose. For example, the Swiss 3R Center defines replacement methods as full or partial replacement methodologies that are based on 1) approaches not entailing the use of living animals such as cultured cells (including primary cells), tissues, and organs as well as the use of testing strategies that take into account existing in vivo data, *in vitro* methods, *in silico* computational methods, physicochemical properties, and nontesting data, and/or on 2) use of animals that, based on current scientific evidence, are not considered capable of experiencing suffering such as some invertebrates and immature forms of vertebrates (https://www.swiss3rcc.org/en/3rs-resources/what-are-the-3rs).

In the case of the EU Directive 2010/63/EU, the following principle of replacement is given in Article 4.1:

> *"a scientifically satisfactory method or testing strategy, not entailing the use of live animals, shall be used instead of a procedure".*

This implies avoiding the use of whole living animals. In this case, using whole living fish embryos or insects like *Drosophila* instead of a mouse would not be considered a replacement alternative but a refinement alternative, since the research still uses whole living animals.

Replacement is more challenging to achieve than reduction and refinement, since it requires a long-term and complex development process that can require multiple regulatory validation steps, while reduction and refinement can be applied in the shorter term within each experiment that will still encompass the use of whole living animals. In article 4, the EU Directive 2010/63/EU defines the principle of reduction as ensuring "*that the number of animals used in projects is reduced to a minimum without compromising the objectives of the project*," and of refinement as "*eliminating or reducing to the minimum any possible pain, suffering, distress, or lasting harm to the animals*."

Before submitting a project authorization request for an animal study, it is expected that a thorough search for the Three Rs is performed. In case Three R alternatives are available, they should be implemented. Furthermore, reformulating the research question may also help in identifying replacement alternatives. For example, by splitting the research question into e.g., the various mechanistic steps and modes of action as described in Replacement methods and testing strategies section of this chapter. Some additional questions that may help in identifying ways to replace animal experimental procedures include (1) what is the target species of interest (human/animal)? (2) what do you want to know/achieve? (3) what is already known? (4) what models are already in use and how applicable are these for answering your question? (5) what other disciplines could help address the question? An example of how to reformulate a research question is the Helpathon organized in the Netherlands (https://www.youtube.com/watch?v=W0r8XBb4WsE). It represents a two-day workshop with multiple stakeholders, in which the goal of the research is separated from the use of animal models. The starting point is to discuss how the research goal can be reached without the use of animals. A new experimental proposal without the use of animals is written, based on the original plan that encompassed animal studies.

Finally, Directive 2010/63/EU attributes a strong role to EU Member States in order to "*contribute to the development and validation of alternative approaches*" and to "*take such other steps as they consider appropriate to encourage research in this field*". (Article 47.1). The Directive also states that "*Member States shall, at national level, ensure the promotion of alternative approaches and the dissemination of information thereon*" (Article 47.4).

Outside the European Union, some national governments are also stepping away from animal testing and moving toward more new approach methodologies (NAMs). For example, in the United States, the Environmental and Protection Agency aims to stop conducting and funding studies on mammals by 2035. Furthermore, the National Toxicology Program for the 21st century makes use of high-throughput *in vitro* screening assays, systems biology, and predictive *in silico* approaches to improve the hazard evaluation of environmental chemicals, based on the predictive value of the investigated assays (Huang et al., 2016).

Replacement methods and testing strategies

New approaches methodologies that can be used to avoid the use of animal testing, also referred to as NAMs, may encompass the following technologies, methodologies, approaches, or combinations thereof.

- In silico *approaches* based on experiments performed via computer simulations as for example, artificial intelligence, machine learning, computational modeling, and (Quantitative) Structure-Activity Relationships (Q) SARs.
- *In chemico assays* that measure the chemical reactivity or other physicochemical properties of substances based on an abiotic assay.

- In vitro *assays* that perform a given experiment in a culture dish, or, more generally, in a controlled environment outside of a living organism.
- *New testing tools* such as high-throughput screening and high-content methods (e.g., genomics, proteomics, and metabolomics).
- *Existing information from human studies* such as epidemiological studies, exposure, when such information is of relevance.
- Methods that aim to improve the *mechanistic understanding* of a certain effect such as adverse outcome pathways (AOPs).
- *Integrated approaches* that combine different sources of information in the form of, e.g., Integrated Approaches to Testing and Assessment (IATA) and Defined Approaches to testing and assessment (DAs).

Individual methods

Individual methods can represent *in silico, in chemico, in vitro*, or other advanced assays which can either fully or partially replace an animal experimentation. A "one-to-one" replacement means that one animal study is replaced by, e.g., an *in vitro* study. An example is the skin irritation test for regulatory purposes, where the irritating properties of a chemical can be studied using reconstructed human skin models cultured *in vitro* (OECD test guideline (TG) 439, 2020a). Another example, in education, is the replacement of live animals used for dissection by alternatives such as synthetic mannequins, simulators, training devices, films, and videos as well as *in silico* alternative teaching modules such as, for example, used for frog dissections (https://norecopa.no/search?q=frog%20dissection*&facet.limit=125). The International Network for Humane Education (InterNICHE) website provides an overview of available alternatives for teaching purposes (http://www.interniche.org/en/alternatives). Finally, in medical research, over the last decade, some gaps in modeling human relevance have been addressed by using complex *in vitro* models. For example, the use of complex architecture including different cell populations, the capacity to challenge cells with biochemical and mechanical stimuli, the introduction of microfluidic components enabling to switch from static to controllable dynamic systems, and the integration with analytical devices to probe the biochemical processes are some of the advances that have been made toward increasing human relevance. One of the many examples is the NC3Rs awarded project on an *in silico* model that predicted the risk of drug-induced heart arrhythmias in humans by using human cardiac cells, more accurately than animal studies had done (https://www.nc3rs.org.uk/news/international-3rs-prize-awarded-computer-modelling-predicts-human-cardiac-safety-better-animal). Nevertheless, even the most recent three-dimensional complex *in vitro* models still have gaps that remain ambitions of the scientific community. These gaps are the current challenges for technology developers (Jeong et al., 2018).

As mentioned previously, it is currently well accepted that more than one method is necessary to allow the full replacement of an animal experimentation. The individual methods making part of a combined approach, also referred to as partial replacement methods, can partially replace an animal experiment by covering, e.g., a certain mechanistic pathway and/or physiological target of relevance. These partial replacement methods can be used in a combined manner with other models and information sources as described here below, as part of a strategy approach for achieving full replacement of the animal test (Manou et al., 2005).

Testing strategies

In mechanistically complex situations, such as systemic toxicity of chemicals or drugs, it is well accepted that one single method might not be sufficient to fully replace an animal experimentation, as the chemical or drug can interact through multiple mechanisms, pathways, and organs. In these cases, usually more elaborate replacement methods are needed due to the complexity of the effects investigated.

Latest advances in biomedical research and bioengineering provide the opportunity to achieve more physiologically relevant human models and reduce the uncertainties associated with interspecies differences. For example, novel *in vitro* methods or approaches that are based on human cells or tissues can provide human-based mechanistic insights that help to unveil the mode of actions or AOPs that may happen in humans (Alliance for Human Relevant Science, 2020; https://www.humanrelevantscience.org/white-papers). This can help speeding up drug development, reduce the need for animal testing, and offer a more ethical model to study disease pathology at organ and tissue level. These models may include *in vitro* human-based tissues and organoids (e.g., patient-specific organoids derived from primary tissues) and human physiological simulations based on microphysiological or organ-on-a-chip technologies (Busquet et al., 2020).

The use of mechanistically relevant models within integrated strategies, combined with the knowledge of the underlying biological process, can contribute to a more mechanistic and predictive human science while decreasing the animal studies that may be necessary. Sistare et al. (2016) report that sophisticated *in vitro* humanized test systems, emerging computational models, and novel translation biomarkers can be used to improve the ability to better predict drug-induced liver injury, and make the development of safer drug candidates less dependent on animal studies. Malloy et al. (2017) highlights the usefulness of using predictive approaches based on, e.g., *in silico* and *in vitro* approaches, computational models, to help assessing the safety of chemicals. Finally, Piersma and colleagues (Piersma et al. (2018); Burgdorf et al., 2019) suggest that a transition to a mechanistically based human-focused safety assessment is needed, which steps away from the traditional animal studies, and defines human biology as a new standard. The ideal at the horizon is the virtual human, where integrated methods come together (Burgdorf et al. 2019). However, further research is necessary to overcome the still existing gaps in scientific knowledge and technological limitations (Dal Negro et al., 2018).

The need to obtain relevant biological and mechanistic information for decision-making processes calls for the use of pathway-based approaches that capture the current understanding and physiological mechanisms (Worth and Patlewicz, 2016). Integrated approaches combine in a logical manner mechanistically based methods, pathways as well as other information sources. The first concepts of integrated approaches that combine different test methods and information sources developed for the regulatory safety assessment of chemicals were described in the late 1990s and 2000s. Since then, the systematic combination of several information sources has been implemented in different areas of regulatory and basic toxicology as described below.

AOPs describe a mechanistic sequence of events that starts with an initial interaction of a stressor with a biomolecule in an organism that causes perturbation in its biology (i.e., molecular initiating event). The sequence of events then progresses through a dependent series of intermediate key events which ultimately culminates in an adverse outcome considered to be relevant to risk assessment or regulatory decision-making (OECD GD 233, 2018a; AOP Wikipedia, https://aopwiki.org).

For example, the AOP used to depict skin sensitization adverse effects (OECD GD 168, 2012) has served as the basis to develop and validate a number of partial replacement non-animal methods that today have gained regulatory acceptance (EU Regulation 2016/1688, 2016; OECD TG 442C, 2018b; OECD TG 442D, 2018c, OECD TG 442E, 2020b). AOP has also been used together with toxicity pathway frameworks to organize and integrate diverse biological data for evaluating the endocrine activity of chemicals (Browne et al., 2017). Browne et al. conclude that without mechanistic data, an endocrine mode of action cannot be determined. A mode-of-action ontology model to support the animal-free evaluation of repeated dose toxicity of chemicals has also been proposed (Desprez et al., 2019). The authors propose critical elements and ways of establishing a mode-of-action ontology model to support the animal-free safety evaluation of chemicals. Finally, Hessel et al. (2018) propose the use of an ontology-driven animal-free testing strategy for developmental neurotoxicity testing. The authors propose the scientific validation of models through coverage of the biological domain, requiring the mining of modern biology, toxicology, and chemical information to feed intelligent designs, in order to define biologically relevant testing strategies for neurodevelopmental toxicity testing.

IATAs integrate and weight the relevant existing evidence collected from multiple information sources. Where required, it also guides the targeted generation of new data to inform regulatory decision-making on potential hazard or risk (OECD GD 256, 2016a; OECD GD 329, 2020c). An IATA includes a degree of expert judgment, for example, in the choice of information sources and their weighting. Nevertheless, some IATA components, such as DAs (see below), can be standardized (i.e., rule-based). IATAs may allow for full or partial replacement of animal testing. For example, the IATA for skin irritation and corrosion represents a full replacement of the animal testing (Eskes and Hofmann, 2017; OECD GD 203, 2014). Furthermore, the IATA for ocular hazard (OECD GD 263, 2019) allows to partially replace (and reduce) animal testing conducted for this purpose (Dal Negro et al., 2018). Finally, the use of IATAs have also been suggested for evaluating the carcinogenic potential of chemicals (Jacobs et al., 2016; Petkov et al., 2016).

DAs consist of a defined set of information sources (e.g., *in silico* predictions, *in chemico*, *in vitro* data) for which a fixed data interpretation procedure (e.g., statistical or mathematical models) is applied to derive a prediction. The results or predictions derived from a DA can either be used on their own (e.g., if they are deemed to be fit for purpose) or be considered together with other sources of information (e.g., in the context of an IATA) within an overall weight-of-evidence approach to contribute to a regulatory decision (OECD GD 255, 2016b). In contrast to the assessment process within IATAs, which involves some degree of expert judgment, results or predictions generated with defined approaches are rule-based. A number of DAs have been proposed, for example, for the assessment of skin sensitization (Kleinstreuer et al., 2018; OECD GD 256, 2016a). Two of those have been adopted in 2021 as full replacement of the animal test by the OECD Test Guidelines Program for predicting skin sensitization hazard and potency (OECD TG 497, 2021). Furthermore, DAs have also been proposed to identify ocular hazards which are currently also under discussions at the OECD level (Alépée et al., 2019a,b).

What is a scientifically satisfactory replacement method?

The use of NAMs in basic, applied, and regulatory sciences has considerably increased and will continue to expand in the coming decennia. The EU Directive 2010/63/EU on the protection of animals used for scientific purposes states that: "*Member states shall ensure that, wherever possible, a scientifically satisfactory method or testing strategy, not entailing the use of a live animal, shall be used instead of a procedure*". The alternative approach not entailing the use of a live animal should therefore be "*scientifically satisfactory*". The characterization of the scientific validity of an alternative approach to animal testing is an important step for gaining acceptance.

For methods aiming to gain regulatory acceptance, this is generally demonstrated through a validation process whereby the scientific validity of the method is assessed. The scientific quality of methods used for other purposes than regulatory applications, is however also important to be considered as to ensure the robustness of the obtained results. This can be achieved by making use of, e.g., GIVIMPs as recommended by the OECD. Finally, a number of guidelines also exist on how to assess the scientific validity of (Q)SARs (see below in section Validation of (Quantitative) Structure-Activity Relationships tools).

The scientific validation of a method

The scientific validation of a method allows to demonstrate the ability of a method to provide reliable and relevant data for a particular purpose, ensuring that sound science-based data are generated. In this context, reliability (or reproducibility) is the extent to which a test method can be performed reproducibly within and between laboratories over time, when performed using the same protocol. Relevance assesses the relationship of the test to the effect of interest and whether it is meaningful and useful for a particular defined purpose. It is the extent to which the test correctly measures or predicts the biological effect of interest. It usually takes into account the biological and mechanistic relevance of the method in relation to the in vivo effect of interest, the predictive capacity of the test method compared to a reference method (e.g., an animal study), its regulatory needs, usefulness, and limitations.

The process from developing a test method to its regulatory acceptance includes a number of steps going from research and development, to test method validation, its peer review, regulatory acceptance, and the final implementation and use of the test method. In each step, if the required criteria to advance to the next step are not met, the method has to be optimized to ensure it meets the necessary requirements to progresses further. The process of scientific validation can lead to different levels of regulatory acceptance/use, depending on the status of scientific validity of the method or approach. For example, annex XI of the EU REACH regulation defines the following levels of acceptance for *in vitro* methods (EU, 2006):

- validated methods might be used for the identification of hazardous and nonhazardous effects;
- suitable methods (i.e., methods having met the criteria to enter the validation process) might be used for the identification of only hazardous effects;
- other methods can be used to gain mechanistic insights.

Collaboration and communication between the different actors involved in the validation process (i.e., test method developers, validation bodies, regulators and end-users) is important to accelerate the acceptance, implementation, and use of a new test method. However, if the scientific validation of a test method can facilitate its regulatory acceptance and standardization, scientific validation shall not to be confused with regulatory acceptance. The regulatory acceptance process assesses whether a (scientifically valid) method is capable of providing the answers that a specific legislation requires. These legislative data requirements may differ between legislation and regions. Information on the progress of alternative, non-animal methods from their validation to their acceptance as a recognized test method for use in various sectors can be found in the European Union Reference Laboratory for alternatives to animal testing (EURL-ECVAM) Tracking System for Alternative methods toward Regulatory acceptance (TSAR, https://tsar.jrc.ec.europa.eu).

A method that is holistically and universally suitable to every scientific purpose does currently not exist; a method can or cannot be the best choice, depending on the scientific question to be answered. This has led to the concept of "qualification" of a technology as opposed/complementary to "validation." The qualification of a model/technology has been embraced not only by the scientific community but also by international regulatory agencies such as the Food and Drug Administration [FDA], and the European Medicines Agency [EMA]. According to the FDA definition in the Predictive Toxicology Roadmap, qualification of a biological and toxicological model and assay is a process that, within a stated context of use, brings to the conclusion that "*the results of an assessment using the model or assay can be relied on to have a specific interpretation and application in product development and regulatory decision-making*" (FDA, 2017; https://www.fda.gov/media/109634/download). The qualification approach has also been embraced by the EMA in 2016 in its Guideline on the principles of regulatory acceptance of Three Rs testing approaches (EMA, 2016). This concept is equally applicable to both toxicology and research, within or outside the regulatory space. It means that, given a specific biological question, the most suitable model/technology that can best answer that question is identified by qualification in that specific context of use. By defining the context of use of a method, its limitations i.e., areas where data generated with that method are not relevant, are also defined.

Good *In Vitro* Method Practices

The OECD Guidance Document on GIVIMPs (GIVIMP, OECD, 2018d) has been developed for *in vitro* test method developers, providers, and end users to help improve the scientific integrity (relevance) and quality (reproducibility) and reduce the uncertainties of the results obtained with *in vitro* methods. When GIVIMP is properly implemented, it increases reproducibility, credibility, reliability, and integrity of the generated data and related reports and publications and will improve the efficiency of method development and applied use.

GIVIMP tackles 10 aspects related to *in vitro* work: (1) roles and responsibilities, (2) quality considerations, (3) facilities, (4) apparatus, material, and reagents, (5) test systems, (6) test and reference/control items, (7) standard operating procedures, (8) performance of the method, (9) reporting of results, and (10) storage and retention of records and materials.

The concepts of GIVIMP are important for both basic research and regulatory science laboratories to help speed up the process of development, ensure a method design of high quality, and facilitate use of the method by industry and other decision-makers. Furthermore, adopting the principles discussed in GIVIMP on a global level can help harmonize the development and execution of NAMs around the world.

Validation of (Quantitative) Structure-Activity Relationships tools

The OECD has developed a number of guidance documents on how to assess the validity and suitability of (Q)SARs models and the adequacy of their predictions (OECD GD 69, 2007; http://www.oecd.org/chemicalsafety/risk-assessment/guidancedocumentsandreportsrelatedtoqsars.htm). Justification for (Q)SARs and Expert Systems can be provided by means of a (Q)SAR Model Reporting Format (QMRF) which proposes the validity of a method by considering the OECD (Q)SAR principles of (1) defining the endpoint, (2) defining the algorithm, (3) defining the AD, (4) defining goodness of fit and robustness, (5) defining predictivity, and (6) providing a mechanistic understanding. The Joint Research Center (JRC) QSAR Model Database provides an inventory of information on available QMRFs that is freely accessible online (https://ec.europa.eu/jrc/en/scientific-tool/jrc-qsar-model-database). In addition, the adequacy and reliability of individual predictions is demonstrated by means of a (Q)SAR Prediction Reporting Format (QPRF).

International acceptance of replacement alternatives to animal experimentation

In the EU, Article 13.1 of Directive 2010/63/EU states that "*Without prejudice to national legislation prohibiting certain types of methods, Member States shall ensure that a procedure is not carried out if another method or testing strategy for obtaining the result sought, not entailing the use of a live animal, is recognised under the legislation of the Union*". As a consequence, where available and recognized under the EU legislation, procedures involving the use of live animals should be replaced with scientifically satisfactory alternative methods.

Furthermore, a number of EU regulations covering different sectors call for the use of alternative methods to animal experimentation as described below.

- A ban on animal testing in the EU exists for both finished cosmetic products (since 2004) and cosmetic ingredients (since 2009) according to EU Regulation 1223/2009 (EU, 2009). Furthermore, since 2013, the EU has imposed a marketing ban on cosmetics tested on animals (finished or containing ingredients) to ensure that imported cosmetics are also not tested on animals.

- The European Regulation for the Registration, Evaluation, Authorization, and Restriction of Chemicals (REACH, EU Regulation 1907/2006) requires that *in vitro* testing be conducted for topical effects such as skin and ocular hazards. It also requires the use of integrated testing strategies that take into account all existing information before undertaking animal experimentation. Finally, it also requires the use of combined mechanistic methods for assessing the skin sensitization potential hazard of chemicals (EU, 2006 updated in 2016).
- The European Regulation for the Classification Labeling and Packaging of substances and mixtures (EU CLP, EU Regulation 1272/2008) encourages the use of tiered weight-of-evidence strategies for assessing the hazard of mixtures. The Regulation also highlights the use of validated alternative methods for specific purposes, as, for example, to confirm the dermal and ocular hazard classification of extreme pH formulations with alkaline or acid reserve that indicates noncorrosion (EU, 2008).
- The European Regulation on the marketing and use of biocidal products requires that vertebrates are used only as a last resort and encourages the use of tiered weight-of-evidence strategies, existing information, reduction and refinement methods (EU Regulation 528/2012). It also describes different acceptance levels, depending on the validation status of the method or approach (EU, 2012).
- Finally, the EMA describes opportunities for the implementation of the Three Rs in human medicinal products (EMA, 2018a), and has created a guideline on the implementation of best practices with regard to the Three Rs in regulatory testing of medicinal products (EMA, 2018b).

At the international regulatory level, the OECD Test Guidelines Program develops and makes available the most relevant internationally agreed test methods and approaches (https://www.oecd-ilibrary.org/environment/oecd-guidelines-for-the-testing-of-chemicals-section-4-health-effects_20745788). In recent years, an increasing number of new OECD test guidelines based on alternative approaches to animal testing has been adopted representing either full replacements, partial replacements, elements of integrated/defined approaches, or full integrated/defined approaches. These Test Guidelines are used by governments, industry, and independent laboratories to assess the safety of chemicals and chemical products, such as medicines, industrial chemicals, and pesticides. They are primarily used in regulatory safety testing and subsequent chemical notification and registration.

In addition, the International Council for Harmonization of Technical Requirements for Pharmaceuticals for Human Use (ICH), the European Pharmacopeia, the International Cooperation on Harmonization of Technical Requirements for Registration of Veterinary Medicinal Products (VICH) and the ISO standards also take into account test methods that are based on alternatives to animal experimentation. For example, a number of Three R methods have been considered in the European Pharmacopoeia including, for example, tests conducted for the quality control of vaccines for human and veterinary use, tests used for blood products and for biological and biotechnological substances (Lang et al., 2018; Milne and Buchheit, 2012). Another example is the adoption of the ICH S10 photosafety evaluation of pharmaceuticals which provides guidance on photosafety testing strategies for pharmaceuticals including the use of chemical and *in vitro* methods (ICH S10, 2015).

Conclusions

Developed over 60 years ago, the principle of the Three Rs has become a hallmark in order to conduct humane experimental procedures. The European Union has been a forerunner in implementing the Three Rs principle in its legislation already since 1986 with the EU directive on the protection of animals used for experimental and other scientific purposes (Directive 86/609/EEC replaced in 2010 by the Directive 2010/63/EU).

The EU directive 2010/63/EU further pioneers in having as a final goal the full replacement of procedures on live animals for scientific and educational purposes as soon as it is scientifically possible to do so. A replacement method or testing strategy is defined as "*a scientifically satisfactory method or testing strategy, not entailing the use of live animals*". A test method or testing strategy is considered to be scientifically valid when it is found to be reliable and relevant for a particular purpose. For methods aiming to gain regulatory acceptance, this is generally demonstrated through a validation process whereby the scientific validity of a test method is assessed. For new developed methods not formally validated, the scientific quality of a method can be assessed through a qualification process to ensure the validity of the obtained results; this assessment can be achieved by making use of GIVIMP as recommended by the OECD.

A replacement method has often been seen as a one-to-one replacement of an animal test. However, with the growing understanding of the underlying biological and physiological mechanistic pathways, and the advancement in sophisticated 3D and human-based physiologically relevant human models, it is currently well accepted that more than one method is necessary to fully replace an animal experimentation.

Methods and approaches that can be used to partially or fully replace the use of animal procedures encompass *in silico* approaches, *in chemico* and *in vitro* assays, as well as integrated approaches that combine different sources of information such as IATA and DAs, new testing tools such as high-throughput screening and high-content methods (e.g., genomics, proteomics, metabolomics) and methods that aim to improve the mechanistic understanding. Although further research is needed to cover current existing knowledge gaps and overcome technological limitations, the latest advances in biomedical research provide the opportunity to contribute to improve the scientific quality of the obtained results in order to achieve more relevant and reproducible outcomes as well as to further the implementation of the Three Rs principle.

References

Alépée, N., Adriaens, E., Abo, T., Bagley, D., Desprez, B., Hibatallah, J., Mewes, K., Pfannenbecker, U., Sala, À., Van Rompay, A.R., Verstraelen, S., McNamee, P., 2019a. Development of a defined approach for eye irritation or serious eye damage for liquids, neat and in dilution, based on cosmetics Europe analysis of *in vitro* STE and BCOP test methods. Toxicol. *In Vitro* 57, 154–163. https://doi.org/10.1016/j.tiv.2019.02.019.

Alépée, N., Adriaens, E., Abo, T., Bagley, D., Desprez, B., Hibatallah, J., Mewes, K., Pfannenbecker, U., Sala, À., Van Rompay, A.R., Verstraelen, S., McNamee, P., 2019b. Development of a defined approach for eye irritation or serious eye damage for neat liquids based on cosmetics Europe analysis of *in vitro* RhCE and BCOP test methods. Toxicol. *In Vitro* 59, 100–114. https://doi.org/10.1016/j.tiv.2019.04.011.

Browne, P., Noyes, P.D., Casey, W.M., Dix, D.J., 2017. Application of adverse outcome pathways to U.S. EPA's endocrine disruptor screening program. Environ. Health Perspect. 125, 096001. https://doi.org/10.1289/EHP1304.

Burgdorf, T., Piersma, A.H., Landsiedel, R., Clewell, R., Kleinstreuer, N., Oelgeschläger, M., Desprez, B., Kienhuis, A., Bos, P., de Vries, R., de Wit, L., Seidle, T., Scheel, J., Schönfelder, G., van Benthem, J., Vinggaard, A.M., Eskes, C., Ezendam, J., 2019. Workshop on the validation and regulatory acceptance of innovative 3R approaches in regulatory toxicology - evolution versus revolution. Toxicol. *In Vitro* 59, 1–11.

Busquet, F., Hartung, T., Palloca, G., Rovida, C., Leist, M., 2020. Harnessing the power of novel animal-free test methods for the development of COVID-19 drugs and vaccines. Arch. Toxicol. 94, 2263–2272. https://link.springer.com/content/pdf/10.1007/s00204-020-02787-2.pdf (Accessed 20 March 2021).

Dal Negro, G., Eskes, C., Belz, S., Bertein, C., Chlebus, M., Corvaro, M., Corvi, R., Dhalluin, S., Halder, M., Harvey, J., Hermann, M., Hoffmann-Dörr, S., Kilian, K., Lambrigts, D., Laroche, C., Louhimies, S., Mahony, C., Manou, I., McNamee, P., Prieto, P., Reid, K., Roggen, E., Schutte, K., Stirling, C., Uhlrich, S., Weissenhorn, R., Whelan, M., 2018. One science-driven approach for the regulatory implementation of alternative methods: a multi-sector perspective. Regul. Toxicol. Pharmacol. 99, 33–49.

Declaration of Bologna, 2009. Reduction, refinement and replacement alternatives in laboratory animal procedures. ATLA 37, 285–289.

Desprez, B., Birk, B., Blaauboer, B., Boobis, A., Carmichael, P., Cronin, M.T.D., Curie, R., Daston, G., Hubesch, B., Jennings, P., Klaric, M., Kroese, D., Mahony, C., Ouédraogo, G., Piersma, A., Richarz, A.N., Schwarz, M., van Benthem, J., van de Water, B., Vinken, M., 2019. A mode-of-action ontology model for safety evaluation of chemicals: outcome of a series of workshops on repeated dose toxicity. Toxicol. *In Vitro* 59, 44–50.

Eskes, C., Hofmann, M., 2017. Overview on current status of alternative methods and testing approaches for skin irritation testing. In: Eskes, C., van Vliet, E., Maibach, H.I. (Eds.), Alternatives for Dermal Toxicity Testing. Springer International Publishing, Cham, Switzerland.

European Medicines Agency (EMA), 2016. Guideline on the Principles of Regulatory Acceptance of 3Rs (Replacement, Reduction, Refinement) Testing Approaches. https://www.ema.europa.eu/en/documents/scientific-guideline/guideline-principles-regulatory-acceptance-3rs-replacement-reduction-refinement-testing-approaches_en.pdf (Accessed 24 March 2021).

European Medicines Agency (EMA), 2018a. Reflection Paper Providing an Overview of the Currently Regulatory Testing Requirements for Medical Products for Human Use and Opportunities for Implementation of the 3Rs. EMA/CHMP/CVMP/JEG-3Rs/742466/2015. https://www.ema.europa.eu/en/documents/scientific-guideline/reflection-paper-providing-overview-current-regulatory-testing-requirements-medicinal-products-human_en.pdf (Accessed 20 March 2021).

European Medicines Agency (EMA), 2018b. Review and Update of EMA Guidelines to Implement Best Practice with Regard to 3Rs (Replacement, Reduction and Refinement) in Regulatory Testing of Medicinal Products – Report on Actions Taken. EMA/CHMP/CVMP/JEG-3Rs/677407/2015. https://www.ema.europa.eu/en/documents/scientific-guideline/review-update-ema-guidelines-implement-best-practice-regard-3rs-replacement-reduction-refinement_en.pdf (Accessed 20 March 2021).

European Union (EU), 2006. Regulation (EC) No 1907/2006 of the European Parliament and of the Council of 18 December 2006 concerning the Registration, Evaluation, Authorisation and Restriction of Chemicals (REACH), establishing a European Chemicals Agency, amending Directive 1999/45/EC and repealing Council Regulation (EEC) No 793/93 and Commission Regulation (EC) No 1488/94 as well as Council Directive 76/769/EEC and Commission Directives 91/155/EEC, 93/67/EEC, 93/105/EC and 2000/21/EC. Off. J. Eur. Union L 396, 1–849. This regulation was amended by i) EC Regulation 2016/863. Off. J. Eur. Union L 144, 27–31; and ii) EC Regulation 2016/1688. Off. J. Eur. Union L 255, 14–16. https://eur-lex.europa.eu/eli/reg/2006/1907/2014-04-10 >. (Accessed 20 March 2021).

European Union (EU), 2008. Regulation (EC) No 1272/2008 of the European Parliament and of the council of 16 December 2008 on classification, labelling and packaging of substances and mixtures, amending and repealing directives 67/548/EEC and 1999/45/EC, and amending regulation (EC) No 1907/2006. Off. J. Eur. Union L 353, 1–1355. This regulation was amended by EC Regulation 2016/918. Off. J. Eur. Union L 156, 1–103. https://eur-lex.europa.eu/legal-content/EN/TXT/HTML/?uri=CELEX:32008R1272&from=EN (Accessed 20 March 2021).

European Union (EU), 2009. Regulation (EC) No 1223/2009 of the European Parliament and of the council of 30 November 2009 on cosmetic products. Off. J. Eur. Union L 342, 59–209. https://eur-lex.europa.eu/legal-content/EN/TXT/HTML/?uri=CELEX:32009R1223&from=EN (Accessed 20 March 2021).

European Union (EU), 2010. Directive 2010/63/EU of the European Parliament and of the council of 22 September 2010 on the protection of animals used for scientific purposes. Off. J. Eur. Union L 276, 33. https://eur-lex.europa.eu/legal-content/EN/TXT/HTML/?uri=CELEX:32010L0063&from=EN (Accessed 20 March 2021).

European Union (EU), 2012. Regulation (EC) No 528/2012 of the European Parliament and of the Council of 22 May 2012 concerning the making available on the market and use of biocidal products. Off. J. Eur. Union L 167, 1–123. https://eur-lex.europa.eu/legal-content/EN/TXT/HTML/?uri=CELEX:32012R0528&from=EN (Accessed 20 March 2021).

European Union (EU), 2016. Regulation (EU) 2016/1688 of 20 September 2016 amending annex VII to regulation (EC) No 1907/2006 of the European Parliament and of the council on the registration, evaluation, Authorisation and Restriction of chemicals (REACH) as regards skin sensitisation. Off. J. Eur. Union L 255, 14–16. https://eur-lex.europa.eu/legal-content/EN/TXT/PDF/?uri=CELEX:32016R1688&from=EN (Accessed 20 March 2021).

Food Drug Administration (FDA), 2017. FDA's Predictive Toxicology Roadmap. https://www.fda.gov/media/109634/download (Accessed 24 March 2021).

Hessel, E.V.S., Staal, Y.C.M., Piersma, A.H., 2018. Design and validation of an ontology-driven animal-free testing strategy for developmental neurotoxicity testing. Toxicol. Appl. Pharmacol. 354, 136–152.

Huang, R., Menghang, X., Sakamuru, S., Zhao, J., Shahane, S.A., Attene-Ramos, M., Zhao, T., Austin, C.P., Simeonov, A., 2016. Modelling the Tox21 10 K chemical profiles for *in vivo* toxicity prediction and mechanism characterization. Nat. Commun. 7, 10425. https://www.nature.com/articles/ncomms10425 (Accessed 20 March 2021).

ICH, 2015. ICH S10 Photosafety Evaluation of Pharmaceuticals. Guidance for Industry. https://www.ema.europa.eu/en/ich-s10-photosafety-evaluation-pharmaceuticals (Accessed 20 March 2021).

Jacobs, M.N., Colacci, A., Louekari, K., Luijten, M., Hakkert, B.C., Paparella, M., Vasseur, P., 2016. International regulatory needs for development of an IATA for non-genotoxic carcinogenic chemical substances. ALTEX 33, 359–392. https://doi.org/10.14573/altex.1601201.

Jeong, C.G., Dal Negro, G., Getsios, S., Ekert, J., 2018. Application of complex *in vitro* models (CIVMs) in drug discovery for safety testing and disease modeling. In: Borenstein, J.T., Tandon, V., Tao, S.L., Charest, J.L. (Eds.), Microfluidic Cell Culture Systems, second ed. Elsevier, Amsterdam, pp. 121–149.

Kleinstreuer, N.C., Hoffmann, S., Alépée, N., Allen, D., Ashikaga, T., Casey, W., Clouet, E., Cluzel, M., Desprez, B., Gellatly, N., Göbel, C., Kern, P.S., Klaric, M., Kühnl, J., Martinozzi-Teissier, S., Mewes, K., Miyazawa, M., Strickland, J., van Vliet, E., Zang, Q., Petersohn, D., 2018. Non-animal methods to predict skin sensitization (II): an assessment of defined approaches. Crit. Rev. Toxicol. 48, 359–374.

Lang, C., Kolaj-Robin, O., Cirefice, G., Taconet, L., Pel, E., Jouette, S., Buda, M., Milne, C., Charton, E., 2018. Replacement, Reduction, Refinement. Animal Welfare Progress In European Pharmacopoeia Monographs: Activities of the European Pharmacopoeia Commission From 2007 to 2017. Pharmaeuropa Bio Sci Notes, pp. 12–36. https://www.edqm.eu/sites/default/files/article-pheur-replacement-reduction-refinement-may2018.pdf (Accessed 20 March 2021).

Malloy, T., Zaunbrecher, V., Beryt, E., Judson, R., Tice, R., Allard, P., Blake, A., Cote, I., Godwin, H., Heine, L., Kerzic, P., Kostal, J., Marchant, G., McPartland, J., Moran, K., Nel, A., Ogunseitan, O., Rossi, M., Thayer, K., Tickner, J., Whittaker, M., Zarker, K., 2017. Advancing alternatives analysis: the role of predictive toxicology in selecting safer chemical products and processes. Integrated Environ. Assess. Manag. 13, 915–925.

Manou, I., Eskes, C., de Silva, O., Renner, G., Zuang, V., 2005. Chapter 2. In: Eskes, C., Zuang, V. (Eds.), Alternative (Non-animal) Methods for Cosmetics Testing: Current Status and Future Prospects, vol. 33 (Suppl. 1). ATLA, pp. 21–26.

Milne, C., Buchheit, K., 2012. EDQM's 3R activities in the field of quality control of vaccines. ALTEX Proceedings 1/12. In: Proceedings of WC8, pp. 65–69. https://www.edqm.eu/medias/fichiers/article_c_milne_kh_buchheit_edqms_3r_activities_in.pdf (Accessed 20 March 2021).

OECD, 2007. Guidance Document on the Validation of (Quantitative) Structure-Activity Relationships [(Q)SAR] Models OECD Series on Testing and Assessment No. 69. OECD, Paris. https://www.oecd-ilibrary.org/docserver/9789264085442-en.pdf?expires=1617647482&id=id&accname=guest&checksum=61D497B9545BE39D77C39C37478D8D4C (Accessed 5 April 2021).

OECD, 2012. Guidance Document No. 168. The Adverse Outcome Pathway for Skin Sensitisation Initiated by Covalent Binding to Proteins. Part 1: Scientific Evidence. Environment, Health and Safety Publications, Series on Testing and Assessment. OECD, Paris. https://www.oecd.org/env/the-adverse-outcome-pathway-for-skin-sensitisation-initiated-by-covalent-binding-to-proteins-9789264221444-en.htm (Accessed 20 March 2021).

OECD, 2014. Guidance Document No. 203 on an Integrated Approached to Testing and Assessment for Skin Irritation/Corrosion. Environment, Health and Safety Publications, Series on Testing and Assessment. OECD, Paris. https://www.oecd.org/publications/guidance-document-on-an-integrated-approach-on-testing-and-assessment-iata-for-skin-corrosion-and-irritation-9789264274693-en.htm (Accessed 20 March 2021).

OECD, 2016a. Guidance Document No. 256 on the Reporting of Defined Approaches and Individual Information Sources to Be Used within Integrated Approaches to Testing and Assessment (IATA) for Skin Sensitization. Series on Testing and Assessment. OECD, Paris. https://www.oecd.org/publications/guidance-document-on-the-reporting-of-defined-approaches-and-individual-information-sources-to-be-used-within-integrated-9789264279285-en.htm (Accessed 20 March 2021).

OECD, 2016b. Series on Testing and Assessment No. 255. Guidance Document on the Reporting of Defined Approaches to Be Used within Integrated Approaches to Testing and Assessment. Series on Testing and Assessment. OECD, Paris, France. http://www.oecd.org/officialdocuments/publicdisplaydocumentpdf/?cote=env/jm/mono(2016)28&doclanguage=en (Accessed 20 March 2021).

OECD, 2018a. Series on Testing and Assessment No. 233. User's Handbook Supplement to the Guidance Document for Developing and Assessing AOPs. Series on Testing and Assessment n. 233. Series on Adverse Outcome Pathways n. 1. OECD, Paris, France. https://one.oecd.org/document/ENV/JM/MONO(2016)12/en/pdf (Accessed 20 March 2021).

OECD, 2018b. Guideline for the Testing of Chemicals No. 442D: In Vitro Skin Sensitisation: ARE-Nrf2 Luciferase Test Method. Organisation for Economic Cooperation and Development, Paris, France. https://www.oecd-ilibrary.org/environment/oecd-guidelines-for-the-testing-of-chemicals-section-4-health-effects_20745788 (Accessed 20 March 2021).

OECD, 2018c. Guideline for the Testing of Chemicals No. 442E: In Vitro Skin Sensitisation Assays Addressing the Key Event on Activation of Dendritic Cells on the Adverse Outcome Pathway for Skin Sensitisation. Organisation for Economic Cooperation and Development, Paris, France. https://www.oecd-ilibrary.org/environment/oecd-guidelines-for-the-testing-of-chemicals-section-4-health-effects_20745788 (Accessed 20 March 2021).

OECD, 2018d. Series on Testing and Assessment No. 286. Guidance Document on Good *In Vitro* Method Practices (GIVIMP). OECD, Paris, France. https://www.oecd-ilibrary.org/docserver/9789264304796-en.pdf?expires=1615743442&id=id&accname=guest&checksum=600D9101AF9E966D18BFDF18B8F13983 (Accessed 20 March 2021).

OECD, 2019. Series on Testing and Assessment No. 263 on an Integrated Approach on Testing and Assessment (IATA) for Serious Eye Damage and Eye Irritation. OECD, Paris. https://www.oecd-ilibrary.org/environment/second-edition-guidance-document-on-integrated-approaches-to-testing-and-assessment-iata-for-serious-eye-damage-and-eye-irritation_84b83321-en (Accessed 20 March 2021).

OECD, 2020a. OECD Guidelines for the Testing of Chemicals No. 439. In Vitro Skin Irritation: Reconstructed Human Epidermis Test Method. Organisation for Economic Cooperation and Development. Originally adopted in 2010, Paris, France. https://www.oecd-ilibrary.org/docserver/9789264242845-en.pdf?expires=1616266731&id=id&accname=guest&checksum=02420FC6454726113DB0D8A88C1B554C (Accessed 20 March 2021).

OECD, 2020b. Guideline for the Testing of Chemicals No. 442C: In Chemico Skin Sensitisation: Direct Peptide Reactivity Assay (DPRA). Organisation for Economic Cooperation and Development, Paris, France. https://www.oecd-ilibrary.org/environment/oecd-guidelines-for-the-testing-of-chemicals-section-4-health-effects_20745788 (Accessed 21 March 2021).

OECD, 2020c. Series on Testing and Assessment No. 329 on Overview of Concepts and Available Guidance Related to Integrated Approaches to Testing and Assessment (IATA). OECD, Paris, France. http://www.oecd.org/env/ehs/testing/series-testing-assessment-publications-number.htm (Accessed 20 March 2021).

OECD, 2021. Guideline for the Testing of Chemicals No. 497: Guideline on Defined Approaches for Skin Sensitisation. Organisation for Economic Cooperation and Development, Paris,

France. www.oecd-ilibrary.org/environment/guideline-no-497-defined-approaches-on-skin-sensitisation_b92879a4-en (Accessed 15 October 2021).

Petkov, P.I., Schultz, T.W., Donner, E.M., Honmam, M., Morita, T., Hamada, S., Wakata, A., Mishima, M., Maniwa, J., Todorov, M., Kaloyanova, E., Kotov, S., Mekenyan, O.G., 2016. Integrated approach to testing and assessment for predicting rodent genotoxic carcinogenicity: IATA for predicting rodent genotoxic carcinogenicity. J. Appl. Toxicol. 36, 1536–1550.

Piersma, A.H., Burgdorf, T., Louekari, K., Desprez, B., Taalman, R., Landsiedel, R., Barroso, J., Rogiers, V., Eskes, C., Oelgeschläger, M., Whelan, M., Braeuning, A., Vinggaard, A.M., Kienhuis, A., van Benthem, J., Ezendam, J., 2018. Workshop on acceleration of the validation and regulatory acceptance of alternative methods and implementation of testing strategies. Toxicol. *In Vitro* 50, 62–74.

Russel, W.M.S., Burch, R.L. (Eds.), 1959. The Principles of Humane Experimental Technique. Methuen, London. http://altweb.jhsph.edu/pubs/books/humane_exp/het-toc (Accessed 20 March 2021).

Sistare, F.D., Mattes, W.B., LeCluysem, E.L., 2016. The promise of new technologies to reduce, refine, or replace animal use while reducing risks of drug induced liver injury in pharmaceutical development. ILAR 57, 186–211.

Tannenbaum, J., Bennett, B.T., 2015. Russell and Burch's 3RS then and now: the need for clarity in definition and purpose. JAALAS 54, 120–132.

Worth, A.P., Patlewicz, G., 2016. Integrated approaches to testing and assessment. In: Eskes, C., Whelan, M. (Eds.), Validation of Alternative Methods for Toxicity Testing, Advances in Experimental Medicine and Biology, vol. 856. Springer, Cham, Switzerland.

Chapter 11

Design of procedures and projects

Derek Fry[1], Adrian Smith[2], Joanne Storey[3], Simon Bate[4], Merel Ritskes-Hoitinga[5,6]

[1]*School of Biological Sciences, University of Manchester, Manchester, United Kingdom;* [2]*Norecopa, c/o Norwegian Veterinary Institute, Ås, Norway;* [3]*GlaxoSmithKline Medicines Research Centre, Stevenage, Hertfordshire, United Kingdom;* [4]*GlaxoSmithKline, Stevenage, United Kingdom;* [5]*SYRCLE, Health Evidence, Radboud University Medical Center, Nijmegen, The Netherlands;* [6]*AUGUST, Clinical Medicine, Aarhus University, Aarhus, Denmark*

Introduction (Derek Fry)

Much of the content of this chapter should be useful to those planning to publish work in journals that hold to the 3Rs principles and some sections are relevant to designing experiments generally. However, the main purpose of the chapter is to consider the design issues a person should think about when preparing a proposal for experimental work on animals which would need authorization under the terms of Directive 2010/63/EU. The first of these is the aim(s) of the work. Does it fall within the list of permitted purposes detailed in Article 5? Could experiments with that aim in mind be done without using vertebrate animals or cephalopods, or completed before the animals reach the critical stage of development? If the answer to the first question is yes and the answer to the second is no, then a project in accord with the Directive would be expected to be needed, and this chapter should help researchers present their proposed experiments for favorable evaluation under the terms of the Directive.

Another preliminary consideration is whether the proposed work is worth doing. Article 18 of the Directive requires an evaluation of "whether the harm to the animals in terms of suffering, pain and distress is justified by the expected outcome." This is a complex matter dealt with later in this chapter, but before going into detailed design and planning a researcher should be fairly confident that studies would pass a "harm-benefit" analysis.

Although the design of experiments contributes principally to "reduction," animal studies should take account of all 3Rs. New possibilities for replacing the animal work previously thought necessary come up for regulatory studies

Practical Handbook on the 3Rs in the Context of the Directive 2010/63/EU
https://doi.org/10.1016/B978-0-12-821180-9.00002-7

and also for drug-receptor interactions, some physiological studies, and disease models. Those judging whether proposed animal-based studies are permissible under the Directive should be aware these may not be permitted if a suitable, practicable alternative exists. When animals that come within the scope of the Directive do need to be used, this should involve least severity and the design of an experiment should take potential refinements into account, including changing to a less-sentient species. Refinement helps in obtaining reliable and reproducible data: results on suffering or stressed animals may be a poor indication of effects in the normal animal, and data on stressed animals is likely to show greater variability than data on unstressed ones.

The recommended websites section at the end of this chapter should help with identifying possible replacement alternatives, additional scope for reduction, and potential refinements. Refinement also comes up in the basic design considerations and animal models sections. It is also important not to exclude replacement and refinement possibilities from the search strategies set up when doing systematic reviews: the section on this topic guides the reader through this, not just for published studies based on animals.

Experimental work starts with the biological insight to formulate a good experimental question, and, for hypothesis-testing experiments, proceeds to devising a testable hypothesis. An experimenter needs to recognize what is feasible and to plan how best to perform the experiment given the constraints. Most animal experiments involve quantitative comparisons, and an appreciation of statistics is important to understand how to organize the experiment so that the results can be properly analyzed. One section of this chapter is specifically to show how different designs can be analyzed, but researchers should consult a statistician for advice on the optimal design for their particular experimental question and local circumstances. It should also be apparent from this chapter that experiments should be designed not only to minimize animal numbers (reduction) but also to minimize severity (refinement), and that housing, administration, and sampling arrangements need to be taken into account in the design so that valid comparisons can be made using the analysis tools available (see also Fry, 2013). This calls for good planning and there is a section covering this.

The Directive envisages work carried out within the framework of a project of up to five years duration. This is a different approach from considering a single experiment. It entails a series of experiments and thinking about the 3Rs throughout the series. This means checking possibilities for using a replacement alternative not just at the start but as the work progresses, using the experience from the early experiments to indicate refinements that could reduce the severity of the later ones, and reconsidering what is the most suitable animal model and what is the best design at each stage in the development of the work, reviewing numbers throughout. Much of the substance of this chapter is related to hypothesis testing, but exploratory, data-gathering and, in particular, pilot experiments could have a place in the series.

In most animal-based experiments which come within the framework of the Directive, the researcher will introduce, in one or more of the experimental conditions ("treatments") being compared, alterations that give the risk of pain, suffering, distress, or lasting harm above the level of the skilled insertion of a hypodermic needle. Even when the experiment only involves behavioral studies or alterations which themselves would be outside the scope of the Directive, the research would need authorization if invasive sampling may be needed. Thus observational studies, such as preference testing or studies on wildlife, which do not involve confinement or exposure to a potentially harmful environment may not of themselves come above the threshold of pain, suffering, distress or lasting harm, but any invasive sampling, or attachments that interfere with an animal's normal behavior, could bring them within the scope of the Directive, and if so the justification for the study would be assessed against the Directive's considerations. Although this chapter does not particularly cover the design of such experiments, it gives fundamental principles that apply across the range of animal experimentation, and could be helpful for gaining a favorable assessment.

Exploratory, pilot or data-gathering experiments which come within the scope of the Directive are justifiable because they are expected to produce sufficient useful data, such as that on which to formulate a hypothesis. There may, for example, be value in including pilot experiments in a developing program. These trials with small numbers allow adverse effects, time to treatment effect, sources of variability, technical problems, and the like, to be noted so that definitive experiments can be better planned and problems avoided. They guard against overoptimism, and judiciously used should mean fewer animals are needed overall. Although they might help estimate variability when no other estimate is available, numbers involved are likely to be too small to give a reliable indication of variability.

In planning a sequence of experiments in accord with the 3Rs, a series of questions can be posed:

What can be done without animals?
What can be done under terminal anesthesia?
What can be done with only mild severity?
What can only be done at more than mild severity?

There may be negatives to all except the last but they are still worth asking, and there may be different answers at different stages of a program of work.

For ongoing studies, it may be that these questions have already been asked, and the work has settled on using a particular animal model for the period of a proposed project. This may be so for scientific studies intended to develop further a previous discovery, and regulatory toxicology and many safety assessment studies are set up for using well-tried models for which there is confidence in the results obtained. However application of the 3Rs principles means the use of the animal studies should be periodically reviewed to see whether nonanimal approaches, less severe animal models, or more effective animal models are available. The section in this chapter on animal models should help in choice of model both initially and at any review stage.

Basic design considerations (Derek Fry)

Well-designed experiments should help the researcher obtain valid results from which safe conclusions can be drawn, know how widely these may apply, and use resources efficiently. They should also minimize severity and be reproducible. And good design starts with a good experimental question or objective.

Objectives

The Directive uses "purpose" and "objective" interchangeably in some cases but with distinctly different meaning in others. Either might apply to the broad area of investigation of a five-year project proposal, but for a particular experiment, the objective needs to be quite specific not only for clarity of design but also to meet the requirements of the directive that numbers are the "minimum without compromising objectives of the project" (Article 4.2). So the objective of each experiment should make clear when it would be achieved and what the scientific outcome would be. It is not uncommon for the objective of an experiment, the question it is looking to answer, not to be clear (Kilkenny et al., 2009), but a good, clearly defined objective is critical to developing the design.

A test of a good objective is whether you would know when you had achieved it, whether it would lead to a clear experimental outcome, and, for a hypothesis-testing experiment, whether it leads to a testable hypothesis. The objective "to determine the effect on radiographic tibial bone density in sheep of different plasma levels of prednisolone" could pass, with perhaps the addition of "to a given degree of confidence," whereas "to study the effect of steroids in sheep" would not, but might be a general aim for the project. For that general aim an evaluator might look for the proposal to answer questions like "why sheep?" and "why steroids?" but that is different from judging whether a particular experimental design is appropriate.

Fundamentals: independent replication, avoiding bias, and suitable comparisons

In all types of experiments, researchers need to be aware of the variability of biological material, the risk of bias, that the best measures of outcome should be used, and that disturbance to the animals should be minimized (Fry, 2013). In behavioral observational or epidemiological studies, investigators need also to be aware of the risk of another linking source of variability contributing to or providing an apparent correlation.

The basic experimental pattern is comparison of groups of units of biological material which are given different treatments. Because of the inherent variability of biological material, parts of the same material treated identically will not give the same response. So a number of the units will be needed to give a good guide to what happens to the particular biological material when under the particular treatment. There is a fundamental need for "replication."

It is crucial that each "replicate" (often called the biological replicate to distinguish it from multiple samples from the same experimental unit) can give an independent indication of the response of the material under the particular condition. The test is "could the biological entity be independently assigned to any of the experimental treatments?" Individual animals grouped together such as fish in a pond or mice in a cage are not independent estimates of effect if the whole group received the same treatment. Local factors like the hierarchy within the group and pond or cage effects mean the pond or cage is the true replicate and the experimental effect has to be judged from the housed groups, not the individuals. If all five mice in a cage are given the same treatment, and two cages receive that treatment, there are two replicates for that effect, not ten. Taking the individual responses as independent when in fact the true measure of the experimental response is, for example, the effect on the group within the cage is termed "pseudo-replication" (Hurlburt, 1984): it can give a very false impression of the ability of the experiment to detect a genuine difference between experimental groups.

Experimental units

The biological replicate has been termed the "experimental unit" and is the element of biological material that is independently and randomly assigned to an experimental group. The key word is "independent," and crucially any two experimental units should be able to receive different treatments. Depending on the type of experiment, the experimental unit may be an animal, a group of animals given the same treatment in the same environment (like fish in a tank receiving the treatment through the water supply), or a part of an animal. When several measurements are made on the same experimental unit, it is important to distinguish the experimental unit from the observational or measurement units, which are not independent estimates of the experimental effect (Lasic et al., 2018). For cross-over designs, in which an animal is exposed successively to different experimental conditions so the treatments being compared are in different time periods, the experimental unit is the animal for a treatment period.

Quantifying variability

The inherent variability of biological material means there is risk of apparent differences between experimental groups being solely due to this, and the design of an experiment should allow for proper quantification of the level of that risk. This is why no definitive experiment should be planned without an understanding of how its results will be analyzed. The section of this chapter on analysis develops this point and shows how different experimental designs can be analyzed.

Avoiding bias by randomization and blinding

In biomedical experiments, variability in whatever is being measured due to the different treatments is seen against a background of many other sources of variation. Some of those sources could affect one group more than the other, giving a "bias" which overlies any experimental effect, and which could indicate an experimental difference when none exists. So the allocation of experimental units to treatment groups should be random to minimize the risk of inherent differences being associated with particular treatments (Fisher, 1960). The same principle of avoiding bias should apply throughout the experiment. The order of administration of compounds to different experimental units should be random, the sequence of sampling random, the positioning of cages in a rack randomized, the treatment sequence within a time period randomized, and so on. Properly carried out, randomization should make inherent differences in biological units or measurement processes equally likely to occur with any treatment and equally likely to affect any response. There is a small chance that a random allocation will happen to assign several experimental units with a particular characteristic to one experimental group, or one part of a varied spatial environment (as in the fully randomized plan view in Fig. 11.1). This is a particular risk with small numbers of experimental units, when a stratified random arrangement may be preferable.

Randomization does not help with subjective bias which may be subtle and unconscious, with the person measuring influenced by what he or she expects. That is dealt with by "blinding," that is arranging that the person measuring is not aware of what treatment the experimental unit has had. If there is risk of subjective bias in allocation then the experiment should be "double-blinded" with the person allocating a biological unit unaware of what treatment it will receive.

Comparisons and controls

When comparing experimental treatments, it is important to include comparisons or controls which avoid risk of misinterpretation of the outcome. If the objective is simply to make direct comparisons, as between different surgical approaches or routes of administration, separate controls, sham-operated or vehicle (compound solvent), for example, would be unnecessary, but in other cases, several separate controls may be needed. This is most clearly seen in batch testing (taking a sample from a batch of vaccine or injectable drug product) to determine whether any toxic agents have been inadvertently introduced. Comparing the effect of the sample against a vehicle control is not enough to give confidence that no toxins are present. If there is no effect, it could be due to extraneous factors. To have confidence that there are genuinely no toxins in the sample, it is important to show that an agent known to be toxic under the same conditions does show an effect.

In such situations, the minimum safe comparison is test agent, negative control, and positive control. A positive control should show an effect and gives confidence that an absence of observable difference between experimental groups is not misinterpreted as a lack of effect when it is actually due to something that has made any effect difficult to observe. Positive controls should really be used whenever such misinterpretation may occur. However, if the objective is simply to show qualitatively that an effect would be detected, only two or three experimental units may be enough for the positive control group, and it can be excluded from the analysis. Positive controls in regulatory toxicology, however, would increase both numbers and severity unnecessarily as the whole testing sequence contributes to confidence that toxicity will be properly assessed.

Basic design issues

Fully randomized designs

Simple comparison of randomized treatment groups is a valid design which is widely applicable and easy to understand. Although commonly used, the fully randomized comparison carries the risk that the experimental effect may be obscured by other sources of variation. A more complex design may well use fewer experimental units to meet the experimental objective.

Randomized block designs

It has been recognized since the 1930s that where there is an influential source of variation other than the experimental treatments, a design which allows the obscuring variabilities to be distinguished from experimental effects is better than the fully randomized one. This involves grouping of experimental units with regard to the unwanted ("nuisance") variability. It is referred to as "blocking," and the design as a "randomized block design."

Blocking may not be an intuitive term for animal experimenters, but is understandable from an agricultural research standpoint. It is explained in Fig. 11.1.

The principle can be readily applied to animal experiments, and Fig. 11.1 gives an example. Instead of plots in a field, think of surgical operations that have to be spread over several time periods. Just as plots in one area can form a block, so can singly housed animals given different operations within a time period. In each time period, the solutions used may differ somewhat, the theater conditions may have changed, the operating team may differ or be performing less well, and so on, but provided the different operative treatments being compared are all done in random sequence in that time period, the

Which of four fertilisers would be best in difficult terrain?

A test field like that on the right has a distinct slope and a gradient of drainage with good drainage at the top and poor drainage and marshy vegetation at the bottom.

Below is a plan view of 16 plots in such a field with the gradient of colouring indicating the variation in drainage from top to bottom of the slope.

Fully randomised design:
In the plan view (left) the 4 fertilisers (1 to 4) are randomly allocated to the 16 plots, and the plot is the experimental unit. No account is taken of the slope and drainage, so an unequal distribution of fertilisers to different drainage areas can result. Crop yield comparisons would be affected by the amount of drainage as well as by the fertiliser.

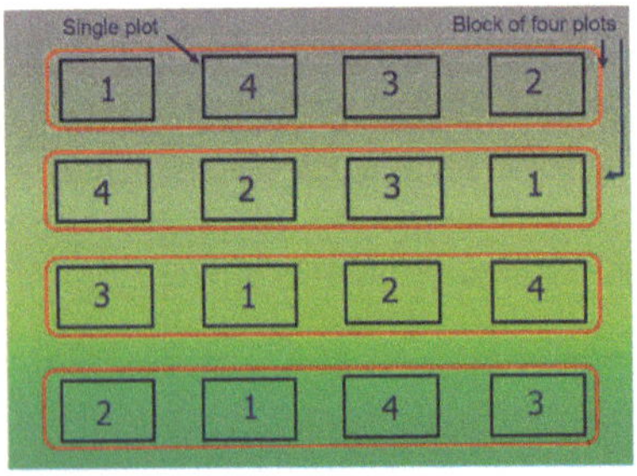

Randomised block design:
Here, the 16 plots are grouped in blocks of four plots, each at a different level down the slope. Each block has a plot with each one of the 4 fertilisers, but distributed in random order across the field. The 4 fertilisers can be compared against each other at every level of the slope, taking out the variability due to drainage from the experimental comparison.

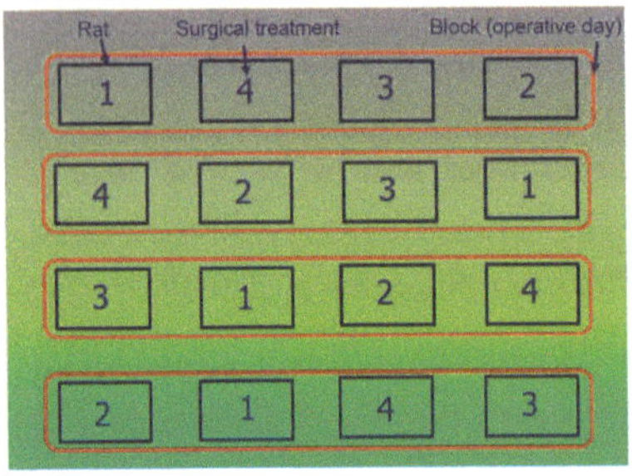

Randomised block design for animal work: comparing surgical treatments:
The "plot" becomes the singly-housed rat on which the surgical treatment is carried out, and the "block" the one day a week the surgeon can do the various operations. The four different treatments being compared are done in random sequence on the operative day. The experimental differences can be separated from any variability in conditions and operative performance from week to week.

FIGURE 11.1 The concept of "blocking."

experimental differences can be compared without the confusion of other sources of variability. Alternatively think of the pattern in the field being applied to any positional arrangement, such as stations in a milking parlor, rodent cages in a cage rack, or fish tanks in a complex water supply arrangement. More broadly, instead of plots arranged in blocks in a field think of pups coming from the same litter: the litter forms a "block" of pups, and provided littermates from each litter are allocated at random to all the treatments being compared there can be within-litter comparison of the effects of the treatments that can be viewed separately from the variability coming from the different parentage.

Blocking can also be used for within-animal comparisons. Supposing a researcher is comparing the local effects of different surface radiations on pig skin. Several small and well-separated skin areas could be marked out on a pig and one of each of the particular radiations being studied applied at random to each. The differing effects of the different radiations could be compared in that pig. But that pig might be unusual, or there might be insufficient replicates of each radiation dose or type to show a clear difference. If several pigs were used, these problems could be overcome: each pig would be regarded as a block and the experiment set up as a randomized block experiment, with each pig contributing one or more replicates to the overall comparison between treatments.

When an experiment has cages or tanks at several levels in the rack, or is spread over a number of time periods, or different laboratories, a randomized block design would likely be the one of choice.

Factorial arrangements

Another approach for reducing the number of experimental units, and therefore animals needed, is to study two or more sources of variability in the same experiment so that the experimental units used contribute information on each of the sources of variability of interest. The variabilities of interest are regarded as factors contributing to the experiment at a number of levels. So if sex difference is of interest that factor would be at two levels, male and female. Other factors might be treatment, strain of rodent, or time of day and these could be at two, three, four, or more levels depending on the number of experimental conditions or strains being compared or the number of time points to be studied.

In describing a factorial approach, the convention is that there is a number for each factor and that number gives the number of levels. So a 2×3 factorial arrangement would mean two factors, one at two levels and the other at three levels. An experiment comparing the effect of a drug against vehicle (solvent) control in two strains of mice would be a 2×2 design, with two levels of the factor "treatment" (drug or control) and two levels of the factor strain. The essential features are given in detail in Festing et al. (2016).

The variability due to random effects and biological variation operating on the experimental units used is estimated from all the units in the experiment so the number of experimental units in a subgroup (such as the strain 1: drug-treated, in the 2×2 design just mentioned) can be quite low. If the effect of one factor depends on the state or level of the other, as when, for example, the responses of males and females to a compound are different, a factorial arrangement can show this "interaction" clearly. If there is no interaction the factorial approach allows more than one experimental question to be asked using the same material, effectively doing more than one experiment on the same experimental units (a point nicely illustrated for a dairy herd by Morris, 1999, p. 64). It is thus efficient in use of resource.

Sizing experiments

Experiments should use the right number of experimental units, neither too few to have a good prospect of obtaining a reliable and reproducible result nor too many so that more animals than necessary are used. The approach to selecting suitable numbers will vary depending on the statistical test to be used and whether the purpose of the experiment is to assess whether treatments are different or equivalent (see the Experimental design and statistical analysis section below for more details on equivalence testing).

When testing for a difference between treatments, there are two main approaches that can be used. Power analysis estimates the number needed for each experimental group being compared (the "sample size"). It is based on the ratio between the size of the effect of biological interest and the estimated variability between experimental units, and takes that ratio and three elements decided by the experimenter to give how many units should be allocated to each experimental group. The experimenter has to decide the acceptable risk of missing a genuine effect, the acceptable risk of finding a false effect generated just from the random variation of the biological material, and whether a difference in one or both directions is important. A power analysis program will then output an estimate of the number of experimental units needed in each of the groups being compared. This approach is simple for fully randomized two-group comparisons but may need modifying for more complex designs. Some pointers on how to estimate the number of experimental units needed for different designs is included in the comparison by Karp and Fry (2021) of the advantages and disadvantages of different designs.

The second main approach is to use the resource equation. This estimates the number of experimental units needed for the whole experiment and is suitable for complex experiments. It does not require estimates of the size of effect and variability but is based on the law of diminishing returns—much information is gained from the first few information-contributing units used but as more units are recruited the amount gained per unit diminishes until an extra unit adds very little to the overall picture. A good balance between units used and information gained could be when between 10 and 20 units contributed. Mead (1988, p. 587) put forward a resource equation, $T + B + E = (N-1)$, suggesting the aim should be for E to be between 10 and 20. From this equation, N (the number needed for the whole experiment) could be determined from T (a treatment component given by the number of comparisons minus 1) and B (a block component, given in most designs by the number of blocks minus 1) and setting E (the number of units on which the estimate of variability between units would be based) between 10 and 20. In a 2×2 factorial arrangement, there would be four factor A/factor B combinations giving 3 degrees of freedom for contribution of the combinations to the total. Putting E at 12 would give N-1 as $12 + 3$, i.e., 15, and so 16 units for the experiment. So 4 units in each of the 4 subgroups should be adequate. More cautiously, setting E to 16 would lead to 5 experimental units in each combination.

Sequential experimental designs provide a quite different way of sizing an experiment. Comparison between the responses of experimental units in different experimental manipulations is made one or two units at a time and continued until a preset significance level or stop-testing point is reached. The number of experimental units and therefore animals used is not based on estimates but is only the number needed to meet the preset criteria. Waterton et al. (2000) took this approach for efficacy testing of potential neuroprotective agents in a rat acute stroke model and provide a good account of how a sequential study should be conducted.

How to plan experiments from day one (Adrian Smith)

> *Achieving reproducible and reliable preclinical research results is a joint responsibility that requires a partnership between the investigative team and the animal care and use program staff.*
>
> Cheleuitte-Nieves and Lipman (2019).

Good planning is essential if experiments are to be both valid and reproducible. A large number of reporting guidelines are now available (e.g., Hooijmans et al., 2010; Percie du Sert et al., 2020a), but no amount of good reporting can improve a poor experiment. A burnt cake will remain burnt, however well it is described. We need to go back to the kitchen and improve the recipe. This section provides a short introduction to the practicalities of planning and conducting animal studies: in short, being prepared.

The first step in planning any animal study should be to determine whether it is necessary at all, or whether it can be replaced by nonanimal methods. This stage should include a comprehensive review of previously published studies. Advice on this topic is available from a number of sources. There may be room for partial replacement even when some animals must be used. For example, there are now a large number of educational and training aids which can be used to help personnel acquire practical skills before they handle live animals (see the NORINA database in the list of recommended websites at the end of the chapter).

Importance of involving animal care staff

Scientists who are not specialists in research animal science, nor conversant with the latest developments in this young field, are likely to be unaware of all the factors that can influence animals and the data collected from them. They may also assume that the animal care staff will attend to all of these. Such assumptions should not be made. Planning must be carried out in close collaboration with the animal care staff from day one, whether the research is to be carried out in a facility or in the field. There are many good reasons for this:

- Animal carers and technicians have a moral right to know what is to be done with the animals in their care.
- They will be more motivated to work creatively and at the best of their ability if they see the potential benefits of the research program.
- They know the limitations of the facility best, and can assist both management and scientists in planning studies which are realistic. They will also be able to inform management about the need for any new equipment, or for allocation of funds for extra staffing.
- Their practical knowledge and ability to think laterally between studies and species may reveal better ways of conducting the research being planned.
- Failure to involve them will create anxiety and opposition to animal research, which may limit their interest in improving the study.

The personnel who have contact with the animals on a regular basis should also be the ones who perform all the procedures on conscious animals. This will minimize one of the major confounding factors in animal research: stress caused by handling and procedures carried out by people in whom the animals have no confidence.

To achieve this level of constructive dialog, there must exist a culture of care in the research facility, in which personnel of all ranks are recognized for the importance of their contributions to the research process. Such a culture should also embrace a culture of challenge, in which anyone can air their concerns about any aspect of a project, without fear of reprisals or ridicule. Constructive questions, even about well-established protocols, will help test their validity, avoiding procedures "which have always been done this way" or "as often as necessary." Standard operating procedures (SOPs) for, at the very least, the most critical steps in a project will help ensure quality. The drafts of these SOPs should be written by those who are familiar with the tasks, and then approved by management.

Detail matters

The importance of attention to detail can be illustrated by an apparently simple procedure such as blood sampling. The procedure can have critical effects on the animals and the validity of the data obtained from them. Blood removal has a definite cost to the animal concerned: not only in the form of blood loss (which may also be internal and therefore not so obvious) but also in terms of stress related to the associated handling, immobilization, and recovery periods. Prior knowledge of the total circulating blood volume of the animal is essential, along with knowledge of how much blood may be withdrawn at specified intervals without causing significant physiological changes. Although there is relatively detailed guidance in the literature, it must be applied sensibly, since animals undergoing procedures may be in a poorer clinical state than normal. The merits of different blood sampling techniques should be judged against these criteria:

- Is the blood vessel visible and therefore relatively easy to puncture?
- Can the amount of blood needed be easily removed?
- Can unwanted bleeding (either internally or externally) be prevented? This extra blood loss must be taken into account when deciding how much to collect.
- Might the technique damage surrounding tissue and/or cause significant pain?
- Can samples be collected rapidly, to avoid artefacts caused by mechanical stress ("milking" the vein), or varying degrees of degradation (metabolic or physical changes in collection tubes which have stood for varying lengths of time prior to centrifugation)?
- Is it feasible to sample all the animals within the stipulated time frame?

Similar questions should be discussed in relation to all other practical procedures, however minor they may appear. For example, when injections are to be given the factors to be considered include:

- the injection site
- the volume and irritability of the injectate
- the risk of deposition outside the target area
- the significance of hepatic metabolism before substances reach the systematic circulation, if given intraperitoneally (Turner et al., 2011)
- the potential for lasting pain
- the feasibility of the technique in the size of animal to be used.

This applies particularly to very small animals, whose relatively high metabolic rate may influence the choice of injection site. For example, subcutaneous injection is likely to provide rapid uptake, thus avoiding the need for attempted intramuscular injection into a tiny muscle, which is likely both to be more painful and difficult, if not impossible, to perform properly.

In field experiments, there may be many additional factors such as the effect of environmental temperature on blood samples and injectable solutions.

Sometimes it is not possible to find definitive answers to technical questions about procedures, or their effect on the animals. If a scientific basis for best practice cannot be found in the literature, a good rule of thumb is to apply "the three Ss" of Carol Newton (Smith and Hawkins, 2016): Good Science,

Good Sense, and Good Sensibilities. In other words, commonsense and critical anthropomorphism (based upon the likelihood of procedures generating pain and discomfort in humans) can be useful adjuncts to good science when scientific evidence is missing.

Planning documents

Many animal studies run for extended periods, encroaching both on weekends and staff holidays. Relief staff may lack full insight into the workings of the facility—and its potential weaknesses. To ensure consistent quality, a Master Plan and contingency plans should also be in place. A Master Plan is an overview of tasks which must be carried out over a period (for example, the project's length, or one year in the life of the facility), and is typically used to ensure that more infrequent procedures, such as service of instruments, are carried out at the prescribed intervals.

Contingency plans describe procedures to be followed in the event of given hazards occurring, such as power outages, flooding, and other threats to the physical buildings and their contents, including health risks to the personnel. These plans should be rapidly reassessed and updated whenever a situation threatening the facility occurs. Imagination should be encouraged: it is better to be accused of being neurotic than to overlook the need for these plans. Contingency plans should be reviewed regularly, following risk assessment of the facility, its various zones and projects. As in other areas of life, two laws tend to apply to animal research facilities: Murphy's Law ("anything that can go wrong, will go wrong") and Sod's Law ("this will happen at the worst possible time").

On a more positive note, a series of even small improvements to a protocol may lead to a significant increase in the quality and robustness of an animal study, the so-called aggregation of marginal gains.

Checklists lie at the heart of quality assurance systems. They reduce the risk of vital functions being forgotten or delayed, they ensure that complex procedures are carried out in the correct sequence, and they encourage collaboration between all parties involved, so that everyone is, literally, on the same page. Checklists are used extensively in the aviation industry, even by experienced pilots on short routine flights, at all stages of operation. It is too late to read them all at the arrival gate. Checklists should be used at all levels of activity in a research establishment.

A thorough check should be made of the facility itself. Many facilities follow the accreditation scheme run by AAALAC International. Part of the accreditation process involves writing a functional description of the facility, known as a Program Description (PD). This forms the backbone of the facility's activities. AAALAC provide a detailed template for writing the PD, covering the four pillars of facility management:

A. The Animal Care and Use Program
B. The Animal environment, Housing and Management
C. Veterinary Care
D. The Physical Plant (infrastructure).

AAALAC's template can be downloaded free of charge from their website. It can be recommended whether there are plans to apply for accreditation or not, since it can be used as a checklist for all aspects of the running of a research facility. The PD can be used actively to inform the facility of elements which should be added to the Master Plan and contingency plans. Many of these elements will involve writing SOPs.

An agreement should also be written between the research team and the animal facility. This should describe the distribution of labor and costs, clarifying all stages of the study, and should define the responsibility for collecting the data to be recorded during the entire project. This reduces the real risk of a situation, long after completion of a study, where data requested by journals or reviewers (for example, environmental parameters in the animal room) cannot be supplied because misunderstandings or unclear divisions of labor led to the data not being collected. In the worst case, it may be impossible to publish the study if crucial data are missing, wasting animal lives, research funds, and human resources.

The PREPARE guidelines

Based upon experiences with research supervision and facility management over a 30-year period, the PREPARE guidelines have been written to give detailed advice on planning animal experiments (Smith et al., 2018). PREPARE consists of a two-page checklist of 15 main topics, supported by a comprehensive website, updated frequently, which provides in-depth information about each topic, including links to the latest peer-reviewed guidelines from across the world: https://norecopa.no/PREPARE.

PREPARE covers the following main topics (see Fig. 11.2):

Formulation of the study

1. Literature searches
2. Legal issues
3. Ethical issues, harm-benefit assessment, and humane endpoints
4. Experimental design and statistical analysis

Dialog between scientists and the animal facility

5. Objectives and timescale, funding and division of labor
6. Facility evaluation

PREPARE

norecopa

The PREPARE Guidelines Checklist

Planning Research and Experimental Procedures on Animals: Recommendations for Excellence

Adrian J. Smith[a], R. Eddie Clutton[b], Elliot Lilley[c], Kristine E. Aa. Hansen[d] & Trond Brattelid[e]

[a]Norecopa, c/o Norwegian Veterinary Institute, P.O. Box 750 Sentrum, 0106 Oslo, Norway; [b]Royal (Dick) School of Veterinary Studies, Easter Bush, Midlothian, EH25 9RG, U.K.; [c]Research Animals Department, Science Group, RSPCA, Wilberforce Way, Southwater, Horsham, West Sussex, RH13 9RS, U.K.; [d]Section of Experimental Biomedicine, Department of Production Animal Clinical Sciences, Faculty of Veterinary Medicine, Norwegian University of Life Sciences, P.O. Box 8146 Dep., 0033 Oslo, Norway; [e]Division for Research Management and External Funding, Western Norway University of Applied Sciences, 5020 Bergen, Norway.

PREPARE[1] consists of planning guidelines which are complementary to reporting guidelines such as ARRIVE[2].
PREPARE covers the three broad areas which determine the quality of the preparation for animal studies:

1. **Formulation of the study**
2. **Dialogue between scientists and the animal facility**
3. **Quality control of the components in the study**

The topics will not always be addressed in the order in which they are presented here, and some topics overlap. The PREPARE checklist can be adapted to meet special needs, such as field studies. PREPARE includes guidance on the management of animal facilities, since in-house experiments are dependent upon their quality. The full version of the guidelines is available on the Norecopa website, with links to global resources, at **https://norecopa.no/PREPARE**.
The PREPARE guidelines are a dynamic set which will evolve as more species- and situation-specific guidelines are produced, and as best practice within Laboratory Animal Science progresses.

Topic	Recommendation
	(A) Formulation of the study
1. Literature searches	☐ Form a clear hypothesis, with primary and secondary outcomes. ☐ Consider the use of systematic reviews. ☐ Decide upon databases and information specialists to be consulted, and construct search terms. ☐ Assess the relevance of the species to be used, its biology and suitability to answer the experimental questions with the least suffering, and its welfare needs. ☐ Assess the reproducibility and translatability of the project.
2. Legal issues	☐ Consider how the research is affected by relevant legislation for animal research and other areas, e.g. animal transport, occupational health and safety. ☐ Locate relevant guidance documents (e.g. EU guidance on project evaluation).
3. Ethical issues, harm-benefit assessment and humane endpoints	☐ Construct a lay summary. ☐ In dialogue with ethics committees, consider whether statements about this type of research have already been produced. ☐ Address the 3Rs (replacement, reduction, refinement) and the 3Ss (good science, good sense, good sensibilities). ☐ Consider pre-registration and the publication of negative results. ☐ Perform a harm-benefit assessment and justify any likely animal harm. ☐ Discuss the learning objectives, if the animal use is for educational or training purposes. ☐ Allocate a severity classification to the project. ☐ Define objective, easily measurable and unequivocal humane endpoints. ☐ Discuss the justification, if any, for death as an end-point.
4. Experimental design and statistical analysis	☐ Consider pilot studies, statistical power and significance levels. ☐ Define the experimental unit and decide upon animal numbers. ☐ Choose methods of randomisation, prevent observer bias, and decide upon inclusion and exclusion criteria.

Topic	Recommendation
	(B) Dialogue between scientists and the animal facility
5. Objectives and timescale, funding and division of labour	☐ Arrange meetings with all relevant staff when early plans for the project exist. ☐ Construct an approximate timescale for the project, indicating the need for assistance with preparation, animal care, procedures and waste disposal/decontamination. ☐ Discuss and disclose all expected and potential costs. ☐ Construct a detailed plan for division of labour and expenses at all stages of the study.
6. Facility evaluation	☐ Conduct a physical inspection of the facilities, to evaluate building and equipment standards and needs. ☐ Discuss staffing levels at times of extra risk.
7. Education and training	☐ Assess the current competence of staff members and the need for further education or training prior to the study.
8. Health risks, waste disposal and decontamination	☐ Perform a risk assessment, in collaboration with the animal facility, for all persons and animals affected directly or indirectly by the study. ☐ Assess, and if necessary produce, specific guidance for all stages of the project. ☐ Discuss means for containment, decontamination, and disposal of all items in the study.
	(C) Quality control of the components in the study
9. Test substances and procedures	☐ Provide as much information as possible about test substances. ☐ Consider the feasibility and validity of test procedures and the skills needed to perform them.
10. Experimental animals	☐ Decide upon the characteristics of the animals that are essential for the study and for reporting. ☐ Avoid generation of surplus animals.
11. Quarantine and health monitoring	☐ Discuss the animals' likely health status, any needs for transport, quarantine and isolation, health monitoring and consequences for the personnel.
12. Housing and husbandry	☐ Attend to the animals' specific instincts and needs, in collaboration with expert staff. ☐ Discuss acclimatization, optimal housing conditions and procedures, environmental factors and any experimental limitations on these (e.g. food deprivation, solitary housing).
13. Experimental procedures	☐ Develop refined procedures for capture, immobilisation, marking, and release or rehoming. ☐ Develop refined procedures for substance administration, sampling, sedation and anaesthesia, surgery and other techniques.
14. Humane killing, release, reuse or rehoming	☐ Consult relevant legislation and guidelines well in advance of the study. ☐ Define primary and emergency methods for humane killing. ☐ Assess the competence of those who may have to perform these tasks.
15. Necropsy	☐ Construct a systematic plan for all stages of necropsy, including location, and identification of all animals and samples.

References
1. Smith AJ, Clutton RE, Lilley E, Hansen KEA & Brattelid T. PREPARE:Guidelines for Planning Animal Research and Testing. *Laboratory Animals*, 2017, DOI: 10.1177/0023677217724823.
2. Kilkenny C, Browne WJ, Cuthill IC *et al.* Improving Bioscience Research Reporting: The ARRIVE Guidelines for Reporting Animal Research. *PloS Biology*, 2010; DOI: 10.1371/journal.pbio.1000412.

Further information
https://norecopa.no/PREPARE | post@norecopa.no | @norecopa

FIGURE 11.2 The PREPARE checklist (available in 25 languages—see the Recommended websites listing at the end of the chapter). *From Smith, A.J., Clutton, R.E., Lilley, E., Hansen, K.E.A., Brattelid, T., 2018. PREPARE: guidelines for planning animal research and testing. Lab. Anim. 52 (2), 135–141. https://doi.org/10.1177/0023677217724823, reprinted with permission from Laboratory Animals.*

7. Education and training
8. Health risks, waste disposal, and decontamination

Quality control of the components in the study

9. Test substances and procedures
10. Experimental animals
11. Quarantine and health monitoring
12. Housing and husbandry
13. Experimental procedures
14. Humane killing, release, reuse, or rehoming
15. Necropsy

Unlike reporting guidelines, PREPARE places emphasis on the need for close collaboration with animal care staff (see in particular topics 5–8 and 10–15). The attention paid to these topics may not be mentioned in the publications coming from the project, but it will have been crucial to their success. The checklist is currently available in 25 languages, and a three-minute cartoon film has been produced to illustrate the similarities to quality assurance in the aviation industry.

PREPARE is designed to be offered to both scientists and animal facilities on a voluntary basis, as an *aide memoire* to avoid forgetting essential tasks. Reporting guidelines can only, at best, improve the next round of studies. Furthermore, compliance with reporting guidelines has, unfortunately, proved to be low, and studies show that many researchers who publish in journals that have endorsed them, have never heard of them (Reichlin et al., 2016). Thus there is good reason to encourage uptake of planning guidelines, without turning them into an additional regulatory hurdle.

Final points

Good planning may reveal opportunities for replacement, reduction, or refinement. These should be highlighted in the publications coming from the research. Since many bibliographic databases only index the title and abstract of a paper, these 3R improvements should wherever possible be flagged in these parts of the paper, or described in a separate methodology paper.

In summary, good planning will lead to:

- *Better science*
- *Improved animal welfare*
- *Advancement of the 3Rs*
- *A safer working environment for both animals and humans*

Animal models (Joanne Storey)

An *animal model* can be defined as a spontaneous or induced (through selective inbreeding or other procedure) replication of physiological mechanism(s), or of a disease state, in a nonhuman animal. The decision to utilize an animal model should include consideration of the biological question under scrutiny. Nonanimal approaches should always be considered, but there are biological systems questions that cannot be answered with a reductionist approach, so need an animal model. However, the selection of the animal model should always be through detailed analysis rather than acceptance that a specific historical model is appropriate.

Importance of selecting the appropriate animal model

A thorough understanding of the literature is necessary to understand both potential nonanimal alternatives and the value of proposed animal models (see also the Preclinical Systematic Reviews section of this Chapter).

Utilization of a well-constructed search string and use of appropriate search tools, see for example EURL ECVAM Search Guide, and bibliographic databases, e.g., Medline, is required to retrieve relevant literature. As mentioned in the Introduction, searching for replacement alternatives should be repeated during a program of work, and results and experience from early experiments looked at for indications of refinements that could reduce the severity of later ones. As work progresses, reconsidering the suitability of an animal model is also important, alongside the experimental design within that given animal model. For example, an exploratory experiment within a given model should inform confirmatory work within the same model.

Consideration of an animal model should explore many factors. In cases when an induction (e.g., via chemical administration or genetic modification) is utilized, the differences with the final disease state are important to consider. Lack of understanding of the pathophysiological events behind disease symptoms can lead to incorrect assumptions and caution is needed when assessing a model on phenotypic symptoms alone. An example is afforded by the animal models of amyotrophic lateral sclerosis (ALS): mouse models expressing a mutant version of the TDP43 protein have been utilized as models for ALS although they differ in key mechanisms from the human disease. Such mice usually die of bowel obstructions, whereas human patients succumb to muscle wasting, often dying from inability to breathe. Thus, for investigation of a number of scientific questions, this particular genetically altered animal model may not be appropriate.

Key measures of disease biology ("biomarkers") can be useful bridges from the human condition to an animal model and provide confidence in the relevance of that model, and should be fully explored when conducting an evaluation. Additionally, consideration should be given to the utility or shortcomings of replicating long-term disease with short-term models or using juvenile animals to investigate diseases of adult patients.

Framework for selecting the appropriate animal model

Points to consider when selecting the appropriate animal model are suggested below but note these are in the context of disease models or models of a biological mechanism thought to contribute to disease (translational relevance).

- **Human or veterinary condition**—How well understood is the human or veterinary condition which is being modeled?/How well understood is the mechanism of interest in the species intended to benefit from the research?
- **Model physiology**—How well is the systemic and molecular physiology of the intended patient replicated in this model?
- **Model pharmacology**—How well does the model replicate the expected pharmacology in patients?

- **Model disease**—How well does the model replicate the etiology and pathogenesis of the patient disease?
- **Reproducibility**—How reproducible are the biological measures of interest in this model?

Periodic retrospective assessment to justify ongoing use of an animal model is important. Additionally, under Directive 2010/63 retrospective assessment is a requirement for work on nonhuman primates and for projects involving severe procedures, and it may also be stipulated at authorization.

The above list focuses on human translation studies, but it is recognized that animal models are required for basic scientific research, disease or symptom characterization, and safety. While the human context should be paramount for human translation studies, a more general approach for models falling outside of that realm should utilize a framework in which, when seeking a new model, the objective is carefully considered. Building from that, a process of understanding the criteria required to meet the objective should be worked through, a search conducted on those scientific criteria, and the models that meet one or more of them evaluated.

Fidelity and discrimination

Consideration of fidelity/discrimination comes among the many criteria for assessing a model. The topic was originally raised by Russell and Burch (1959), with "discrimination" being the extent to which the model reproduces one particular property of the target, and fidelity the general similarity to what is being modeled. For a model to be predictive of the effect of a compound in humans there is a balance to achieve between having an ability to discriminate between treatments while maintaining fidelity or relevance to what is being modeled, which in many cases is the human mechanism. Defining a clear experimental hypothesis will aid in consideration of fidelity and discrimination. The concepts are useful both for animal and nonanimal model assessment.

General points

Bringing the above considerations together should occur with specific reference to Article 13.2 of the Directive: "In choosing between procedures, those which to the greatest extent meet the following requirements shall be selected: (a) use the minimum number of animals; (b) involve animals with the lowest capacity to experience pain, suffering, distress or lasting harm; (c) cause the least pain, suffering, distress or lasting harm; and are most likely to provide satisfactory results."

Methodology and detailed protocols for animal models have further and wider reaching 3Rs impact when made fully accessible, so should be fully published if possible. This would aid, for example, sharing efficiencies in methodology, linking up investigators with experience of a specific methodology, and preventing repeat assays. It is also important to share refinements made with a model, such as welfare improvements and steps to minimize adverse effects.

Standardization of the animal model

Scientists are faced with a paradox: obtaining data with little variability and working toward improved reproducibility requires standardizing many factors within experiments, but this results in a reduction of the generalizability of the results. This was made clear by Fisher in 1960 “The exact standardization of experimental conditions, which is often thoughtlessly advocated as a panacea, always carries with it the real disadvantage that a highly standardized experiment supplies direct information only in respect of the narrow range of conditions.”

For a particular model to be useful across several laboratories, the process of induction and the characteristics of the model, and perhaps other features, may need to be standardized. Yet, the impact of biological heterogeneity on reproducibility of results is increasingly recognized and warrants further consideration at the time of experimental design.

Within animal models, there is scope for biological heterogeneity and this is being discussed increasingly in the scientific literature. Sources of heterogeneity include but are not limited to age, sex, and strain. A current paradigm shift is being proposed (Voelkl et al., 2020), namely deliberate heterogenization, including in an experiment, for example, different housing/environmental conditions, ages, sex, test times, and alternate test systems. If such sources of variability are included within the experimental design, the appropriate statistical analysis is essential, and the design should be discussed with statisticians.

It is additionally important to consider how reliance on a particular genotype may not adequately reflect the genetic diversity of natural populations. Also in immunology studies, reliance on pathogen-free conditions for the animals may mean the immune responses seen in experiments are limited.

A further point is that maximizing a signal window (large effect size) within an animal model, while improving the signal/noise (effect size to variability) ratio, may be constraining other interesting scientific results. Consider, for example, how anesthetic choice may artificially dampen one physiological response allowing focus on another, e.g., the anesthetic effect on cell types within the cerebellum (Bengtsson and Jörntell, 2007). So in a study of mossy fibers in the cerebellum, using an anesthetic which depressed responses in other cerebellar fibers could enhance the responses of interest and improve effect size, but this would be at the cost of being unrepresentative of general cerebellar function, and with the risk of having less biological relevance. New advances/technological limitations should also be recognized.

Surrogate markers for monitoring may in some cases be useful, but we should watch that we are not opting for the cleanest signal or easiest surrogate marker to monitor rather than challenging ourselves to look for new advances/ technologies to really understand what is biologically relevant.

Animal studies have been increasingly criticized for lack of reproducibility (Baker, 2016; Begley and Ellis, 2012; Prinz et al., 2011) but detailed discussion on the topic is outside of the scope of this chapter. Clearly concerns around reproducibility raise significant ethical and economic questions. In order to support reproducible science, good methodological documentation is essential to enable conclusions to be drawn on the generalizability of the results. For example, the extent to which bias has been minimized in a study and the use of power calculations to estimate numbers needed are important pieces of information for an informed judgment on how likely the results are to be generalizable. There are several guidelines and templates available to support the reporting of animal experiments that should be consulted e.g., ARRIVE v2.0 (Percie du Sert, 2020a), "Landis 4 criteria" (Landis et al., 2012), Gold standard publication checklist (Hooijmans et al., 2010).

Scientists should ultimately consider both standardization and variability from both a biological and statistical viewpoint. Your model should allow measurement of a statistically significant result that has biological relevance. It may be beneficial to achieve more generalizable results, or to put it another way to achieve the greatest amount of knowledge per individual study, likely gained by utilizing heterogenized study designs. In turn this can be factored into the Harm—Benefit Analysis (see below) in terms of the benefit being obtained from an individual animal experiment.

Experimental design and statistical analysis (Simon Bate)

It has long been recognized that the correct use of statistics and experimental design is important when implementing the 3Rs. In their seminal work, Russell and Burch (1959) state that "Every time any particle of statistical method is properly used, fewer animals are employed than would otherwise have been necessary." Whether it be designing more efficient experiments that use fewer animals but generate reliable conclusions, or using more appropriate statistical methods that provide better insights into the objectives of the experiment, in animal-based research the correct implementation of statistics is often a crucial part of the experimental process (see Bate and Clark, 2014; Gosselin, 2019).

Research conducted using animals usually involves significant biological variability, and yet the sample sizes employed can be relatively small. These issues increase the risk that experimental conclusions may be unreliable (Button et al., 2013). However, the researcher has a lot of control over the experimental design employed. They must take advantage of this if they are to improve the reliability of the experimental results. For example, they can decide the bodyweight range of the animals employed, their genetic status, and the environment they are housed in. If the results generated are to be reliable then this researcher control, and the information it provides, must be made use of when designing the experiment and analyzing the data generated.

In this section, some useful statistical analysis techniques are described, with particular focus on when and where they should be applied. We follow the principle that the statistical analysis should reflect the experimental design employed. For example, if nuisance sources of variability were taken into account when designing the study, then this should be reflected in the analysis (Bate and Clark, 2014; Festing et al., 2016). These texts provide detailed illustrations of how these tools are used in practice, which is outside the scope of this chapter. Instead of stepping through an example analysis, for some points reference is made to the section of the freely available statistics program InVivoStat, which readers can use with their own data. However, if available we strongly recommend seeking out the support of a professional statistician. They can advise on both the experimental design and statistical analysis so that the animals are used in the most efficient way and the conclusions of the study are reliable and reproducible.

Experimental objectives and analysis strategies

As already said above in this chapter, when setting up an experiment, the first thing to consider is the objective of the experiment. Is the experiment to test a predefined hypothesis (a "confirmatory" experiment), or to generate knowledge that will lead to a testable hypothesis (an "exploratory" experiment). The latter include "pilot" studies to refine an existing animal model or gain a better understanding of the experimental variability (Huang et al., 2019). As commented above, the Directive considers work conducted over a five-year period, and this can include the initial work carried out to set up/refine an animal model, experiments to generate hypotheses to test, and then confirmatory experiments to assess a predefined hypothesis (Sorzano and Parkinson, 2019).

Exploratory and confirmatory experiments require different experimental designs, each with their own particular statistical analysis strategy. At the risk of overgeneralizing, there are two main types of statistical analysis, "estimation" and "hypothesis testing." The purpose of the former is to estimate the size (and/or variability) of the response, under different experimental conditions, and to put a statistically defined interval around these estimates. The hypothesis testing approach, which generates *P*-values and conclusions regarding statistical significance, is a more formal analysis approach and can be used when testing predefined hypotheses.

In practice experiments can have multiple objectives. For example, an experiment designed to test a predefined hypothesis regarding the effect of a novel compound on the hippocampus (using MRI imaging) can also be used as an exploratory experiment to investigate the effect of the compound on other brain regions. Results from such exploratory analyses should be treated with caution and positive findings verified in follow-up confirmatory experiments. It is important to acknowledge that confirmatory experiments test for particular hypotheses and these should not be adjusted after the experimental results have been generated, a practice known as "HARKing" (Hypothesizing After the Results are Known) Munafo et al. (2017).

Design and analysis of pilot and exploratory studies

Pilot and exploratory studies are performed to gain a better understanding of the animal model or obtain an estimate of the variability. The designs used may be relatively small and yet can involve assessing many factors of interest. Bate and Clark (2014) define a type of design (the "large factorial" design) that can be used to identify which effects influence the response and crucially how they interact with each other (see also Shaw et al., 2002). The statistical analysis of such studies can generate estimates of the effects along with measures of the reliability of these estimates; for example, group means with either standard error of means (SEMs) or confidence intervals.

Design and analysis of confirmatory experiments

Confirmatory experiments are performed to assess specific predefined hypotheses. They should be designed using a formal framework that includes identifying a suitable sample size before running the experiment (Huang et al., 2019).

To begin with the researcher should consider whether they are "testing for a difference" (i.e., trying to show that the intervention causes a change in the response) or whether they are "testing for equivalence" (i.e., trying to show that the intervention has no effect on the response). These are different scenarios and require different analysis strategies.

"Testing for a difference" is a commonly employed analysis strategy and begins with the definition of the "null" hypothesis (i.e., that the treatment has no effect). The researcher then attempts to disprove this by generating data. They also need to consider the "alternative" hypothesis: this can be one-sided (i.e., the treatment is bigger or smaller than the comparator) or two-sided (the treatment effect can be in either direction). For a given biologically relevant effect, and an estimate of the variability of the response, the researcher should choose a suitable sample size (Ruxton and Colegrave, 2011; Lazic et al., 2018). The statistical analysis will generate, among other things, P-values. The P-value is the probability of observing data as extreme as that observed, or more extreme, when the null hypothesis is true. When the P-value is less than a predefined significance level (usually .05 or 5%), then it is standard practice to conclude the observed results do not support the null hypothesis, and hence that the alternative is true. There are some things to note with this approach:

- The *P*-value is only related to the null hypothesis—it is not directly related to the alternative hypothesis and hence does not provide any information about the alternative hypothesis. Additionally, when the null is rejected the *P*-value is no longer meaningful (as it is generated using the discredited null hypothesis).
- A *P*-value of .04 is almost the same as a *P*-value of .06, even though it is standard practice to declare the former "statistically significant" but not the latter.
- The choice of sample size is critical if the *P*-values are to be reliable and informative, otherwise observed effects that are biologically important may be not declared statistically significant (or vice versa).

Although not as popular, if the researcher is trying to prove that experimental interventions are equivalent then a "test of equivalence" should be used, for example, the Two One-Sided test or TOST approach (Limentani et al., 2005). Before performing a TOST test, the researcher must define the equivalence bounds (i.e., the region where results are considered equivalent). As long as the confidence interval around the observed difference between the experimental intervention means is within this region, then equivalence can be declared.

Statistical analysis tools

In this section, some parametric and nonparametric analyses are described. The parametric tests rely on making "parametric" assumptions, but as long as these assumptions hold they provide a powerful set of tools that can take account of the experimental design used.

Parametric analysis: t-test

The t-test is perhaps the most commonly employed test in animal research. By t-test we imply that the dataset analyzed consists of only two groups. The observed difference between the means of the two groups is compared to a pooled estimate of the within-group variability. We contend this test should rarely, if ever, be used if the experiment involves more than two groups. Using a t-test to compare only two of several groups does not make use of all the information collected, and hence is not the most powerful test available, if the experiment (1) has one factor at more than two levels (e.g., increasing doses of a treatment and control); (2) involves multiple factors of interest, each with two or more levels (i.e., is a factorial design); (3) has additional measurements taken on the animal (baseline bodyweight, age, or pretreatment response); (4) involves measuring the animals repeatedly over time (Bate and Clark, 2014). Using less powerful tests imply more animals will be required to achieve the same level of confidence.

Parametric analysis: analysis of variance

Increasing the number of experimental units in a group, all other things being equal, will improve the reliability of the estimate of the group mean. The same is true of the estimate of the variability. If responses from all experimental units in the experiment are used to estimate the underlying variability (rather than taking two groups only, as you would do when performing a t-test) then the estimate of the variability will be more reliable, resulting in more reliable/reproducible conclusions. This is one of the benefits of performing Analysis of Variance (ANOVA). All of the data are used to estimate the variability, and this estimate is then employed to generate the statistical results. The ANOVA approach can be used when (1) the factor of interest has more than two levels; (2) the experiment involves multiple factors at two or more levels (i.e., factorial designs); (3) additional factors are included in the design to account for nuisance sources of variability (i.e., block designs). While the "ANOVA table" provides tests of overall effects, we can also make pairwise comparisons between group means. The crucial difference between an ANOVA-based pairwise test and the t-test is the estimate of the variability used. In the ANOVA-based pairwise test, the variance estimate is obtained from all experimental units, not just two groups in isolation, and hence is more reliable.

An ANOVA analysis can be used to perform hypothesis tests (i.e., generation of *P*-values) but can also be employed to estimate group means (and differences between the means) with confidence intervals, where the pooled estimate of variability is used to calculate the confidence intervals. These latter results are useful when performing exploratory analyses. A detailed description of the ANOVA analysis approach can be found in many textbooks, for example, Bate and Clark (2014, Chapter 5.4) and Festing et al. (2016, Chapter 4).

All modern statistical software packages can perform ANOVA, and without the need for all the groups to be the same size. For example, InVivoStat's "Single Measures Parametric Analysis" module (www.invivostat.co.uk) performs multiway ANOVA and has the ability to include blocking factors (Clark et al., 2012).

Parametric analysis: analysis of covariance

If a block design has been employed by the researcher, then a categorical blocking factor can be included in the ANOVA analysis to account for any variability associated with the blocking effect. It may be the case, however, that continuous variables are available that, while not of direct interest themselves, can help account for some of the nuisance variability in the response. Examples included baseline bodyweight or a pretreatment measure of the response. These continuous variables are known as covariates, and the analysis that includes them is called Analysis of Covariance (ANCOVA). If a suitable covariate is included in the analysis, then it can reduce the variability and hence improve the sensitivity and reliability of the results. Many statistical packages, such as InVivoStat, allow the user to include covariates in the statistical analysis. When including covariates in the statistical analysis certain assumptions are made (Bate and Clark, 2014). InVivoStat's output provides information about whether the covariate should be included in the analysis.

Parametric analysis: linear and nonlinear regression

These tools tend to be used for exploratory analyses when the researcher investigates the effect of a continuous variable on a response. There are strong links between regression and ANCOVA. The former involves including in the analysis continuous variables that are "of interest," and hence the purpose of the statistical analysis is to estimate the relationship between the response and the continuous variable. The latter involves including nuisance continuous variables in the statistical analysis to solely account for nuisance sources of variability, hence estimating any relationships is not of direct interest.

Sometimes the relationship between the response and the continuous variable is nonlinear. For example, when modeling a dose-response relationship (Liao and Liu, 2009). Such analyses are used mainly in exploratory experiments, where summary measures of the dose-response relationship (perhaps the maximum concentration, CMax, or half-maximal inhibitory concentration, IC_{50}) are of interest. Many modern packages provide the user with the ability to fit nonlinear curves to data, for example, InVivoStat's "Dose-Response" module, that fits the Hill equation to data.

Parametric assumptions

All of the above tests rely on certain parametric assumptions—the responses should be continuous (i.e., responses can take any numerical value), normally distributed, and the variability should be the same across the experimental groups (the "homogeneity of variance" assumption). Additionally the individual responses should be independent, and this can be assumed if the randomization has been performed correctly.

As discussed above, most parametric analyses rely on a pooled estimate of the variability. For the results of such analyses to be reliable, then the pooled estimate of variability should reflect the variability across all the experimental groups. Where this is not the case (and this can be informally assessed by considering the error bars on the "Means with SEMs" plots often used by researchers) then the results can be unreliable.

The parametric assumptions can be assessed using formal statistical tests and many software packages offer various options. For example, tests for normality include the D'Agostino-Pearson and Shapiro-Wilkes tests, and tests for homogeneity of variance include the Bartlett and Brown-Forsythe tests (Zar, 2019). InVivoStat uses graphical tools (the QQ-plot for normality and the Residuals vs. Predicted plot for homogeneity of variance). These latter tools are flexible and can be applied to any ANOVA, ANCOVA, and/or more complicated analyses.

If the assumptions do not hold then the researcher should try to transform the response, prior to analysis, to address any issues regarding the parametric analysis assumptions (Bate and Clark, 2014). Transformations, such as log, square root or arcsine, can be performed manually before the start of the analysis or within the analysis package itself (InVivoStat has an option to transform the data during the analysis process).

Parametric test: repeated measures analysis

It may not be possible to perform the randomization that is required for the parametric assumption of independence to be satisfied. For example, if animals are measured repeatedly over time, then the testing will be performed, by definition, in a nonrandom order (day 1 must come before day 2). This lack of randomization implies that the results for each animal will be related over time and hence the independence of the responses cannot be assumed. To account for the lack of independence either (1) a summary measure of each animal's responses should be analyzed (for example, the mean response, the CMax or the area under the curve) or (2) a repeated measures analysis should be employed. The former approach is recommended if a suitable summary measure exists, as it can address the scientific question that is being assessed. The repeated measures analysis makes certain assumptions about the within-animal measurements that may not hold. However, such an analysis can account for any relationships between the within-animal responses and hence allow the independence assumption to be relaxed.

Many statistical packages have repeated measures tools, for example, repeated measures ANOVA and/or repeated measures mixed models. InVivoStat performs a repeated measures mixed model within the "Repeated Measures Parametric Analysis" module (Clark et al., 2012). One benefit of the mixed model over the ANOVA-based approach is that it deals with missing data. With the ANOVA-based approach, any animal with a missing single response is excluded from the analysis, whereas they will still be included in the mixed model analysis.

Nonparametric analyses: Mann–Whitney, Kruskal–Wallis and Friedman's tests

Nonparametric tests tend to avoid investigating model parameters and focus instead on assessing the nature of the distribution. Such tests do not depend on assumptions about the distribution of the data. They should be used when the data are not normally distributed and no transformation can be applied that would make it valid to use a parametric test. As with parametric tests, nonparametric tests depend on the assumption that the measurements are all independent. They tend to be less powerful when compared to the parametric test equivalents, if the parametric assumptions hold. Commonly used nonparametric tests include the Mann–Whitney test (equivalent to the t-test), the Kruskal–Wallis test (equivalent to the one-way ANOVA), and Friedman's test (equivalent to the two-way ANOVA without interaction), see McKillup (2011).

Multiple comparison tests

When performing multiple statistical tests (be that multiple t-tests or multiple pairwise comparisons following ANOVA), then the more tests the researcher performs, the greater the risk of finding a false positive result. There are many analysis tools available to reduce this risk; common examples include Bonferroni, Dunnett's, and Tukey HSD. Generally these tests make it more difficult to declare a result statistically significant, hence reduce the false positive risk; however, these tests also reduce (to a greater or lesser extent) the ability to identify true positive results. For a discussion of these tests, including when/where to apply them and the controversies surrounding them, see Bate and Clark (2014).

Summary

In conclusion, the statistical analysis should reflect the experimental design used and should be chosen to maximize the statistical power. This strategy will help to ensure that the impact of any sources of variability, which were identified when constructing the design, can be accounted for in the statistical analysis. This should give more reliable/reproducible results, while allowing the researcher to reduce animal use.

Preclinical systematic reviews (Merel Ritskes-Hoitinga)

Systematic reviews are very helpful when preparing new projects and procedures, as they lead to a synthesis of already available knowledge, thereby also identifying knowledge gaps and preventing unnecessary duplication of animal studies. By obtaining a critically appraised complete body of evidence of what has already been published before and by identifying knowledge gaps, a suitable basis for designing new projects is reached about what will be needed to study these gaps. This applies not only to animal studies but also to projects with nonanimal alternatives and novel methods, or a combination of animal and nonanimal studies. As such this helps meet the requirements of EU Directive 2010/63EU, contributing to the implementation of the 3Rs. A thorough and complete literature search is aimed for in one of the steps in systematic reviews, thereby also safeguarding the 3Rs. As well as the implementation of the 3Rs, systematic reviews also contribute to better science by creating transparency on the quality and translational value of published data. Since 1992 the methodology has been well-known in the clinic, and systematic reviews of clinical studies have been performed routinely within the global Cochrane Collaboration, leading to evidence-based medicine. As many animal studies are aiming to inform about and improve human health, it seems only logical to also perform systematic reviews for preclinical investigations, so they become more evidence-based. Moreover, by combining clinical and preclinical systematic reviews an analysis of the translational evidence is provided. The number of preclinical systematic reviews is still well below the number of clinical ones, and for many reasons, it is desirable that it increases.

What are systematic reviews?

A systematic review is commonly defined as a literature review focused on a single question which tries to identify, select, appraise, and synthesize all high-quality research evidence relevant to that question. As developed by the Cochrane Collaboration, a systematic review comprises a number of standard steps: (1) Phrasing the research question, (2) Searching for all evidence, (3) Selecting the relevant studies for the question, (4) Extracting the study characteristics (species, sex, age, etc.), (5) Assessing the study quality and/or doing a risk of bias assessment, (6) Synthesizing the outcome data and if possible, doing a statistical analysis of all results found (meta-analysis). By collecting and synthesizing the information already available, new results and insights can be obtained without doing new studies. As the Cochrane methodology is specifically oriented toward clinical studies, adaptations have been necessary to use these methods for the preclinical field.

Why do systematic reviews?

The specific search for the 3Rs appears to be quite challenging, as there is a huge number of 3R databases and websites available, each with a different content, structure, and search options. By performing questionnaires and interviews on 3R searches among researchers, animal welfare officers, and members of ethical review committees in the Netherlands, it was concluded that existing 3R possibilities are not always found (van Luijk, 2017). Strictly speaking under EU Directive 2010/63EU, you must use existing 3R methods, but what if it is so difficult to do effective 3R searches? The advantage of systematic reviews is that they require a thorough literature search in at least two databases, and for that reason this method is considered quite effective for achieving better science and finding and implementing 3R possibilities at the same time. Effective literature searches can be performed by using search filters for animal studies in Pubmed and Embase, which have been developed by SYRCLE (SYstematic Review Center for Laboratory (animal) Experimentation; www.syrcle.nl). By using these filters combined with the particular area of interest, all animal studies published in that area and included in Pubmed and Embase can be found. SYRCLE has also developed several tools adapted to preclinical systematic reviews, e.g., a Risk of Bias tool, and has published guidance on how to perform a meta-analysis (Hooijmans et al., 2014; de Vries et al., 2014). More information is also available on the SYRCLE website and through the free e-learning module (https://syrcle.ekphost.nl; login code: syrcle). Ideally, before embarking on new projects, a systematic review should have been done to analyze all the evidence already available on a certain topic.

An unfortunate benefit of preclinical systematic reviews is that they show the inadequate quality of publications of animal studies: 50%–80% of all publications included in systematic reviews do not report essential details such as method of randomization, blinding procedures, power calculation and mention of sex, strain designation, diet composition, etc. This makes a valid scientific interpretation of the study results biased (compromised internal validity) and compromises the generalizability of the results (to another species like the human; external validity). It needs to be stressed that use of PREPARE and ARRIVE guidelines is of the utmost importance for experimental studies. Next to publication quality, many systematic reviews also demonstrate the low translational value of animal studies for humans. At present we cannot tell whether this low translation is caused by low internal validity (poor study quality) or low external validity (whether in general research results from animals are relevant for the human situation), or both.

A 2020 impact study in the Netherlands has shown that an educational program financed by the Dutch Health Funding Agency ZonMW, aimed at stimulating the use of preclinical systematic reviews, has greatly influenced research and researchers in a positive way (https://www.zonmw.nl/fileadmin/zonmw/documenten/Fundamenteel/MKMD_2020/Report_ZonMw_Preclinical_impact_study_SR_final.pdf). In the Netherlands, over a period of 10 years, more than 20 one-day workshops have been organized with about 400 participants, resulting in 88 workshop participants starting their own preclinical systematic review. The operation was conducted under the supervision of SYRCLE co-workers to guarantee quality and keep the process flowing. Questionnaires and interviews among the researchers involved revealed that they thought they had learned new skills and understood the value of systematic reviews, and indicated motivation toward commitment to changing their laboratory practice and field of science for the better. In some cases, researchers chose not to go down the road of animal studies any more, choosing alternative methods or a clinical study instead.

Because systematic reviews have been developed in the clinic for randomized clinical trials, systematic review is sometimes considered a useful method only for preclinical (translational) research, and not for fundamental science. However, for any topic, a systematic review is a useful exercise, especially because the step of searching for all evidence makes sure a good literature search is performed, preventing unnecessary duplication. Such an in-depth comprehensive literature search is also helpful for doing scoping or mapping reviews in order to get a good overview of available data without going into too much depth (Leenaars et al., 2020). In fundamental science, systematic reviews can lead to new results and insights, and the formulation of truly novel hypotheses (Hollyer et al., 2019).

How do SRs help to implement the 3Rs?

The extensive searches done for systematic reviews can also make unnecessary duplication transparent. In a systematic review on intestinal anastomosis models, Yauw et al. (2015) found that 88 out of a total of 1342 studies had duplicated the adverse effect of chemotherapy on wound healing. This adverse effect is already well-known, and Yauw et al.'s systematic review can hopefully prevent new animal studies on this particular topic (Reduction). Moreover, in the same systematic review, it was discovered that in 83% of the publications included there was no mention of whether surgery had been performed under sterile or nonsterile conditions, and 91% did not mention whether postoperative analgesics were used. If it is not mentioned in a publication, it is not clear whether an intervention was applied or not. Cochrane rules are that when it has not been mentioned, it is presumed not to have been done. If postoperative analgesics were not given, this raises serious concern about both animal welfare and possible adverse effects on the experimental results. It is essential that sterile surgery and postoperative analgesia are used and reported (Refinement).

Currie et al. (2013) performed a systematic review on animal experiments studying bone cancer pain. This study provides a good example of how a systematic review can be done and the value of doing it. They 'searched three electronic databases ... using the terms [bone cancer OR osteosarcoma] AND [pain OR analgesia]," limited results to animals, and then selected for further analysis publications that described studies of bone cancer animal models that measured a pain-related behavioral outcome, and that described this outcome in control or sham-injured animals or from the contralateral side of a tumor-bearing animal. Titles and abstracts were assessed, with translation if needed, and fuller details obtained of studies meeting the prespecified inclusion criteria. Although 1156 studies were initially identified, removal of duplicates, reviews, and studies not meeting the inclusion criteria reduced this to 112. Data from these were subjected to a quality assessment and meta-analysis.

This exercise gave useful information on the level of pain seen in different animal models, providing evidence on which a researcher could question whether a high-pain model was necessary for a particular study. It also indicated that the best effect size was seen with the least invasive method of assessing pain, termed "spontaneous pain", meaning that this was assessed from incidence of unprovoked vocalization, guarding, change of gait, and so on. Although the greatest differences in results occurred when studying spontaneous pain behaviors, mechanically evoked pain behaviors were more often reported. The review highlighted that spontaneous pain behaviors are important outcome measures for clinical relevance and for the development of effective therapeutic targets. Measuring spontaneous behaviors is associated with lower degrees of discomfort as compared to evoked pain behavior, and this review illustrates that systematic reviews can be useful for planning procedures, helping in assessing their severity and the animal welfare aspects of reaching the goal of the study.

For selecting a suitable (animal) model system, systematic reviews lead to more evidence-based choices. De Vries et al. (2012) looked at tissue engineering experiments for cartilage defects. Rodents and rabbits turned out to be frequently used, however, the pathophysiological process in these species differs significantly from the human process. Human cartilage defects are often limited to the cartilage, causing healing to be problematic as there is no blood vessel ingrowth. Experimental cartilage defects in rodents and rabbits will always turn out to be osteochondral, as the cartilage layer is so thin. This implies that lesions extend into the bone underneath the cartilage, always leading to blood vessel ingrowth and successful healing. As this process is so different from humans, larger animal species are therefore more suitable. Veterinary patients as models for human medicine have gained popularity (one health/one medicine concept), as veterinary patients such as horses, have similar cartilage defects as humans. By studying these veterinary patients, more relevant therapeutic results for humans can be obtained too.

Systematic reviews can also be done for studies on alternatives, such as in vitro studies (Golbach et al., 2016). By reviewing the evidence from studies that are based on the use of human materials, a systematic review can contribute to preventing and replacing animal studies. Systematic reviews hold great promise for the future, as they can provide the evidence for the validation of alternative *replacement* tests (Stephens et al., 2018).

Harm-benefit assessment (Joanne Storey)

Article 18 of the Directive requires an evaluation of "whether the harm to the animals in terms of suffering, pain and distress is justified by the expected outcome." Researchers should specifically consider whether the results that will be obtained are worth the level of harm or lasting suffering to the animals and the Harm-Benefit Assessment (HBA) is an important aspect in achieving this.

Determining harms

A procedure that results in pain, suffering, distress, or lasting harm can be counted within the harms of an HBA. As detailed in Article 13.2 in choosing between procedures, those which to the greatest extent meet the following requirements shall be selected: (a) use the minimum number of animals; (b) involve animals with the lowest capacity to experience pain, suffering, distress, or lasting harm; and (c) cause the least pain, suffering, distress, or lasting harm. Chapter 9 details harm to experimental animals and should be cross-referenced.

An AALAS-FELASA working group on HBA has provided an excellent overview of the main concepts and thinking (Brønstad et al., 2016) including a detailed review of the HBA literature and key documents. In the area of harms, the many factors relevant were grouped into three. The largest harms are within the subgroup "animal welfare harms" and include discomfort, duration of procedure relative to lifespan, housing, handling, etc. The second subgroup is "animal rights harm" in which the intrinsic value and rights of the animal are included. The last subgroup entitled "quality harms" is tied to scientific quality and includes the nonpublication of negative results. It is important to recognize that harms are greater than simply pain and suffering: all sources that may negatively impact on animals should be considered. Refer to Chapter 9 for more detailed information.

In the context of the 3Rs, consideration of harm may result in paradoxes. An opinion is not offered here on these, but the reader should be aware of them. Firstly, when considering *reduction*, harms may be limited to a few animals, such that the total number of animals is lower, yet from a *refinement* perspective a lower level of harm per individual animal may be considered utilizing a greater total number of animals. These ethical dilemmas require careful thought. The reader should also consider contemporary definitions of the 3Rs that may help in thinking through the dilemmas (e.g., the NC3Rs' contemporary definition of Reduction: "Appropriately designed and analyzed animal experiments that are robust and reproducible, and truly add to the knowledge base"). Secondly, the many factors defined as harms are often dependent on the sentience and consciousness of the animals. In the original definition of *replacement* by Russell and Burch (1959), replacement should consider the replacement or substitution of higher animals by animals of lower sentience. Note that in EU Directive 2010/63 Article 8, experiments in nonhuman primates are called out specifically as being more harmful compared to other species.

In projects written to comply with the Directive, harms are often well described and documented with appropriate humane endpoints defined. When novel compounds or novel procedures are being tested it is important to clearly define humane endpoints and to consider risk mitigation plans ahead of a new procedure. Additionally, harms can be minimized through appropriate use of analgesia and anesthesia, and these should be considered with veterinary input appropriate for the species and strain. The reader is referred to Chapter 9 for an in-depth review of humane endpoints.

Determining benefits

A requirement of Article 13.2 is that procedures chosen should be ones "most likely to provide satisfactory results." This underlines the importance of good experimental design to ensure valid results, and a priori decisions on robust statistical analysis. The benefit should go beyond extrapolation of benefit to human health and should relate directly to the likelihood of success of the experiment achieving an answer to its hypothesis, and the appropriateness of the model to answer the biological question.

The AALAS-FELASA working group on HBA (Brønstad et al., 2016), utilized literature reviews to group factors that impact benefit. The groups were defined as "scientific quality" (e.g., aim and objectives, originality and feasibility), "promise or potential outcome" (e.g., predicted benefit for humans, animals, or the environment and long-term/secondary benefits), and finally the "actual outcome" or benefit that includes primary/direct benefit, safety, knowledge, or education benefits. It is of note that, in general, benefits may be less well defined than harms. Again in terms of the 3Rs, the reader may find the contemporary definitions of the 3Rs offered by the NC3Rs helpful (website: The 3Rs | NC3Rs). In particular, improving the scientific quality of experiments is an opportunity to improve benefit and key to this is good experimental design.

In the context of translational research, it is useful to consider the potential human benefit within the context of the known translational relevance of the animal model being utilized and then consider the "likelihood of success" of a specific, individual experiment in a given model. There should be a clearly defined plan of work, which may include the methodology, an explanation for the choice of method or technique and experimental design, and justification of both species to be used and the animal models. Furthermore, the plan for making the results known, by publication, further dissemination or communication, is important to ensure the work is adding to the knowledge base. The Directive states that realistic objectives should be a factor in assessing likelihood of success and specifically when articulating objectives they should be specific, i.e., have a focused question or aim and be measurable, and note taken of how success or failure of the work will be measured. Further questions to consider are whether the aim is achievable and realistic. An element of time is key to assessing realistic goals; can the work be delivered in the outlined time frame? But also consider timing in relation to other scientific work, i.e., how will the results inform future work and how has prior knowledge informed the current hypothesis?

Resourcing of the work is also important for success and should include considerations of finances, access to required facilities appropriate for the work and of course ensuring availability of both scientific and care personnel for the duration of the proposed work. Considerations should also include any planned work following the in vivo ("in life") phase, so, for example, the planned histological processing and analysis should be clear. Experience and prior track record can give confidence that success will be achieved, and publication record may be useful evidence in supporting this, as can retrospective assessments from previous projects.

Assessing harms against benefits

The 3Rs should have been evidently addressed and consideration for each R been demonstrated prior to the HBA. Importantly, the step of the HBA brings the harms and the benefits together to be assessed against each other, and therefore goes beyond the application of the 3Rs. The HBA itself should be a systematic approach to compare both the harms and benefits. The aim of the HBA is that the potential benefits should always outweigh the likely adverse effects, or the harms minimized, and the potential benefits maximized. The literature contains a range of approaches to HBA, from the use of algorithms, checklists/generic process, and visual representations such as the "Bateson cube" (Bateson, 1986). Regardless of methodological approach, it is useful to have varied representation of skill set and also mind sets within the group of individuals charged with conducting an HBA, and transparency of the process is key.

Additionally, it is worth recognizing that the HBA should be repeated during a program of work, due to its dependence on context. Scientific advances may alter the balance and reduce the benefit achievable, weighting the harms greater than the benefit. Nonanimal alternative methods are increasing rapidly in availability and potential application (Low et al., 2020) and should be an early consideration to check the continued requirement for a given animal experiment.

Further reading

Several publications referred to above for particular points merit wider reading and there are others suitable for further reading not mentioned above. Festing et al. (2016) and Ruxton and Colegrave (2011) presume little mathematical or statistical understanding and cover many aspects of experimental design at a level suitable for new researchers. McKillup's (2011) introduction to statistical analysis is at a similar level. Festing's website, www.3Rs-reduction.co.uk, covers similar material to sections of this chapter and includes interactive elements and self-tests. For refinement aspects of experimental design see Morton (1998), and for agricultural experiments and their analysis see Morris (1999). A comprehensive treatment of experimental design and statistical analysis, including designs not covered here, is found in Quinn and Keough (2002) and Bate and Clark (2014). Both the PREPARE and the ARRIVE Guidelines have useful explanatory texts with them (see Percie du Sert 2020b for the ARRIVE Guidelines explanations). A broad consideration of harm-benefit analysis is given in Brønstad et al. (2016), and Laber et al. (2016) provides useful guidance on how a harm-benefit analysis may be carried out.

References

Baker, M., 2016. Is there a reproducibility crisis? Nature 533, 452–454. https://doi.org/10.1038/533452a.

Bate, S.T., Clark, R.A., 2014. The Design and Analysis of Animal Experiments. Cambridge University Press, Cambridge.

Bateson, P., 1986. When to experiment on animals. New Sci. 109, 30–32.

Begley, C.G., Ellis, L.M., 2012. Raise standards for preclinical cancer research. Nature 483, 531–533.

Bengtsson, F., Jörntell, H., 2007. Ketamine and xylazine depress sensory-evoked parallel fiber and climbing fiber responses. J. Neurophysiol. 98 (3), 1697–1705.

Brønstad, A., Newcomer, C.E., Decelle, T., Everitt, J.I., Guillen, J., Laber, K., 2016. Current concepts of harm–benefit analysis of animal experiments – report from the AALAS–FELASA working group on harm–benefit analysis – Part 1. Lab. Anim. 50 (1S), 1–20.

Button, K.S., Ioannidis, J.P., Mokrysz, C., Nosek, B.A., Flint, J., Robinson, E.S., Munafò, M.R., 2013. Power failure: why small sample size undermines the reliability of neuroscience. Nat. Rev. Neurosci. 14 (5), 365–376.

Cheleuitte-Nieves, C., Lipman, N.S., 2019. Improving replicability, reproducibility, and reliability in preclinical research: a shared responsibility. ILAR J. 60 (2), 113–119. https://doi.org/10.1093/ilar/ilaa009.

Clark, R.A., Shoaib, M., Hewitt, K.N., Stanford, S.C., Bate, S.T., 2012. A comparison of InVivoStat with other statistical software packages for analysis of data generated from animal experiments. J. Psychopharmacol. 26 (8), 1136–1142.

Currie, G.L., Delaney, A., Bennett, M.I., Dickenson, A.H., Egan, K.J., Vesterinen, H.M., et al., 2013. Animal models of bone cancer pain: systematic review and meta-analyses. Pain 154 (6), 917–926.

de Vries, R.B., Buma, P., Leenaars, M., Ritskes-Hoitinga, M., Gordijn, B., 2012. Reducing the number of laboratory animals used in tissue engineering research by restricting the variety of animal models. Articular cartilage tissue engineering as a case study. Tissue Eng. B Rev. 18, 427–435.

de Vries, R.B., Wever, K.E., Avey, M.T., Stephens, M.L., Sena, E.S., Leenaars, M., 2014. The usefulness of systematic reviews of animal experiments for the design of preclinical and clinical studies. ILAR J. 55 (3), 427–437.

Festing, M.F.W., Overend, P., Cortina Borja, M., Berdoy, M., 2016. The Design of Animal Experiments: Reducing the Use of Animals in Research through Better Experimental Design (Revised and Updated Edition). Sage Publications Ltd., London.

Fisher, R.A., 1960. The Design of Experiments, seventh ed. Oliver and Boyd, Edinburgh.

Fry, D., 2013. Experimental design: reduction and refinement in studies using animals. In: Bayne, K., Turner, P.V.((Eds.), Laboratory Animal Welfare. Academic Press, London, pp. 95–113.

Golbach, L.A., Portelli, L.A., Savelkoul, H.F.J., Terwel, S.,R., Kuster, N., de Vries, R.B.M., Verburg-van Kemenade, B.M.L., 2016. Calcium homeostasis and low-frequency magnetic and electric field exposure: a systematic review and meta-analysis of in vitro studies. Environ. Int. 92–93, 695–706.

Gosselin, R.D., 2019. Guidelines on statistics for researchers using laboratory animals: the essentials. Lab. Anim. 53 (1), 28–42.

Hollyer, T.R., Bordoni, L., Kousholt, B.S., Judith van Luijk, J., Ritskes-Hoitinga, M., Østergaard, L., et al., 2019. The evidence for the physiological effects of lactate on the cerebral microcirculation: a systematic review. J. Neurochem. 148 (6), 712–730.

Hooijmans, C.R., Leenaars, M., Ritskes-Hoitinga, M., 2010. A gold standard publication checklist to improve the quality of animal studies, to fully integrate the Three Rs, and to make systematic reviews more feasible. Altern. Lab. Anim. 38, 167–182.

Hooijmans, C.R., IntHout, J., Ritskes-Hoitinga, M., Rovers, M.M., 2014. Meta-analyses of animal studies: an introduction of a valuable instrument to further improve healthcare. ILAR J. 55 (3), 418–426.

Huang, W., Percie du Sert, N., Vollert, J., Rice, A.S.C., 2019. General principles of preclinical study design. In: Handbook of Experimental Pharmacology. Springer, Berlin, Heidelberg.

Hurlburt, S.H., 1984. Pseudoreplication and the design of ecological field experiments. Ecol. Monogr. 54, 187–211.

Karp, N.A., Fry, D., 2021. What is the optimum design for my animal experiment? BMJ Open Sci. 5, e100126. https://doi.org/10.1136/bmjos-2020-100126.

Kilkenny, C., Parsons, N., Kadyszewski, E., Festing, M.F.W., Cuthill, I.C., Fry, D., Hutton, J., Altman, D.G., 2009. Survey of the quality of experimental design, statistical analysis and reporting of research using animals. PLoS One 4 (11), e7824. https://doi.org/10.1371/journal.pone.0007824.

Laber, K., Newcomer, C.E., Decelle, T., Everitt, J.I., Guillen, J., Brønstad, A., 2016. Recommendations for addressing harm–benefit analysis and implementation in ethical evaluation – report from the AALAS–FELASA working group on harm–benefit analysis – Part 2. Lab. Anim. 50 (1S), 21–42.

Landis, S.C., Amara, S.G., Asadullah, K., Austin, C.P., Blumenstein, R., Bradley, E.W., et al., 2012. A call for transparent reporting to optimize the predictive value of preclinical research. Nature 490, 187–191.

Lazic, S.E., Clarke-Williams, C.J., Munafo, M.R., 2018. What exactly is 'N' in cell culture and animal experiments? PLoS Biol. 16, e2005282.

Liao, J.J., Liu, R., 2009. Re-parameterization of five-parameter logistic function. J. Chemometr. J. Chemom. Soc. 23 (5), 248–253.

Leenaars, C., Tsaioun, K., Stafleu, F., Rooney, K., Meijboom, F., Ritskes-Hoitinga, M., Bleich, A., 2020. Reviewing the animal literature: how to describe and choose between different types of literature reviews. Lab. Anim. https://doi.org/10.1177/0023677220968599. Nov 1:23677220968599 (Epub ahead of print).

Limentani, G.B., Ringo, M.C., Ye, F., Bergquist, M.L., McSorley, E.O., 2005. Beyond the t-test: statistical equivalence testing. Anal. Chem. 77, 221A–226A.

Low, L.A., Mummery, C., Berridge, B.R., Austin, C.P., Tagle, D.A., 2020. Organs-on-chips: into the next decade. Nat. Rev. Drug Discov. 20, 345–361. https://doi.org/10.1038/s41573-020-0079-3.

McKillup, S., 2011. Statistics Explained. An Introductory Guide for Life Scientists. Cambridge University Press, Cambridge.

Mead, R., 1988. The Design of Experiments. Cambridge University Press, Cambridge.

Morris, T.R., 1999. Experimental Design and Analysis in Animal Sciences. CABI Publishing, Wallingford, Oxon (UK).

Morton, D.B., 1998. The Importance of Non-statistical Design in Refining Animal Experimentation. In: ANZCCART Facts Sheet. ANZCCART News, vol. 1. ANZCCART, Osmond, Australia, p. 12.

Munafò, M.R., Nosek, B.A., Bishop, D.V., Button, K.S., Chambers, C.D., Percie du Sert, N., et al., 2017. A manifesto for reproducible science. Nat. Hum. Behav. 1 (1), 1–9.

Percie du Sert, N., Hurst, V., Ahluwalia, A., Alam, S., Avey, M.T., Baker, M., et al., 2020a. The ARRIVE guidelines 2.0: updated guidelines for reporting animal research. PLoS Biol. 18 (7), e3000410. https://doi.org/10.1371/journal. pbio.3000410.

Percie du Sert, N., Ahluwalia, A., Alam, S., Avey, M.T., Baker, M., Browne, W.J., et al., 2020b. Reporting animal research: explanation and elaboration for the ARRIVE guidelines 2.0. PLoS Biol. 18 (7), e3000411. https://doi.org/10.1371/journal.pbio.3000411.

Prinz, F., Schlange, T., Asadullah, K., 2011. Believe it or not: how much can we rely on published data on potential drug targets? Nat. Rev. Drug Discov. 10, 712–713.

Quinn, G.P., Keough, M.J., 2002. Experimental Design and Data Analysis for Biologists. Cambridge University Press, Cambridge.

Reichlin, T.S., Vogt, L., Wurbel, H., 2016. The researchers' view of scientific rigor - survey on the conduct and reporting of in vivo research. PLoS One 11, e0165999. https://doi.org/10.1371/journal.pone.0165999.

Russell, W.M.S., Burch, R.L., 1959. The Principles of Humane Experimental Technique. Methuen, London.

Ruxton, G., Colegrave, N., 2011. Experimental Design for the Life Sciences. Oxford University Press, Oxford.

Shaw, R., Festing, M.F., Peers, I., Furlong, L., 2002. Use of factorial designs to optimize animal experiments and reduce animal use. ILAR J. 43 (4), 223–232.

Smith, A.J., Clutton, R.E., Lilley, E., Hansen, K.E.A., Brattelid, T., 2018. PREPARE: guidelines for planning animal research and testing. Lab. Anim. 52 (2), 135–141. https://doi.org/10.1177/0023677217724823.

Smith, A.J., Hawkins, P., 2016. Good science, good sense and good sensibilities: the three Ss of Carol Newton. Animals 6 (11), 70. https://doi.org/10.3390/ani6110070.

Sorzano, C.O.S., Parkinson, M., 2019. Statistical Experiment Design for Animal Research.

Stephens, M.L., Akgun-Olmez, S.G., Hoffmann, S., de Vries, R., Flick, B., Hartung, T., et al., 2018. Adaptation of the systematic review framework to the assessment of toxicological test methods: challenges and lessons learned with the Zebrafish embryotoxicity test. Toxicol. Sci. 171 (1), 56–68.

Turner, P.V., Brabb, T., Pekow, C., Vasbinder, M.A., 2011. Administration of substances to laboratory animals: routes of administration and factors to consider. J. Am. Assoc. Lab. Anim. Sci. 50 (5), 600–613.

van Luijk, J., 2017. The Next Steps towards Responsible Animal-Based Research (Ph.D. thesis). Radboud University, Nijmegen, The Netherlands.

Voelkl, B., Altman, N.S., Forsman, A., Forstmeier, W., Gurevitch, J., Jaric, I., et al., 2020. Reproducibility of animal research in light of biological variation. Nat. Rev. Neurosci. 21, 384–393.

Waterton, J.C., Middleton, B.J., Pickford, R., Allott, C.P., Checkley, D., Keith, R.A., 2000. Reduced animal use in efficacy testing in disease models by the use of sequential experimental designs. In: Balls, M., van Zeller, A.M., Halder, M. (Eds.), Progress in the Reduction, Refinement and Replacement of Animal Experimentation. Elsevier, Amsterdam, pp. 737–745.

Yauw, S.T., Wever, K.L., Hoesseini, A., Ritskes-Hoitinga, M., van Goor, H., 2015. Systematic review of experimental studies on intestinal anastomosis. Br. J. Surg. 102 (7), 726–734.

Zar, J.H., 2019. Biostatistical Analysis. Pearson Education, London.

Recommended websites

3Rs-Reduction.co.uk. An interactive course on experimental design produced by Dr. Michael Festing. http://www.3rs-reduction.co.uk. An overview of Dr. Festing's publications is available at: https://norecopa.no/Festing.

AAALAC International Program Description. https://www.aaalac.org/program-description.

Alternatives and Searching the Literature for Animal Research, UC Davis Library. https://www.library.ucdavis.edu/guide/alternatives.

Conducting a pilot study. https://nc3rs.org.uk/conducting-pilot-study.

ECVAM-EURL Search Guide. https://ec.europa.eu/jrc/en/scientific-tool/eurl-ecvam-search-guide.

EURL ECVAM dataset on alternative methods to animal experimentation (DB-ALM). http://data.europa.eu/89h/b7597ada-148d-4560-9079-ab0a5539cad3.

Experimental design/statistics. https://nc3rs.org.uk/experimental-design.

Fry, D.J., 2014. Teaching experimental design. ILAR J. 55, 457–471. https://doi.org/10.1093/ilar/ilu031. https://academic.oup.com/ilarjournal/article/55/3/457/643598.

Gosselin, R.-D., 2019. Guidelines on statistics for researchers using laboratory animals: the essentials. Lab. Anim. 53 (1), 28–42. https://journals.sagepub.com/doi/full/10.1177/0023677218783223.

InVivoStat (Statistical Software Package). http://invivostat.co.uk.

Lakens, D. Sample Size Justification. https://doi.org/10.31234/osf.io/9d3yf.

SYRCLE (SYstematic Review Center for Laboratory animal Experimentation). https://www.syrcle.network.

The 3R Guide database of guidelines for planning animal research and testing. https://norecopa.no/3R-Guide.

The Experimental Design Assistant – EDA. https://www.nc3rs.org.uk/experimental-design-assistant-eda.

The International Culture of Care Network. https://norecopa.no/CoC.

The National Centre for the Replacement. Refinement and Reduction of Animals in Research, UK NC3Rs. https://nc3rs.org.uk.

The NORINA database of alternatives and supplements to animal use in education and training. https://norecopa.no/NORINA.

The PREPARE checklist. https://norecopa.no/PREPARE/prepare-checklist.

The TextBase database of textbooks and other literature within Laboratory Animal Science and related disciplines. https://norecopa.no/TextBase.

Index

Note: 'Page numbers followed by "b" indicate boxes, those followed by "f" indicate figures and those followed by "t" indicate tables.'

B

C

D

E

I

K

L

T

V

W

Z

Kolibri 360 GmbH
Ferdinand-Jühlke-Straße 7
99095 Erfurt, Deutschland
produktsicherheit@kolibri360.de